Encyclopedia of Petroleum Exploration and Production

Volume I

Encyclopedia of Petroleum Exploration and Production

Volume I

Edited by **Andy Margo and**
Michael Dedini

NY RESEARCH PRESS
New York

Published by NY Research Press,
23 West, 55th Street, Suite 816,
New York, NY 10019, USA
www.nyresearchpress.com

Encyclopedia of Petroleum Exploration and Production: Volume I
Edited by Andy Margo and Michael Dedini

International Standard Book Number: 978-1-63238-151-4 (Hardback)

Printed in China.

Contents

Preface

With the rate of dwindling resources in our world, there is always a constant and never ending need for newer energy sources, especially petroleum reservoirs. This pushes scientists and technicians all over the world to improve on various existing technologies. Petroleum exploration is one such field where constant innovation is needed. It refers to the search for petroleum deposits or rather oil and natural gas resources in the earth's crust. This is a process that relies on a methodical and systematic application of technology by geoscientists that eventually leads towards viable prospects for drilling as well as the actual drilling of these sites. A lot of risk capital is needed in this field to explore prospective sites, especially ones that could have an uncertain outcome. This process of exploration is constantly evolving as it can change a company's fortunes overnight. Petroleum production technology is a field closely related to petroleum exploration. Since working in the petroleum industry can provide a well paid career and exciting opportunities for worldwide employment, it is always advisable to stay up- to- date with the current technologies that envelop exploration. The need for trained operators with knowledge of the current state and new methods of petroleum production is always there in this industry.

I would like to thank those whose hard work has gone behind these researches as well as those who support the concept and goal of this book.

Editor

1

Empirical correlations for viscosity of partially hydrolyzed Polyacrylamide

Changhong Gao

Abstract During water flooding operations, polymer can be added to injected water to improve sweep efficiency and retard viscous fingering. This technique, namely polymer flooding has achieved successful large-scale field applications in China. Polymer flooding is also being tested in several heavy oil fields. For oil field applications, partially hydrolyzed polyacrylamide (HPAM) is the most widely-used polymer. HPAM solution must maintain high viscosity to effectively mobilize oil. In reality however, polymer is very sensitive to salinity and hardness of base water and underground water. This phenomenon is studied in this paper. The two HPAM test samples have molecular weight of 10 million and 20 million Dalton. The first group of tests was conducted to study the effect of salinity on HPAM viscosity. The second group of tests was carried out to evaluate the effect of hardness on HPAM viscosity. It was discovered that both sodium and calcium ions can severely damage polymer viscosity. Data analysis shows that the empirical non-Newtonian parameters are strong functions of water salinity and hardness. Based on test data, empirical formulas are proposed to calculate HPAM viscosity influenced by polymer concentration, temperature, water salinity and water hardness. These formulas are useful tools for predicting HPAM viscosity in field applications.

Keywords HPAM · Viscosity · Salinity · Hardness · Degradation · Power law

C. Gao
Sinopec Shengli Oil Field Company, Dongying, China

Introduction

During oil and gas production, water is often injected into reservoirs to sweep oil to the production well. Because oil viscosity is often much higher than water viscosity, injected water moves faster than oil, leaving oil behind. This phenomenon is referred to as viscous fingering. Polymer can be added to injected water to retard viscous fingering. Polymer increases the viscosity of injected water, reduces water mobility, and therefore achieves a more stable displacement (Littmann 1988).

Polymer flooding has become the most successful chemical method for enhanced oil recovery (EOR). For the Daqing oil field in China, polymer flooding has contributed to more than 10 % of EOR (Wang et al. 2009). For field applications, partially hydrolyzed polyacrylamide (HPAM) is the most widely used polymer, thanks to its relatively cheap price and good solubility in water. The molecular structure of HPAM is given in Fig. 1 (Sheng 2010).

Polymer solution must maintain high viscosity to mobilize oil in reservoir. However, high shearing, high temperature, high salinity and hardness often exist in oil reservoirs and production systems. The polymer can severely degrade under these harsh conditions, which is the major issue facing polymer flooding operation (Gao 2011).

HPAM degradation has been studied for many years. In an early study, it was observed that NaCl, $CaCl_2$ and $MgCl_2$ depressed HPAM viscosity, and HPAM was very sensitive to divalent ions (Mungan 1972). These findings were confirmed by other studies (Maerker 1975; Ward and Martin 1981; Ryles 1988). Therefore it is not a surprise that most successful field applications of polymer flood were carried out in low-salinity reservoirs (Needham and Doe 1987). Moreover, shearing and thermal degradation also have negative impacts on viscosity of HPAM (Gao 2013).

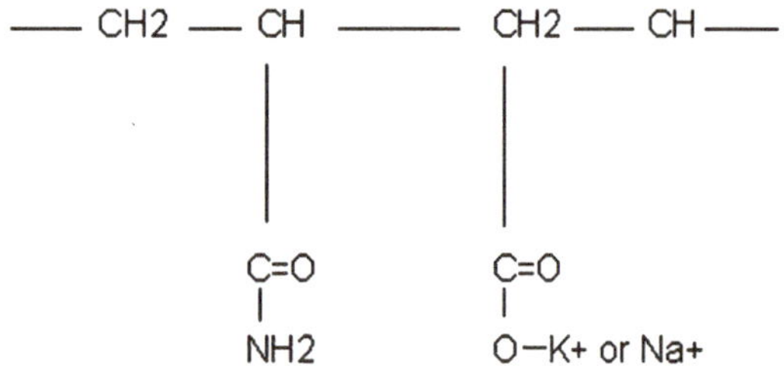

Fig. 1 Structure of HPAM

Polymer solution is a non-Newtonian fluid that follows the power law given in Eq. 1, where μ is the viscosity (cP), γ is the shear rate (1/s), k is the consistency index, and n is the flow behavior index (Rabia 1985). More complex models were also developed to describe polymer viscosity more accurately, such as the Ellis model and Carreau model (Sochi 2009).

$$\mu = k\gamma^n \tag{1}$$

It has to be pointed out that the previous models failed to quantify the influence of individual factors on polymer viscosity. As a result, it is still a challenge to correctly predict the viscosity of HPAM under the complex underground environment. In recent years, polymer flood attracted large-scale field applications in China (Sheng 2010). However, the change in polymer viscosity while HPAM flows through injection facilities and underground rocks has to rely on frequent sampling. There is no reliable formula to predict the viscosity behavior of HPAM in oilfield applications.

This paper obtains polymer viscosity data through laboratory testing. The author also attempts to build numerical correlations for viscosity of HPAM solution. The proposed correlations take into account the various factors that affect polymer viscosity. This work improves the prediction of polymer viscosity in underground environment.

Materials and methods

HPAM is the most widely used polymer product for EOR projects. Recent EOR projects employed HPAM products with high molecular weights. Two HPAM samples, coded as 3,100 and 3,200, were supplied by a polymer producer in the form of powders. The molecular weight of product 3,100 was around 10 million Dalton, and that of product 3,200 was 20 million. For both the products, the degree of hydrolysis was around 25 %.

For the first group of tests, polymer powder was added to NaCl brine to result in a concentration of 1,500 ppm. The concentration of NaCl ranged from 0.1 to 3 % by weight. The polymer solution was stirred for 48 h to ensure good mixing. Polymer viscosity was then measured at 50 °C under varied shear rates with a torque-type viscometer.

The second group of tests was to study the effect of calcium ions on polymer viscosity. The base brine contains sodium chloride and calcium chloride. The concentration of NaCl was controlled at 1 % by weight. The concentration of $CaCl_2$ ranged from 0.05 to 0.15 % by weight. Polymer powder was added to NaCl and $CaCl_2$ brine to result in a concentration of 1,500 ppm. The polymer viscosity was then measured at 50 °C under varied shear rates.

Experimental results

The viscosity data for the product 3,100 in NaCl brine are given in Fig. 2. The viscosity data for the product 3,100 in $CaCl_2$ brine are given in Fig. 3. These trends comply with the power law. Similar test results on the product 3,200 were obtained, but not presented here. It is obvious that both NaCl and $CaCl_2$ severely damage the HPAM viscosity, while HPAM is more sensitive to calcium concentration.

Model formulation and validation

By curve fitting, the power law parameters for the two groups of tests are obtained, as presented in Tables 1, 2. Figures 4, 5 plot the relationships between the power–law parameters and the NaCl concentration (C_{Na}, wt %). It can be seen that the parameter k is depressed at high salinity, while the parameter n slightly increases at elevated salinity. It is obvious that the two HPAM products follow similar trends. The generalized relationship can be expressed as Eqs. (2), (3), where a, $a1$, α, and $\alpha1$ are empirical parameters.

$$\ln(k) = a \times \ln(C_{Na}) + a1 \tag{2}$$

$$n = \alpha \times \ln(C_{Na}) + \alpha1 \tag{3}$$

Figures 6, 7 plot the relationships between the power-law parameters and the $CaCl_2$ concentration (C_{Ca}, wt %).

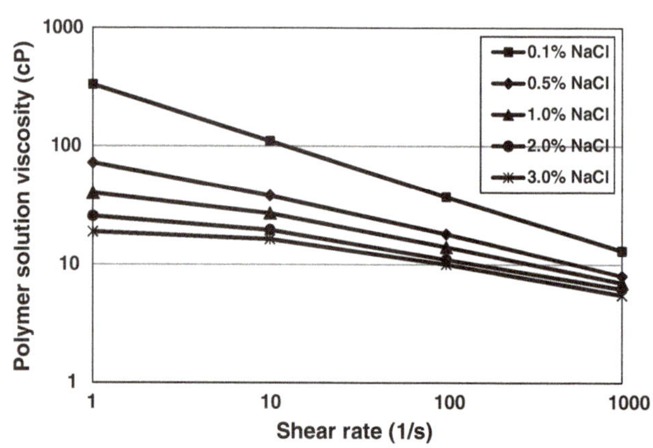

Fig. 2 Viscosity data for the product 3,100 in NaCl solutions

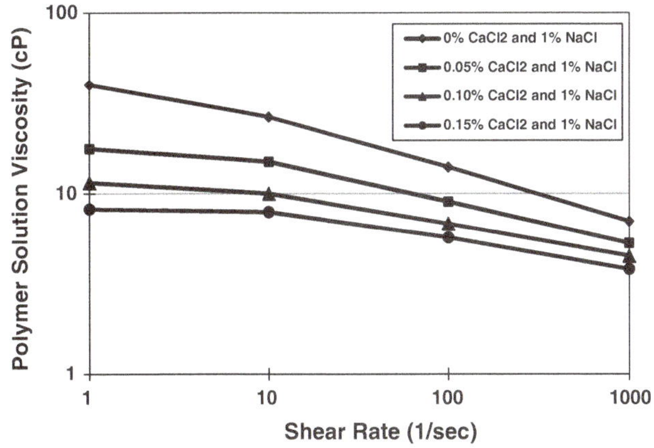

Fig. 3 Viscosity data for the product 3,100 in $CaCl_2$ solutions

Table 1 Power law parameters for tests in NaCl

Polymer product	NaCl (wt%)	*k* (tested)	*k* (calculated)	*n* (tested)	*n* (calculated)
3,100	0.5	75.33	73.88	−0.319	−0.315
	2.0	27.94	30.85	−0.209	−0.221
	3.0	21.32	19.93	−0.181	−0.173
3,200	0.5	189.80	171.83	−0.384	−0.405
	2.0	65.18	71.65	−0.350	−0.361
	3.0	50.72	55.48	−0.331	−0.339

Table 2 Power law parameters for tests in $CaCl_2$

Polymer product	$CaCl_2$ (wt%)	*k* (tested)	*k* (calculated)	*n* (tested)	*n* (calculated)
3,100	0.05	19.70	18.50	−0.179	−0.179
	0.10	12.45	12.47	−0.139	−0.142
	0.15	9.10	9.22	−0.114	−0.113
3,200	0.05	40.99	43.22	−0.265	−0.277
	0.10	30.48	29.24	−0.236	−0.233
	0.15	21.88	21.67	−0.198	−0.199

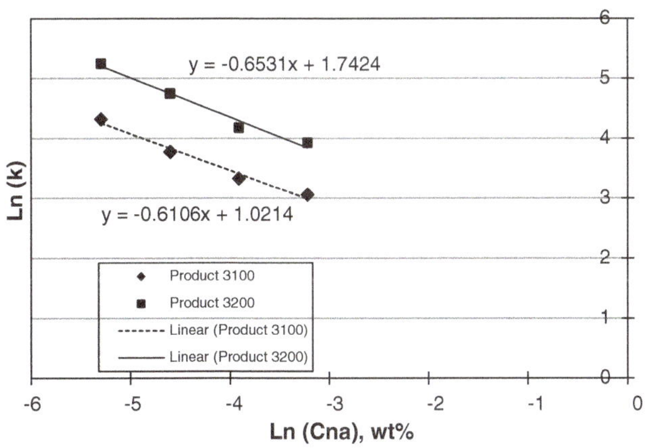

Fig. 4 Relationship between NaCl concentration and parameter *k*

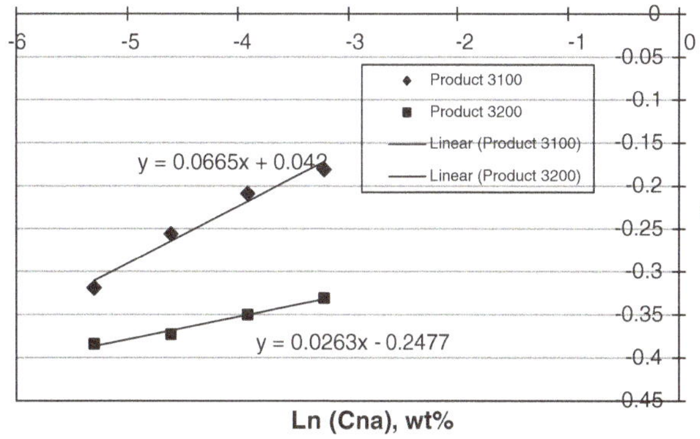

Fig. 5 Relationship between NaCl concentration and parameter *n*

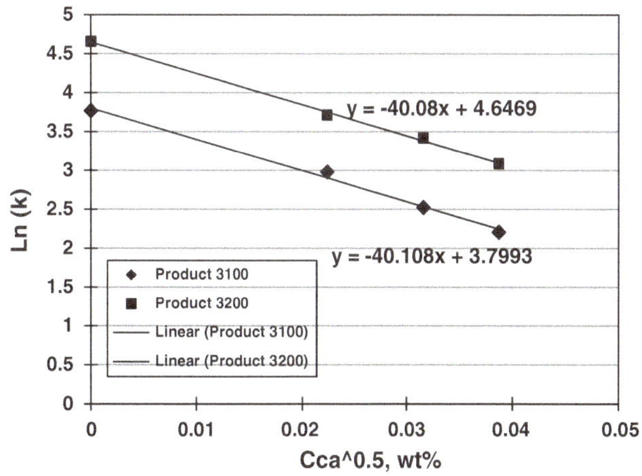

Fig. 6 Relationship between $CaCl_2$ concentration and parameter *k*

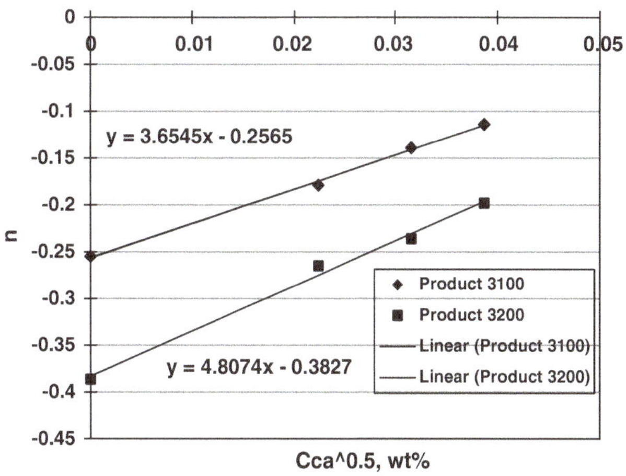

Fig. 7 Relationship between $CaCl_2$ concentration and parameter *n*

It can be seen that the parameter *k* is depressed at high salinity, while the parameter n slightly increases at elevated salinity. These trends are similar to those in Figs. 4, 5. The generalized relationship can be expressed as Eqs. (4), (5), where *b*, *b*1, *β*, and *β*1 are all empirical parameters.

$$\ln(k) = b \times (C_{Ca})^{1/2} + b1 \quad (4)$$

$$n = \beta \times (C_{Ca})^{1/2} + \beta 1 \quad (5)$$

In a previous paper (Gao 2013), the following equations were proposed to correlate the power–law parameters with polymer concentration (PC, ppm) and temperature (T, °C), where c, $c1$, d, $d1$, λ, $\lambda 1$, ω, $\omega 1$ are all empirical parameters.

$$\ln(k) = c \times \ln(\text{PC}) + c1 \quad (6)$$

$$\ln(k) = d \times T + d1 \quad (7)$$

$$n = \lambda \times \ln(\text{PC}) + \lambda 1 \quad (8)$$

$$n = \omega \times T + \omega 1 \quad (9)$$

Combining Eqs. (2), (4), (6), and (7), the following formula is obtained, where e is an empirical parameter:

$$\ln(k) = a \times \ln(C_{Na}) + b \times (C_{Ca})^{1/2} + c \times \ln(\text{PC}) + d \times T + e \quad (10)$$

Combining Equations (3), (5), (8), and (9), the following formula is obtained, where ε is an empirical parameter:

$$n = \alpha \times \ln(C_{Na}) + \beta \times (C_{Ca})^{1/2} + \lambda \times \ln(\text{PC}) + \omega \times T + \varepsilon \quad (11)$$

Based on the test data in this paper and in a previous paper (Gao 2013), the empirical values are obtained with Levenberg–Marquardt method. For the HPAM product with molecular weight of 10 million Dalton, the formulas are given in Eqs. (12), (13):

$$\ln(k) = -0.63 \times \ln(C_{Na}) - 42.457 \times (C_{Ca})^{1/2} + 1.848 \times \ln(\text{PC}) - 0.02 \times T - 12.582 \quad (12)$$

$$n = 0.068 \times \ln(C_{Na}) + 3.984 \times (C_{Ca})^{1/2} - 0.143 \times \ln(\text{PC}) + 6.944 \times 10^{-4} \times T + 1.115 \quad (13)$$

For the HPAM product with molecular weight of 20 million Dalton, the formulas are presented in Eqs. (14), (15):

$$\ln(k) = -0.631 \times \ln(C_{Na}) - 42.169 \times (C_{Ca})^{1/2} + 2.094 \times \ln(\text{PC}) - 0.016T - 13.713 \quad (14)$$

$$n = 0.032 \times \ln(C_{Na}) + 4.743 \times (C_{Ca})^{1/2} - 0.192 \times \ln(\text{PC}) + 6.313 \times 10^{-4}T + 1.208 \quad (15)$$

These formulas are then validated against the test data. The k and n values are calculated with these formulas and compared with the respective values obtained from experimental data, as presented in Tables 1, 2, 3 and 4. It can be seen that the calculated data match the test data relatively well. The advantage of these formulas is that the effects from all known factors are considered, while the previous models failed to include these influences (Zheng et al. 2000).

Table 3 Power law parameters for tests at different polymer concentrations

Polymer product	Cp (ppm)	T (°C)	k (tested)	k (calculated)	n (tested)	n (calculated)
3,100	500	30	4.28	3.70	−0.077	−0.069
	1,000		11.33	13.26	−0.160	−0.0169
	2,000		42.64	47.74	−0.257	−0.268
	3,000		123.81	101.00	−0.339	−0.326
3,200	500	30	6.45	6.09	−0.123	−0.117
	1,000		24.35	26.00	−0.244	−0.250
	2,000		103.20	110.96	−0.390	−0.383
	3,000		275.94	259.36	−0.480	−0.461

Table 4 Power law parameters for tests at different temperatures

Polymer product	Cp (ppm)	T (°C)	k (tested)	k (calculated)	n (tested)	n (calculated)
3,100	2,000	50	28.90	28.95	−0.260	−0.250
		70	19.00	19.40	−0.233	−0.236
		90	14.00	13.01	−0.220	−0.294
3,200	2,000	50	73.70	74.38	−0.373	−0.367
		70	57.42	54.00	−0.360	−0.355
		90	39.43	39.22	−0.331	−0.342

Conclusion

This paper presents the test data for HPAM viscosity in sodium chloride and calcium chloride solutions. It was observed that both monovalent and divalent ions have negative effects on HPAM viscosity. Data analysis shows that for the two HPAM products tested, the related power–law parameters are strong functions of water salinity and hardness.

This paper also presents a novel approach to model viscosity of HPAM solution. Previous models are failed to incorporate the factors that influence polymer viscosity. The proposed empirical correlations quantify the effects of salinity, hardness, temperature, and polymer concentration on HPAM viscosity. In future work, these correlations can be employed by pore-network model to study the polymer viscosity in macroscopic porous media.

References

Gao C (2011) Scientific research and field applications of polymer flood in heavy oil recovery. J Petroleum Explor Prod Technol 1(2):65–70

Gao C (2013) Viscosity of partially hydrolyzed polyacrylamide under shearing and heat. J Pet Explor Prod Technol.

Littmann W (1988) Polymer flooding. Elsevier, London, pp 3–4

Maerker JM (1975) Shear degradation of partially hydrolyzed polyacrylamide. SPE J 15(4):311–322

Mungan H (1972) Shear viscosity of ionic polyacrylamide solutions. SPE J 12(6):469–473

Needham R, Doe P (1987) Polymer flooding review. J Petrol Technol 39(12):1503–1507

Rabia H (1985) Oil well drilling engineering principles and practice. Graham & Trotman, London, pp 96–98

Ryles RJ (1988) Chemical stability limits of water soluble polymers used in oil recovery processes. SPE Reserv Eng J 3(1):23–34

Sheng J (2010) Modern chemical enhanced oil recovery. Gulf Professional Publishing, Houston, Texas, pp 101–102

Sochi T (2009) Single-phase flow of non-Newtonian fluids in porous media. Thesis of University College London, London, UK, pp 8–11

Wang D, Dong H, Lv C, Fu X, Nie J (2009) Review of practical experience by polymer flooding at Daqing. SPE Reserv Eval Eng 12(3):470–476

Ward JS, Martin FD (1981) Prediction of viscosity for partially hydrolyzed polyacrylamide solutions in the presence of calcium and magnesium ions. SPEJ 21(5):623–631

Zheng CG, Gall BL, Gao HW, Miller AE, Bryant RS (2000) Effects of polymer adsorption and flow behaviour on two phase flow in porous media. SPE Reservoir Eval Eng 3(3):216–222

A new approach of splitting C6+ composition of pipeline gas for hydrocarbon dew point determination

Livinus Aniefiok · Obah Boniface

Abstract Hydrocarbon dew point (HCDP) is a critical consideration for pipeline operations. Equation of state (EOS) method combined with Gas Chromatograph analysis is one of the methods used for HCDP determination. Most of the GCs on pipeline gas and in end-user installations are generally C6+ design and a few are C9+ design. In applying a HCDP limit using C6+ data, it is prudent to use an appropriate split of the C6+ composition. Though several fixed ratios of C6+ composition splits have been published in a Gas Processors Association standard and by leading chromatograph manufacturers for application, but may not truly reflect the C6+ composition of a particular gas. This work therefore presents a very simple procedure for splitting C6+ component into C6/C7/C8 for any particular pipeline gas provided that the specific gravity of the C6+ data is known. The method involves; determination of the molecular weights of C6/C7/C8 using three well-established hydrocarbon physical properties correlations, solution of the algebraic equations of the apparent molecular weight of C6+ using matrix notation, and application of logarithmic distribution to the calculated "quasi-mole fraction" of the individual C6+ components. An application of this approach to a field C6+ data is presented. In order to show the capabilities of the new approach, results comparison of calculated HCDP as well as cricondentherm (using HYSYS® with Peng Robinson EOS) between the new approach and some common C6+ composition splits is made.

Keywords C6+ splits · Pipeline gas · Hydrocarbon dew point

L. Aniefiok
Department of Chemical/Petroleum Engineering, University of Uyo, Uyo, Akwa-Ibom State, Nigeria

O. Boniface
Federal University of Technology, Owerri, Imo State, Nigeria

Introduction

Hydrocarbon dew point (HCDP) defines whether the natural gas stream in a pipeline at a given pressure and temperature consists of a single gas phase or two phases, gas and liquid. Figure 1 shows a phase diagram depicting the HCDP for a typical natural gas.

When natural gas is processed, the heavier components are removed in order to supply high quality gas to the market and also to ensure the safety and reliability of the pipeline system. For example, gas turbine generator plants have a written requirement in their warrantees that the fuel gas must be totally gaseous. In order to comply with this requirement, the gas is superheated to a minimum of 50 °F/28 °C above the highest dew point of the gas at the pressure regulators located ahead of the burner section.

The two methods currently in use for determining HCDP are; the manual visual dewpoint approach and the equation of state (EOS) method using Gas Chromatograph (GC) analysis. The manual method was developed by the US Department of Interior, Bureau of Mines and has been codified into a standard test method by the American Society of Testing and Materials (ASTM D 1142-95 1995). It uses a chilled mirror or dew point tester. This approach is labour intensive. Although automated continuous units are available, they are expensive and, unlike GCs, are currently not part of most existing gas transmission facilities

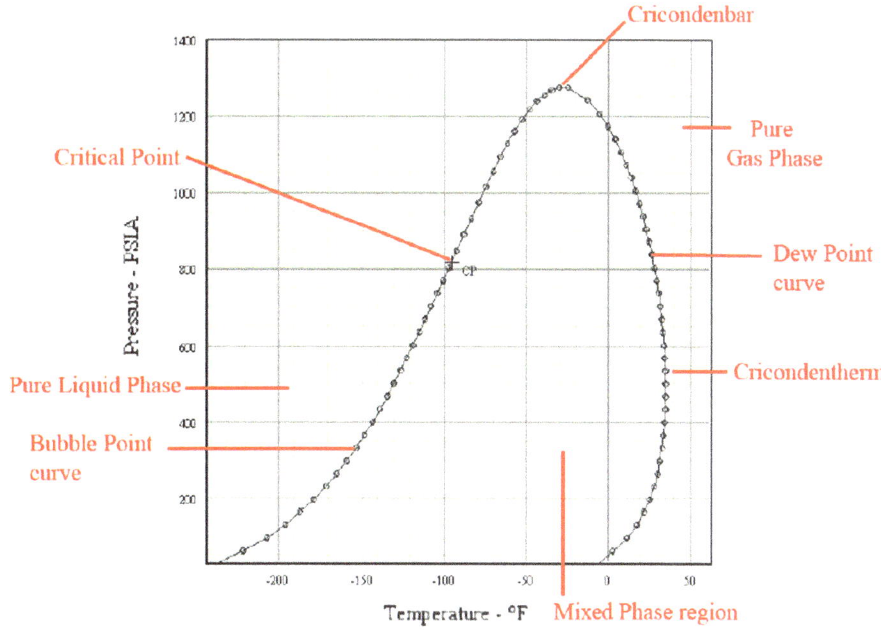

Fig. 1 A phase diagram for a typical natural gas (source: Shane 2000, Emerson Process Management, class # 5300)

(NGC-GPA 2005). In addition, the determination of the phase envelope or the cricondentherm would be difficult using this method since it is a periodic spot checking only. The indirect method uses a GC for compositional analysis in conjunction with an EOS to estimate the dew point as well as the cricondentherm and phase envelope if desired. As the HCDP is the condition when the heavy components begin to drop out into the liquid phase, the accurate measurement of the heavier components is critical for meaningful determination of the HCDP. The composition of the pipeline gas for custody metering is determined by the use of GCs, most of these analyzers are generally C6+ and a few are C9+ design (Jack 2010). The GC C6+ design measures the individual hydrocarbons up to normal-pentane using GPA 2261 (2000) procedure and then report the heavier components as a combined C6+ measurement.

In the absence of detailed analytical data for hydrocarbon plus fractions in a hydrocarbon mixture, erroneous predictions and conclusions can result if the plus fraction is used directly as a single component in the mixture phase behaviour calculations. Numerous authors have indicated that these errors can be substantially reduced by splitting or breaking down the plus fraction into a manageable number of fractions for equation of state calculations.

For energy calculation and HCDP calculation, many gas processors and end-users are now applying equation of state, either a Peng–Robinson (PR) or Soave–Redlich–Kwong (SRK) equation of state, to pipeline gas composition; the C6+ component is split into a fixed ratio of *n*-hexane, *n*-heptane and *n*-octane. Some commonly used percentage C6+ composition splits, based on empirical studies of most pipeline gases, are published in a GPA standard (NGC-GPA 2005): GPA 60 % C6, 30 % C7, 10 % C8; Daniels 47 % C6, 36 % C7, 17 % C8; GPA 50 % C6, 25 % C7, 25 % C8. Application of any of these fixed ratios for a particular delivery point of natural gas may not be a truly representative of the measured gas because the specific gravity of C6+ generated with the fixed ratios may likely deviate from that of the measured gas. Even with the most recent approach, used by gas processors and end-users to determine the percentage characterization of C6+ for a given pipeline, which is by taken the weighted average compositions of the regional supply on that pipeline, the ability of the average characterization to reflect the true composition of a particular gas within a region greatly depends on the variance of the individual components of all gases throughout the region (NGC-GPA 2005). Though, the traditional C6+ analysis provides insufficient data for a valid HCDP calculation (Ernst and Pettigrew 2005), improving on the repeatability of the prediction capability of the C6+ characterisation is essential.

This work presents a very simple and practical procedure for extending the percentage distribution of C6+ composition of pipeline gas into C6/C7/C8 in order to improve HCDP determination. An illustrative example in which the new procedure is applied to field C6+ data is also presented. In order to show the capabilities of the new approach, results comparison of calculated HCDP as well as cricondentherm (using HYSYS® with Peng Robinson EOS) between the new approach and some common C6+ composition splits is made.

Using C6+ data for HCDP limit application on a particular pipeline gas, it is prudent that an appropriate split be employed. If we consider the specific gravity, molecular weight, mole fraction of C6+ of a gas are known, the challenge facing the gas processor or end-user becomes one of splitting the C6+ fraction into C6/C7/C8 that can be used to predict the HCDP and phase behaviour of the gas by equation of state. Based on the observation reported by several researchers (Katz 1983; Lorenz et al. 1964; Pedersen et al. 1982; Ahmed et al. 1985) that lighter hydrocarbon systems exhibit exponential molar distribution, an equation is therefore formulated which could be used to appropriately split the C6+ data after the calculation of the "quasi-mole fraction" of the individual C6+ pseudocomponents from the application of three well-established hydrocarbon physical properties correlations.

The procedure employed to generate an appropriate percentage distribution of C6+ composition of a gas is summarized below:

- The apparent molecular weight of the C6+ composition, if characterized into C6/C7/C8, is defined mathematically by the following equation;

$$(M_a)_{C6+} = \sum_{i=6}^{8} y_i M_i \tag{1}$$

- If the specific gravity of C6+ (γ_{C6+}) is considered, Eq. (1) can then be expressed as:

$$\gamma_{C6+} = \frac{(M_a)_{C6+}}{M_{air}} = \sum_{i=6}^{8} y_i M_i$$

$$\gamma_{C6+} M_{air} = \sum_{i=6}^{8} y_i M_i \tag{2}$$

- Expanding the equation and inserting the value, 28.96, of the apparent molecular weight of air (M_{air}) gives;

$$28.96\gamma_{C6+} = y_6 M_6 + y_7 M_7 + y_8 M_8 \tag{3}$$

M_i, molecular weight of the ith component in the C6+; y_i, "quasi-mole fraction" of the component i in the C6+.

Though there are many correlations for estimating the molecular weight of petroleum fractions, most of which use specific gravity and boiling point as correlation parameters, three well-established hydrocarbon physical properties correlations (Katz and Firoozabadi 1978; Ahmed 1985; Robinson and Peng 1978) have been chosen for the determination of the molecular weights of C6/C7/C8; in so doing three forms of Eq. (3) are generated. Table 1 presents the approximate molecular weight values of C6/C7/C8 from the chosen correlations.

The three forms of Eq. (3) are thus;

$$28.96\gamma_{C6+} = 84y_6 + 96y_7 + 107y_8 \tag{4}$$

$$28.96\gamma_{C6+} = 85y_6 + 94y_7 + 106y_8 \tag{5}$$

$$28.96\gamma_{C6+} = 86y_6 + 100y_7 + 114y_8 \tag{6}$$

Note: Katz and Firoozabadi (1978) proposed a set of tabulated properties. Ahmed (1985) correlated Katz–Firoozabadi-tabulated physical properties with the number of carbon atoms of the fraction. The generalized equation is of the form:

$$\theta = a_1 + a_2 n + a_3 n^2 + a_4 n^3 + (a_5/n) \tag{7}$$

where $\theta =$ any physical properties, but for this work, it represents the molecular weight; $n =$ number of carbon atoms, i.e. 6,7, ..., 45; $a_1 - a_5 =$ coefficients of the equation given Table 2.

Robinson and Peng (1978) put forward a correlation for the determination of the molecular weight of the paraffinic group of hydrocarbon system. The equation is of the form;

$$\text{For Paraffinic group:} \quad MW = 14.026n + 2.016 \tag{8}$$

MW = molecular weight; $n =$ number of carbon atoms.

Equation (4) through (6) can be written in matrix form as follows;

$$\begin{bmatrix} 84 & 96 & 107 \\ 85 & 94 & 106 \\ 86 & 100 & 114 \end{bmatrix} \begin{bmatrix} y_6 \\ y_7 \\ y_8 \end{bmatrix} = \begin{bmatrix} 28.96\gamma_{C6+} \\ 28.96\gamma_{C6+} \\ 28.96\gamma_{C6+} \end{bmatrix} \tag{9}$$

- To solve the matrix expression requires that the specific gravity of the C6+ must be known.
- The calculated "quasi-mole fraction", that is $\gamma_6, \gamma_7, \gamma_8$, of the individual C6+ pseudocomponents from the application of three well-established hydrocarbon physical properties correlations are then expressed in an exponential molar distribution percentage pattern. The equation is of the form;

Table 1 Approximate values of the molecular weight of C6/C7/C8

Katz and Firoozabadi (1978)	Component	C6	C7	C8
	Molecular weight, *M*	84	96	107
Ahmed (1985)	Component	C6	C7	C8
	Molecular weight, *M*	85	94	106
Robinson and Peng (1978)	Component	C6	C7	C8
	Molecular weight, *M*	86	100	114

Table 2 Coefficients of Ahmed for calculating hydrocarbon molecular weight

θ	a_1	a_2	a_3	a_4	a_5
Molecular weight, M	−131.11375	24.96156	−0.34079022	2.4941184×10^{-3}	468.32575

$$\% \text{ mole distribution,} \quad z_i = \frac{\mathrm{Exp}(y_i)}{\sum_{i=6}^{8}(\mathrm{Exp}(y_i))} \times 100 \quad (10)$$

Equations (9) and (10) are therefore the proposed equations to be used for the splitting of C6+ composition of any given pipeline gas provided that the C6+ specific gravity is known.

Application to field C6+ data

The first example uses a gas sample from Queensland Gas Pipeline reported by Jemena Asset Management (Queensland Gas Pipeline Measurement Manual, GTS-199-PR-GM-001 2011). The gas sample was analyzed online by a C6+ gas chromatography for its bulk molecular compositions. The sample information, gas molecular compositions and sample properties are presented in Table 3. The C6+ has a 0.0304 mol%. If we assume the specific gravity of 0.6245 reported to be that of the C6+ data, then the molar percentage distribution of the C6+ data can be determined using Eqs. (9) and (10). Equation (9) therefore becomes;

$$\begin{bmatrix} 84 & 96 & 107 \\ 85 & 94 & 106 \\ 86 & 100 & 114 \end{bmatrix} \begin{bmatrix} y_6 \\ y_7 \\ y_8 \end{bmatrix} = \begin{bmatrix} 28.96(0.625) \\ 28.96(0.625) \\ 28.96(0.625) \end{bmatrix}$$

$$\begin{bmatrix} 84 & 96 & 107 \\ 85 & 94 & 106 \\ 86 & 100 & 114 \end{bmatrix} \begin{bmatrix} y_6 \\ y_7 \\ y_8 \end{bmatrix} = \begin{bmatrix} 18.085 \\ 18.085 \\ 18.085 \end{bmatrix}$$

Using MATLAB, the above matrix algebra can be solved. Thus, the results of the 'quasi-mole' fraction, i.e.

Table 3 Queensland gas pipeline gas sample

Components	Composition, mol%
N_2	2.4910
CO_2	0.9960
C1	89.9510
C2	4.9800
C3	0.9940
i-C4	0.3080
n-C4	0.3090
i-C5	0.1001
n-C5	0.1002
C6+	0.0304
Specific gravity (air = 1)	0.6245

$\gamma_6, \gamma_7, \gamma_8$ are 0.2133, 0.1919, −0.1706 respectively. To obtain the actual percentage molar distribution of the C6+ composition, the 'quasi-mole' fraction are expressed in exponential pattern using Eq. (10);

$$z_i = \frac{\mathrm{Exp}(y_i)}{\sum_{i=6}^{8}(\mathrm{Exp}(y_i))} \times 100$$

$$\sum_{I=6}^{8}(\mathrm{Exp}(y_i)) = \mathrm{Exp}(y_6) + \mathrm{Exp}(y_7) + \mathrm{Exp}(y_8)$$
$$= \mathrm{Exp}(0.2133) + \mathrm{Exp}(0.1919) + \mathrm{Exp}(-0.1706)$$
$$= 1.2378 + 1.2115 + 0.8432$$
$$= 3.2925$$

Thus, the percentage split of the C6+ components are;

$$z_6 = \frac{1.2378}{3.2925} \times 100 = 37.593\ \% \approx 37.59\ \%$$

$$z_7 = \frac{1.2115}{3.2925} \times 100 = 36.798\ \% \approx 36.80\ \%$$

$$z_8 = \frac{0.8432}{3.2925} \times 100 = 25.609\ \% \approx 25.61\ \%$$

The percentage distribution of the C6+ component of the gas sample from Queensland Gas Pipeline with specific gravity of 0.6245 is approximately; 37.59 % C6, 36.80 % C7, and 25.61 % C8. Thus, Table 4 illustrates the resulting distributions of the concentration of hexanes, heptanes, octanes using common composition splits (of the C6+ percentage mole of 0.0304) similar to using a company, contract or historical characterization assumption, and the new method.

For the C6+ data compositions in Table 4, calculation of the HCDPs as well as cricondentherm, using HYSYS® with Peng Robinson EOS, were performed. (Note that: in the case of C6+ data with no split, the C6+ is treated as a single cut; its normal boiling point of 311.98 °F R was estimated using the correlation by Riazi and Daubert (1987).The predicted HCDPs, at a pipeline pressure of 215 psia, for the five C6+ data are presented in Table 5. A comparison of the HCDP results, at the pipeline operating pressure, shows a slight variance between the new method and the traditional C6+ split ratios (Daniels 47/35/17, GPA 50/25/25 and the GPA 60/30/10). The predicted HCDP ranges from as low as −40 °F for the C6+ data with no split ratio to 2.861 °F for the new split method. The HCDP results for Daniels 47/35/17, GPA 50/25/25 and the new split method are −2.151, 1.038 and 2.861 °F respectively; this shows that the new split method has the maximum HCDP value.

Table 4 Illustration of the resulting distribution of C6/C7/C8 for various common splits

Component	C6+ data with no split (composition, mol%)	C6+ data with the new method split ratio (composition, mol%)	C6+ data with 47/35/17 split ratio (composition, mol%)	C6+ data with GPA 2261 split ratio (50/25/25) (composition, mol%)	C6+ data with GPA 2261 split ratio (60/30/10) (composition, mol%)
N_2	2.4910	2.4910	2.4910	2.4910	2.4910
CO_2	0.9960	0.9960	0.9960	0.9960	0.9960
C1	89.9510	89.9510	89.9510	89.9510	89.9510
C2	4.9800	4.9800	4.9800	4.9800	4.9800
C3	0.9940	0.9940	0.9940	0.9940	0.9940
i-C4	0.3080	0.3080	0.3080	0.3080	0.3080
n-C4	0.3090	0.3090	0.3090	0.3090	0.3090
i-C5	0.1001	0.1001	0.1001	0.1001	0.1001
n-C5	0.1002	0.1002	0.1002	0.1002	0.1002
C6+	0.0304	–	–	–	–
C6		0.01143	0.014432	0.0152	0.01824
C7		0.01130	0.010747	0.0076	0.00912
C8		0.00786	0.00522	0.0076	0.00304
Total	100	100	100	100	100

Table 5 Comparisons of hydrocarbon dew point predictions and cricondentherm along with cricondenbar

	C6+ data with no split	C6+ data with 47/35/17 split ratio	C6+ data with the new method split	C6+ data with GPA 2261 split ratio (50/25/25)	C6+ data with GPA 2261 split ratio (60/30/10)
Hydrocarbon dew point prediction at 215 psia (°F)	−40	−2.151	2.861	1.038	−8.0140
Cricondentherm (°F)	−22.78	4.627	9.204	7.334	−0.5866
Cricondenbar (psia)	1,047	1,104	1,114	1,107	1,091

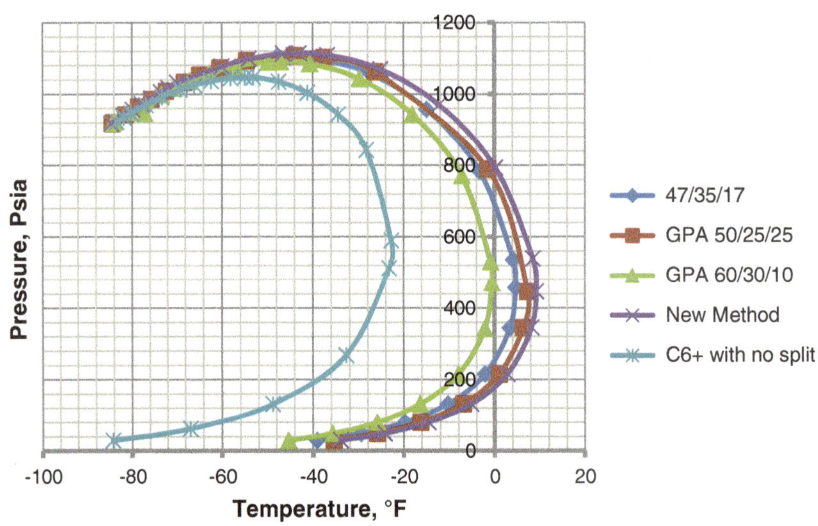

Fig. 2 Different phase envelopes, dew point curve portions, using different C6+ characterisation methods of a pipeline gas with specific gravity of 0.6245

Also, comparison of the predicted cricondentherm using the various C6+ characterisation approaches shows significant variation. Table 5 also presents the comparison of the predicted cricondentherm values. Figure 2 shows the comparison of the phase envelopes (of the dew point curve portions) for the various C6+ data compositions. The

Table 6 Typical union gas sample

Components	Composition, mol%
N_2	2.4910
CO_2	0.9960
O_2	0.02
C1	95.2
C2	2.50
C3	0.20
i-C4	0.03
n-C4	0.03
i-C5	0.01
n-C5	0.01
C6+	0.01
Specific gravity (air = 1)	0.58

phase envelope of the new split ratio is above those of the traditional C6+ split ratios (Daniels 47/35/17, GPA 50/25/25, and GPA 60/30/10). The worst case seems to be the C6+ with no split which is far below all split methods (Table 5).

Performing similar calculations, using Eqs. (9) and (10), for the typical pipeline gas on the Union Gas system given in Table 6 (Chemical Composition of Natural Gas (2005, http://www.uniongas.com) with specific gravity of 0.58 gives the following C6+ percentage distributions; (37.303 % C6, 36.568 % C7, 26.130 % C8. The ratio of the split of the C6+ component of the typical pipeline gas with specific gravity of 0.58 is approximately; 37.3 % C6, 36.6 % C7, and 26.1 % C8. Thus, Table 7 presents the resulting distributions of the concentration of hexanes, heptanes, octanes and nonanes using common composition splits (of the C6+ percentage mole of 0.010) similar to using a company, contract or historical characterization assumption, and the new method.

Figure 3 shows the comparison of the phase envelopes (of the dew point curve portions) for the various split methods of the C6+ data compositions of the Union Gas system. The phase envelope of the new split ratio gives the maximum cricondentherm value. There is a slight difference in the phase envelope predicted using GPA 50/25/25, Daniels 47/35/17 and the new split ratio. However, for this example, the GPA 50/25/25 dew point curve tends to be closer to that of the new split ratio than the dew point curve predicted by the Daniels 47/35/17 split ratio. The phase envelope of the C6+ data with no split appears to be far away from that of the other split methods.

Conclusions

Monitoring of HCDP of a pipeline gas is very important in order to ensure that high quality gas is supplied by gas processor/pipeline operators to the end-users. It is prudent to use an appropriate split of the C6+ composition when applying a HCDP limit using C6+ data. Therefore, a very simple and practical procedure for the estimation of the

Table 7 Illustration of the resulting distribution of C6/C7/C8 for various common splits

Component	C6+ data with no split (composition, mol%)	C6+ data with 47/35/17 split ratio (composition, mol%)	C6+ data with the new method split ratio (37.3/36.6/26.1) (composition, mol%)	C6+ data with GPA 2261 split ratio (50/25/25) (composition, mol%)	C6+ data with GPA 2261 split ratio (60/30/10) (composition, mol%)
N_2	2.4910	1.3	1.3	1.3	1.3
CO_2	0.9960	0.70	0.70	0.70	0.70
O_2	0.02	0.02	0.02	0.02	0.02
C1	95.2	95.2	95.2	95.2	95.2
C2	2.50	2.50	2.50	2.50	2.50
C3	0.20	0.20	0.20	0.20	0.20
i-C4	0.03	0.03	0.03	0.03	0.03
n-C4	0.03	0.03	0.03	0.03	0.03
i-C5	0.01	0.01	0.01	0.01	0.01
n-C5	0.01	0.01	0.01	0.01	0.01
C6+	0.01				
C6		0.004747	0.00373	0.005	0.006
C7		0.003535	0.00366	0.0025	0.003
C8		0.001717	0.00261	0.0025	0.001
Total	100	100	100	100	100

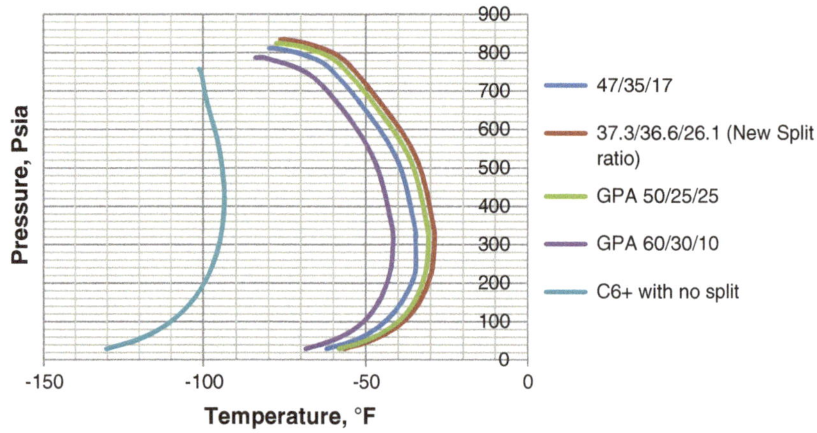

Fig. 3 Different phase envelopes, dew point curve portions, using different C6+ characterisation of the union gas system with specific gravity of 0.58

percentage distribution of C6+ composition for any particular pipeline gas with a known C6+ specific gravity in order to improve HCDP determination has been formulated. The appropriate percentage distribution of C6+ pipeline gas composition may hovers around 37–38 % C6, 36–37 % C7 and 24–26 % C8. Based on the field C6+ data examples, the new split method has the maximum HCDP value when compared to the HCDP predictions of some common C6+ split ratios. Application of this new method to split the C6+ data may therefore represent and also enhance the prediction of HCDP of any particular gas.

References

Ahmed T (1985) Composition modelling of Tyler and mission canyon formation oils with CO_2 and lean gases, final report submitted to the Montana's on a new track for science (MONTS) program. Montana National Science Foundation Grant Program

Ahmed T, Cady G, Story A (1985) A generalised correlation for characterizing the hydrocarbon heavy fractions. In: SPE paper 14266 presented at the SPE 60th annual technical conference, Las Vegas, 22–25 September 1985

ASTM D 1142-95 (1995) Standard test method for water vapor content of gaseous fuels by measurement of dew-point temperature. American society for testing and materials, Philadelphia

Chemical Composition of Natural Gas (2005) http://www.uniongas.com

Ernst K, Pettigrew D (2005) Hydrocarbon dew point monitoring of natural gas using field-mounted on-line gas chromatographs. Pipeline Gas J

GPA 2261 (2000) Analysis for natural gas and similar gaseous mixtures by gas chromatography, 2000. Gas Processors Association, Tulsa

Jack H (2010) Hydrocarbon dew point is a critical consideration for pipeline operations. Pipeline Gas J 237(7)

Katz D et al. (1983) Overview of phase behaviour of oil and gas production. JPT 1205–1214

Katz DL, Firoozabadi A (1978) Predicting phase behaviour of condensate/crude-oil systems using methane interactions coefficients. JPT 1649–1655

Lorenz J, Bray BG, Clark CR (1964) Calculating viscosities of reservoir fluids from their compositions. JPT Trans AIME 231:1171

NGC-GPA (2005) White paper on liquid hydrocarbon dropout in natural gas infrastructure. Natural Gas Council for the Federal Energy Regulatory Commission, NGC+ Liquid Dropout Task Group (Feb)

Pedersen K, Thomassen P, Fredenslud A (1982) Phase equilibria and separation processes. In: Report SEP 8207. Institute for Kemiteknit, Denmark Tekniske Hojskole, Denmark

Queensland Gas Pipeline Measurement Manual (2011) Document number: GTS-199-PR-GM-001

Riazi MR, Daubert TE (1987) Characterization parameters for petroleum fractions. Ind Eng Chem Res 26(24):755–759

Robinson DB, Peng DY (1978) The characterization of the heptanes and heavier fractions. In: Research report 28, GPA, Tulsa

Shane H (2000) Determination of hydrocarbon dew point using a gas chromatograph, class # 5300. Emerson Process Management, Gas Chromatograph Division, Houston, USA

3

Experimental investigation of residual saturation in mixed-wet porous media using a pore-scale approach

Ruthut Larpudomlert · Victor A. Torrealba · Zuleima T. Karpyn · Phillip Halleck

Abstract Wettability is an important factor in terms of flow distribution and the amount of oil left behind in a petroleum reservoir after primary, secondary, and tertiary recovery processes. In fact, most alkaline flooding operations are aimed at wettability reversal. Therefore, a good understanding of how wettability, and other relevant factors (e.g., fluid saturation and porosity), affects displacement efficiency and fluid distribution at the pore-scale can lead to successful predictions of flow properties at the macroscale. In this study, pore-scale two-phase fluid flow in a synthetic mixed-wet granular media was investigated using X-ray microCT. Analysis of residual fluid structures in granular media composed of a mixture of glass and plastic beads (ranging from 0.4 to 0.6 mm in diameter) showed that the number of trapped water blobs was 2.4 times greater than that of oil, whereas most of the blobs were between 0.0001 and 0.001 mm^3 in size (i.e., 100 times smaller than in the case of uniform wettability), and were smaller than the mean pore volume (0.03 mm^3). The ratio of surface area to blob volume showed a slight tendency of water blobs to wet a higher surface area of the grains than oil blobs, for blob volumes larger than the mean pore size; this phenomenon can be attributed to stronger wetting affinity of the glass grains to the water phase than that of the plastic grains to the oil phase. Furthermore, statistical analysis of the distance between residual oil and water blobs to each solid surface confirms preferential wetting affinity of oil and water to plastic and glass surfaces, respectively.

Keywords Wettability · Fluid trapping · Micro computed tomography · Multiphase · Pore-scale

R. Larpudomlert · V. A. Torrealba ·
Z. T. Karpyn · P. Halleck
John and Willie Leone Family Department of Energy and Mineral Engineering, EMS Energy Institute, The Pennsylvania State University, University Park, PA 16802-5000, USA

Introduction

The study of fluid transport in subsurface is important in earth sciences applications, like geochemistry, ground water remediation, and petroleum engineering. In the area of hydrocarbon recovery, the industry has turned to enhanced oil recovery techniques in which complex chemical techniques are used to alter the original fluid and/or rock properties to increase the production of the remaining oil and gas in hydrocarbon reservoirs. However, these processes depend on pore-level heterogeneities and the ability of the injectant to contact the rock and in situ fluids. Thus, it is essential to understand factors such as interfacial tension, capillary forces, and wettability to correctly predict fluid mobility and recovery efficiency Speight (2009).

Over the last several decades, various research studies have been conducted with the purpose of quantifying and validating flow properties and transport mechanisms. However, given the difficulty of obtaining direct pore-scale measurements, pore-level models have been used to predict both flow properties and transport mechanisms. These include: network models, designed as simplified representations of a pore system consisting of pore bodies and pore throats; Lattice-Boltzmann models (LB), used to calculate single-phase and two-phase flow patterns van Dijke and Piri (2007); and mesh-free Lagrangian particle methods, such as smooth particle hydrodynamics numerical models

(SPH), developed to simulate multicomponent immiscible and miscible flow in porous media Tartakovsky and Meakin (2006).

Network modeling has been successfully used in various studies. Suicmez et al. (2008) studied the effects of wettability for water-wet, weakly water-wet, oil-wet, and weakly oil-wet media and pore-level displacement on hydrocarbon trapping, using three-phase-flow pore-scale network modeling. They found that as the medium becomes more oil-wet, the amount of oil trapped by water in the presence of gas increases. Zhao et al. (2010) also used network modeling to study the effect of wettability on oil recovery after water flooding, focusing on the effects of initial water saturation, contact angle distribution and oil-wet fraction on oil recovery. Similarly, Jackson et al. (2003) studied the effect of wettability on water flooding at the reservoir scale using a network model in conjunction with reservoir scale conventional simulations. The three-dimensional network model was derived from Oren et al. (1997) to represent a sample of Berea sandstone in a study suggesting that the assumption of uniform wettability leads to an underestimation of oil recovery. Oren et al. (1997) constructed a 3-D sandstone model as an input for a two-phase model to simulate primary drainage and water injection for water-wet and mixed-wet systems with the purpose of predicting relative permeability and electrical conductivity. Yet, there is still concern with regard to extending a simplified network model technique to the field scale due to the difficulty of adequately describing the complex heterogeneous nature of porous systems.

Recently, X-ray microCT (MCT) has made it possible to obtain high-resolution three-dimensional maps of porous structures, and thus provides better insight into the pore-scale processes taking place during fluid flow experiments from direct three-dimensional imaging and resource data, which allows the development of better network models. MCT is a non-invasive technique that reveals the internal structure of an object by capturing variations in density and atomic composition. Thus, MCT images give a representation of the pore geometry of a system Chaouki et al. (1997) and Karpyn et al. (2010). In dry granular samples, the strong contrast between the solid phase and the pore space provides information for studying the porosity distribution of the system Ketcham and Carlson (2001).

Additionally, MCT images have been used to construct accurate three-dimensional networks for predicting the transport behavior and characteristics of porous media. Jiang et al. (2007) used network-extraction techniques based on MCT images to extract a geometrical and topological network representing the pore structures of three sandstone core samples. The set of algorithms developed by Jiang was designed to improve the efficiency of common thinning algorithms; the method comprised calculating a three-dimensional Euclidean distance map, clustering voxels, and extracting the network of the pore space. They also described partitioning the pore space into nodes and bonds and determining geometrical features, such as radii and shape factors. Prodanović et al. (2007) studied fluid displacement in a Berea core sample, and used MCT to obtain three-dimensional images of the pore structure. The high-resolution images provided detailed pore geometry characteristics (e.g., pore volume, and throat area) as input to Lattice-Boltzmann models, which were used to relate pore-scale parameters to bulk flow properties (e.g., absolute permeability).

Properties such as residual trapping, interfacial area, and capillary forces have also been defined based on high-resolution images from synchrotron X-ray CT. Culligan et al. (2006) suggested that quantifying the nonaqueous phase liquid–water interfacial area provides a measure of the expected area available for mass transfer dissolution in porous media. Al-Raoush and Willson (2005) used synchrotron X-ray tomographic images of a multiphase porous media system to capture the entrapment of the nonwetting phase at residual saturation. This study provided a framework for using high-resolution three-dimensional imaging to extract the pore network and quantify fluid distributions. The results showed the non-wetting phase trapped primarily in the largest pore spaces and in the pore bodies with the highest aspect ratios. In addition, the use of high-resolution X-ray CT provided datasets of a sufficiently high quality to observe the representative elementary volume. However, the size of these images was not sufficient to provide estimates of some of the bulk properties, such as residual saturations and interfacial areas.

Previous investigations of trapped oil clusters in water-wet glass bead packs using X-ray microCT by Karpyn et al. (2010) were performed at larger dimensions (approximately 25.4 mm in diameter and 90.53 mm long). The granular system used in this experiment consisted of spherical glass beads ranging from 0.40 to 0.60 mm in diameter. A spatial distribution of the trapped oil clusters for the entire bead pack was presented to explain fluid distribution. Likewise, Landry et al. (2011) presented quantitative evidence of immiscible fluid distribution in an oil-wet acrylic bead pack, thereby offering a direct comparison with the results presented by Karpyn et al. (2010). Landry's experimental results showed how wettability and saturation history affect fluid distribution. In addition, it was found that the specific surface area of the wetting phase of the acrylic bead pack was less than for the glass bead pack, whereas the specific area of the nonwetting phase of the acrylic bead pack was greater than that of the glass bead pack. Thus, suggesting that the affinity of acrylic beads- to oil was weaker than that of glass beads to water.

With the notable exceptions of Lebedeva and Fogden (2011), Delshad et al. (2006) and Anderson (1987), the vast majority of the work found in the literature focuses on uniformly wetted systems (e.g., water-wet and oil-wet systems). However, it is generally accepted that most geologic porous media are not uniformly wetted. In addition, crude oil can alter the rock state by adsorbing or depositing polar components, thus resulting in a mixed-wet state of relatively oil-wet subareas where oil makes contact combined with unaltered subareas covered by brine. The term "mixed-wettability" was introduced by Salathiel (1973) to describe systems with large pores that are predominantly oil-wet and smaller ones that are water-wet, and in which the residual oil saturation is much lower than that obtained in uniformly wetted systems. Donaldson et al. (1969) reported that changes in wettability affect water flood behavior. They treated the core samples with various amounts of organochlorosilane to change the wettability of the cores from water-wet to oil-wet. The results showed that less oil was recovered after a water injection as the system became more oil-wet. Moreover, according to Morrow Norman and Geoffrey (2001) an infinite number of possible wetting states exist between the strongly water-wet and the strongly oil-wet, and similarly there are many varied flow behaviors Tiab and Donaldson (2004). Understanding the effect of wettability on fluid recovery requires information pertaining to pore-scale fluid distribution, wetting preferences, and pore geometry Kumar (2008). Multiple pore-level models have been proposed to explain the effect of mixed wettability. Heiba et al. (1983) used network models to study the distribution of oil under mixed-wettability conditions. Raeesi and Piri (2009) used a pore-scale network model to study the effect of wettability and trapping. Radke et al. (1992) developed a pore-level illustration to explain how mixed-wettability forms and how it affects the reservoir rock.

The purpose of this study is to analyze a mixed wettability granular pack, and to provide a comparative framework for previous uniform wettability experiments reported in the literature, focusing on the interplay of porosity distribution, fluid trapping behavior, and the effect of wettability. The study offers comprehensive information about the pore structure and immiscible fluid spatial distribution. Such information can be used to generate a basis for pore-level modeling and macroscopic-level predictions. A bead pack with a mixture of glass and acrylic beads was used to represent mixed wettability conditions. The dimensions of the system are representative to those of previous experiments reported in the literature. The results were obtained using the following approach: MCT was used to map the residual fluids (oil, water) and the solid beads (glass, plastic); the resulting image sets were segmented to identify the plastic and glass beads, the oil phase, and water phase. From these data, the porosity, fluid blob size, and saturation distributions were obtained. Finally, the obtained results for the mixed wettability system were compared to uniform wettability data.

Methodology

Summary of data set protocol

For the mixed wettability system, glass and plastic beads, approximately 0.40–0.60 mm in diameter, were packed into a cylindrical plastic core holder of 103.32 mm in length with an inner diameter of 25.4 mm. The proportion of glass and plastic beads in the mix was 66.61 and 33.39 %, respectively.

Flow distributors were added to the upper and lower ends of the bead pack to achieve a uniform flow profile. In addition, a metal screen was placed between the flow distributors and the bead pack. The core holder was placed in the X-ray microCT scanner in the vertical position.

The dry bead pack was first evacuated and pre- saturated with brine. Kerosene was then injected into the pack to achieve irreducible water saturation, and finally brine was injected into the pack until residual oil saturation was achieved. The lighter phase, kerosene was injected from the top of the core holder to minimize buoyancy effects, whereas the heavier brine phase was injected from the bottom to minimize gravitational segregation due to the density difference of the fluids. At the end of each stage of fluid injection, the core sample was scanned using MCT. The experimental sequence is shown in Fig. 1.

In this project, MCT was used to determine the fluid structures in the porous medium. The MCT system consists of an X-ray source, a detector, a translation system, and a computer system that controls motion and data acquisition Karpyn et al. (2010). MCT imaging is a visualization technique used for examining internal structure of opaque three-dimensional objects by producing stacks of two-dimensional images (slices) that reveal the interior of the objects in a nondestructive manner Denison et al. (1997). The scans of this core sample were composed of 3,900 slices, each of which consisted of 1,024 × 1,024 pixels. The MCT system was tuned to a voxel resolution of 0.026 × 0.026 × 0.027 mm, in which 0.027 mm represented the height of the voxel or the thickness of each slice.

Data sets

Four sets of images were obtained using MCT at different stages: dry sample, 100 % brine saturated sample, irreducible brine saturation (S_{wirr}), and residual oil saturation (S_{or}).

Fig. 1 Schematic diagram of the experimental procedure showing the sequence of the kerosene (oil) and brine (water) injections

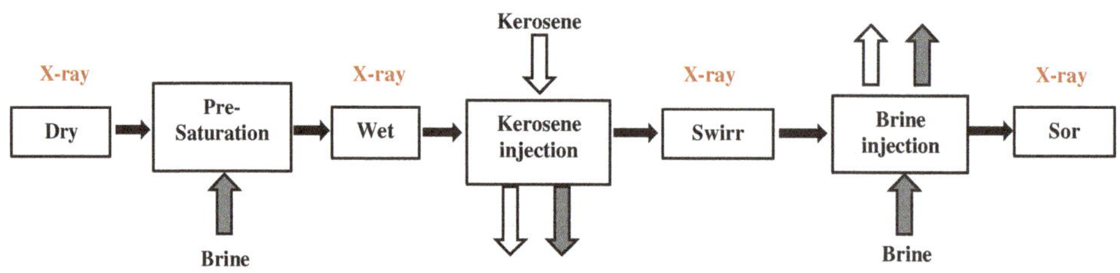

Dry beads

The vacuumed bead pack was scanned dry to obtain an accurate map of the beads and pore structure. Differences in CT numbers for beads and pore space is large due to the difference in density and atomic number. High-density materials characteristically have a high CT number (represented by the light gray regions), whereas low-density materials characteristically have a low CT number (represented by the dark gray regions). Therefore, at this stage, it was easy to segregate the beads from the pore structure.

Wet beads

After the dry sample was scanned, the core sample was presaturated with brine and scanned again. The resulting images were used in the segmentation process to obtain the fluid distribution in the pore space. Compared with the plastic bead, the glass beads and the brine are denser and are shown in light gray, whereas the plastic beads are shown in dark gray in the images.

Irreducible water saturation (S_{wirr})

At this stage, kerosene was injected into the core sample to displace the brine, until it reached irreducible water saturation. Then, the core sample was scanned. For this stage, most of the pore space was filled with kerosene with only a small amount of brine. As the plastic beads and the kerosene are less dense than the glass beads and brine, the regions occupied by kerosene and plastic beads are darker than the regions occupied by glass beads and brine.

Residual oil saturation (S_{or})

The last stage was to inject brine into the core sample to displace the fluid in the previous stage, until it reached residual oil saturation. Then, the core sample was scanned. In contrast to the previous stage, most of the pore space was filled with brine with only a small amount of kerosene. Again, as the plastic beads and the kerosene are less dense than the glass beads and brine, the regions occupied by glass beads and brine are brighter than the regions occupied by kerosene and plastic beads. Thus, the major difference between S_{wirr} and S_{or} is that the pore space in S_{wirr} appears to be dark (dark gray), whereas the pore space in S_{or} appears to be bright (light gray).

Segmentation and quantification

In image processing, the most important step is to separate two distinct materials from each other, i.e., foreground (object) and background, using a proper thresholding method. The simplest segmentation techniques employ global thresholding in which one threshold value is applied to the whole system. Values below the threshold value are designated as one material, and values above are designated to the other material Al-Raoush and Willson (2005).

Various techniques have been used to obtain the most accurate threshold value, such as k-means cluster analysis used by Porter and Wildenschild (2010); local thresholding criteria based on indicator kriging mentioned in Prodanović et al. (2007); and the Otsu method used by Otsu (1979).

Additionally, Mehmet (2004) reported a survey of over 40 image-thresholding techniques to categorize and compare the performance of each. The thresholding techniques were categorized into six major groups according to the information explored such as histogram-shape-based methods, clustering-based methods, entropy-based methods, object-attribute-based methods, spatial methods, and local methods. The performance criteria were based on the misclassification error, edge mismatch, relative foreground area error, modified Hausdorff distance, and region non-uniformity. The inclusive details are described in the survey of image-thresholding techniques and quantitative performance evaluation (2004). The final evaluation performance scores of each method were calculated and ranked from highest to lowest. For the nondestructive testing image datasets, clustering-based methods are among the seven highest-ranking techniques.

One of the most effective clustering-based methods is the Otsu thresholding, which works by iterating between all the possible threshold values to find a particular value where the sum of the foreground and the background is at its minimum. Assume that the number of pixels at the gray level i from all the L levels is denoted by n_i and the number of pixels (N) is the summation of n_i, which is

$N = n_1 + n_2 + n_3 \ldots n_L$. The gray level probability distribution (p_i) is

$$p_i = \frac{n_i}{N} \quad (1)$$

At threshold level k, the pixels are divided into two classes: C_0 [background, (1, 2,..., k)] and C_1 [foreground, ($k+1$, $k+2$,..., L)]. Thus, the probability, mean, and class variances for each class can be described as follows:

Class C_0:

Background probability

$$\omega_0 = \sum_{i=1}^{k} p_i \quad (2)$$

Background mean value

$$\mu_0 = \sum_{i=1}^{k} \frac{ip_i}{\omega_0} \quad (3)$$

Background class variance

$$\sigma_0^2 = \sum_{i=1}^{k} \frac{(i-\mu_0)^2 p_i}{\omega_0} \quad (4)$$

Class C_1:

Foreground probability

$$\omega_1 = \sum_{i=k+1}^{L} p_i \quad (5)$$

Foreground mean value

$$\mu_1 = \sum_{i=k+1}^{L} \frac{ip_i}{\omega_1} \quad (6)$$

Foreground class variance

$$\sigma_1^2 = \sum_{i=k+1}^{L} \frac{(i-\mu_1)^2 p_i}{\omega_1} \quad (7)$$

where the total mean is

$$\mu_T = \sum_{i=1}^{L} ip_i \quad (8)$$

To evaluate the correctness of the threshold value, three parameters are calculated,

$$\lambda = \frac{\sigma_B^2}{\sigma_W^2},\ \kappa = \frac{\sigma_T^2}{\sigma_W^2},\ \eta = \frac{\sigma_B^2}{\sigma_T^2},$$

for within-class variance:

$$\sigma_W^2 = \omega_0\sigma_0^2 + \omega_1\sigma_1^2 \quad (9)$$

between-class variance:

$$\sigma_B^2 = \omega_0(\mu_0 - \mu_T)^2 + \omega_1(\mu_1 - \mu_T)^2 \quad (10)$$

and, total variance:

$$\sigma_T^2 = \sum_{i=1}^{L} (i-\mu_T)^2 p_i \quad (11)$$

To obtain an accurate threshold value, these three parameters must be optimized. In other words, the threshold that provides the minimum within-class variance (σ_w^2) and the maximum between-class variance (σ_B^2) will be the best threshold value.

In this study, the Otsu method was implemented via the *graythresh* algorithm in MATLAB. Each pixel at the different gray level i was counted and stored to construct a histogram, and the number i was used as the number of bin. The advantages of this method are its implementation simplicity; its stability is based on the integration of the histogram, and it covers a wide scope of unsupervised decision procedures.

This method can be applied to separate only two different materials at a time. To segment the three-phase or four-phase systems using Otsu thresholding, a two-step approach is applied. First, the pore space is segmented from the solid (glass and plastic) beads by subtracting the dry sample from the wet sample to remove the solid beads. Thus, the two-phase system is obtained. Then, the *graythresh* MATLAB algorithm was applied again to the extracted sample set, such that all the voxels with a CT number lower than the thresholding value (k) represented the subtracted solid and all the voxels with a CT number higher than the thresholding value (k) represented the pore space (higher CT number). Once the pore space and the subtracted solid space have been segmented, their discretized 3D grids are used to remove the bead space or pore space from the subsequent core sample scans. The next step was to segment the kerosene from the brine and to segment the glass beads from the plastic beads. The 3D grid of the pore space binary 1 and the 3D grid of the solid space binary 2 images were multiplied to the S_{wirr} or S_{or}, and dry sample sets to isolate the fluid in the pore space area, and the mixed beads in the solid area, respectively. Similarly, the *graythresh* algorithm was applied to the multiplied samples to obtain their thresholding values. Once the residual oil and water were segmented from the S_{or} and S_{wirr} samples, respectively, their volume and the surface area for each of the disconnected blobs were analyzed using the three-dimensional visualization software Avizo 3D version 6.3. The stack of images was processed to obtain volume and surface area individually for each separate body. The volume of each segmented body, X, was approximated by the number of voxels forming it:

$$V(X) = \sum_{i,j,k} I(x_i, y_j, z_k), \quad (12)$$

where $I(x_i, y_j, z_k)$ is the intensity of the pixels of the coordinates x_i, y_j, and z_k. The volume of the pixels was scaled to give the true area in the unit associated with the image. Note that, when applying this to binary images, I is equal to 1, or 0. Moreover, the application area 3D was approximated using the intercept, defined as the number of entries in an object along a given direction to account for the exposed surface of the outer pixels.

In addition to the quantification of the fluid saturation, the residual fluid volume, and the surface area, the explicit imaging of the plastic and glass beads was used to evaluate the proximity of the residual fluid structure to the define solid surface via MATLAB. This was achieved by computing the shortest distance from each fluid pixel (blob) to a particular solid type (plastic or glass) to describe the tendency of the fluid trapping with respect to wettability. To determine the shortest distance, the boundary pixels were first defined for both the blobs and the beads by considering the closest 26 neighboring pixels to each. Any pixels surrounded by 26 neighboring pixels were not counted as boundary pixels. The distance between the fluid blob pixels and any solid bead pixel was calculated using the Euclidian distance (D_E):

$$D_E = \sqrt{[((X_s - X_f) \times x)^2 + ((Y_s - Y_f) \times y)^2 + ((Z_s - Z_f) \times z)^2]} \quad (13)$$

where X_s, Y_s, and Z_s represent the solid pixel coordinates; X_f, Y_f, and Z_f represent the fluid pixel coordinates; and x, y, and z represent the voxel size of x, y, and z directions, respectively. Once a fluid voxel was found, then the shortest distance to a solid surface is computed. In this analysis, 800 slices were considered in the mid-region, which was approximately one-third of the core sample.

Results and discussion

Porosity distribution

The average porosity of the bead pack was found to be 42.04 %, with local axial porosity values at different heights along the z-axis varying from 39.99 to 44.55 %, as shown in Fig. 2. The average porosity is in agreement with the theoretical porosity for rhombohedral and cubic packing of regular-sphere grains, which is 25.96 and 47.64 %, respectively. This suggests that the packing is mostly cubic. Moreover, a small variation in percent porosity is good evidence that the bead packing is relatively homogeneous, with the greatest variation at 10 mm in height.

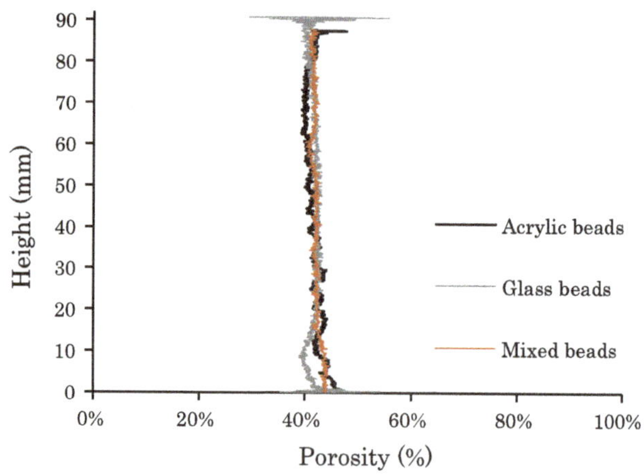

Fig. 2 Vertical profile of bead pack porosity

The average porosity for the mixed-bead pack is comparable to that of the glass bead and plastic bead packs where their average porosity was 41.6 and 42.1 %, respectively. All of the three bead packs show a drop in porosity at the bottom part, at 20 mm in height that may have resulted from localized compaction.

Fluid phase distribution

In addition to pore space and solid region segmentation, each fluid phase within the pore space at S_{wirr} and S_{or} was defined. The segmentation procedure was similar to that used for the previous two sample sets.

For S_{wirr}, kerosene was injected to displace the brine from the pore space, resulting in a small proportion of brine being trapped inside the pore space (irreducible water), whereas the rest of the pore space was filled with kerosene. For S_{or}, brine was injected to displace all the mobile kerosene from the previous stage, resulting in a small proportion of kerosene being trapped inside the pore space (residual oil). The rest of the pore space was filled with brine.

Trapped fluid blobs segmented from CT data ranged in size from 1.9×10^{-5} mm^3 (approximately one single voxel) to 1.18×10^3 mm^3 for water blobs, and from 1.9×10^{-5} to 1.95×10^3 mm^3 for oil blobs. The total volume of the water blobs and the oil blobs was 2.7×10^3 and 3.1×10^3 mm^3, respectively. Moreover, the single largest oil blob found at residual oil saturation was larger than any trapped water blob, suggesting that water remained trapped at the crevices of grain-to-grain contacts and therefore a more dominant wetting phase in this mixed-wet system.

The blob cumulative frequency in Fig. 3 shows that most of the blobs fell in a size between 0.0001 and 0.001 mm^3, making them 100 times smaller in size than those in the single-wettability glass and plastic bead packs

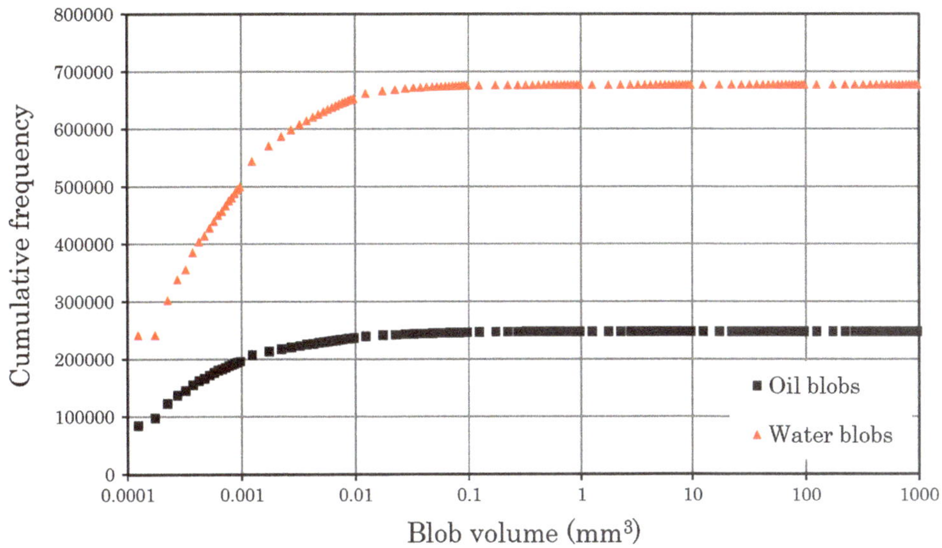

Fig. 3 Cumulative water blob frequency after the oil injection and oil blob frequency after the water injection

reported by Karpyn et al. (2010) and Landry et al. (2011). This difference in sizes of the blobs is expected since the wetting fluid tends to be imbibed by the small pores in the uniformly wet system, causing the nonwetting fluid to be trapped in the larger pores Raeesi and Piri (2009), Kumar et al. (2008), Zhao et al. (2010) and Al-Futaisi and Patzek (2003).

Moreover, Fig. 3 demonstrates a larger quantity of small water blobs as compared to oil blobs. This observation can be explained by the higher proportion of glass beads (water-wet surfaces) compared to acrylic beads (oil-wet surfaces), given that water will remain trapped in the pendular rings around the water-wet grain contacts. Interesting analogies from natural systems can be found in Tiab and Donaldson (2004), Sahimi (1995) and Kumar et al. (2008), whereby water blobs formed from the pre-saturation of brine, as the small water-wet pores are spontaneously imbibed by water due to capillary suction.

Figure 4 shows the cumulative volume percent of each oil blob group to the total residual oil trapped in the pore space and each water blob group to the total residual water trapped in the pore space. Disregarding the largest blob volume, the steepest slope of the oil blobs plot and the water blobs plot was found in the range of 0.01–1 mm^3 and 0.001–0.1 mm^3, respectively, suggesting that the blobs in these groups account for an important portion of the total residual fluid trapped in the pore space. This examination is consistence with the findings of Salathiel (1973), according to which smaller pores are filled with water, whereas the oil preferentially wets larger interconnected pores Anderson (1987).

The spatial distribution along the *z*-direction of all the disconnected water blobs and oil blobs is presented in

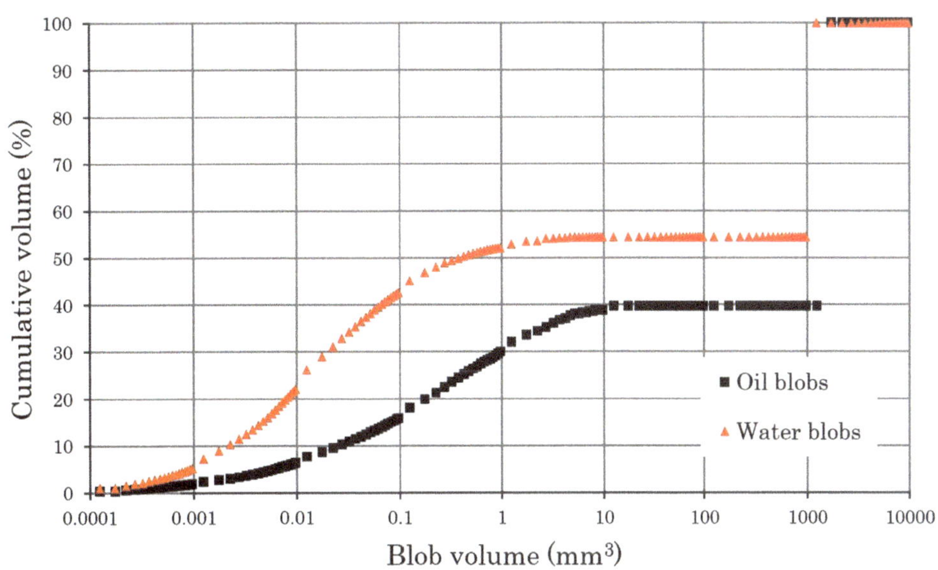

Fig. 4 Cumulative volume percent of water blob group to the total residual water and oil blob group to the total residual oil

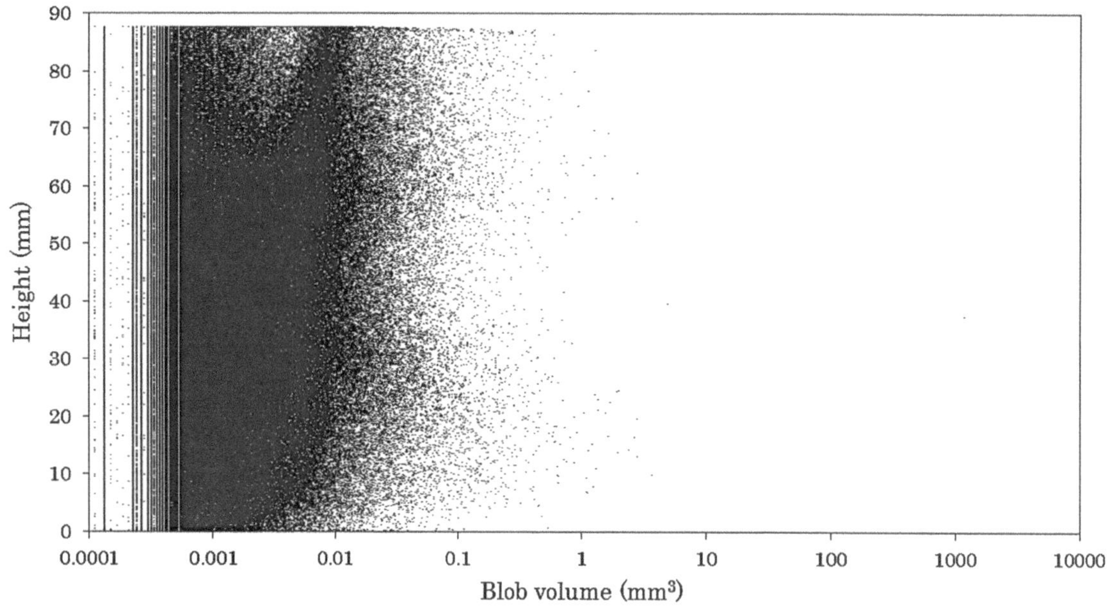

Fig. 5 The spatial distribution along the *z*-direction of the disconnected water blobs after oil injection

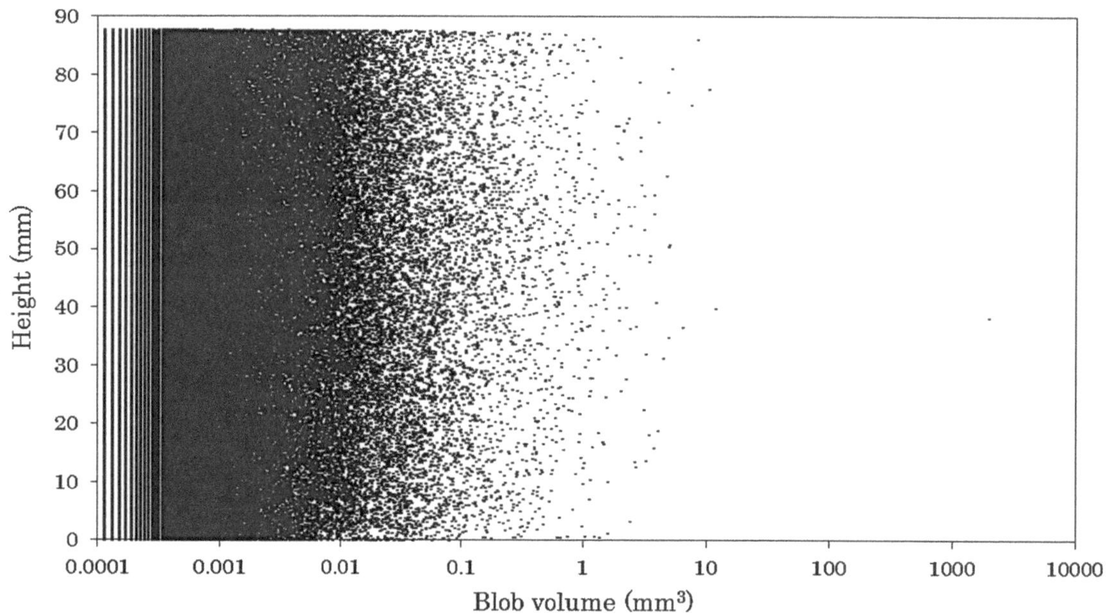

Fig. 6 The spatial distribution along the *z*-direction of the disconnected oil blobs after water injection

Figs. 5 and 6, respectively, wherein each spot represents the vertical location of the centroid of each blob. In the water blob distribution plot, there are smaller amount of medium size blobs on the upper parts and smaller amount of large size blobs on the lower parts of the bead pack. With regard to oil blob distribution, the plot shows that the oil blobs were distributed homogeneously throughout the bead packs. This suggests that oil blobs size is more uniform than that of water blobs.

Individual water blob volumes and oil blob volumes were plotted against their surface area in which each spot in the plot represents one disconnected blob, as shown in Fig. 7. The surface to volume ratio of a sphere was also plotted using a solid line as a reference. The plots show that the blobs were located close to the reference line when their volume is small, and then start to deviate from the reference line as their volume increases. The deviation started at a volume of 0.02–0.03 mm^3, which is

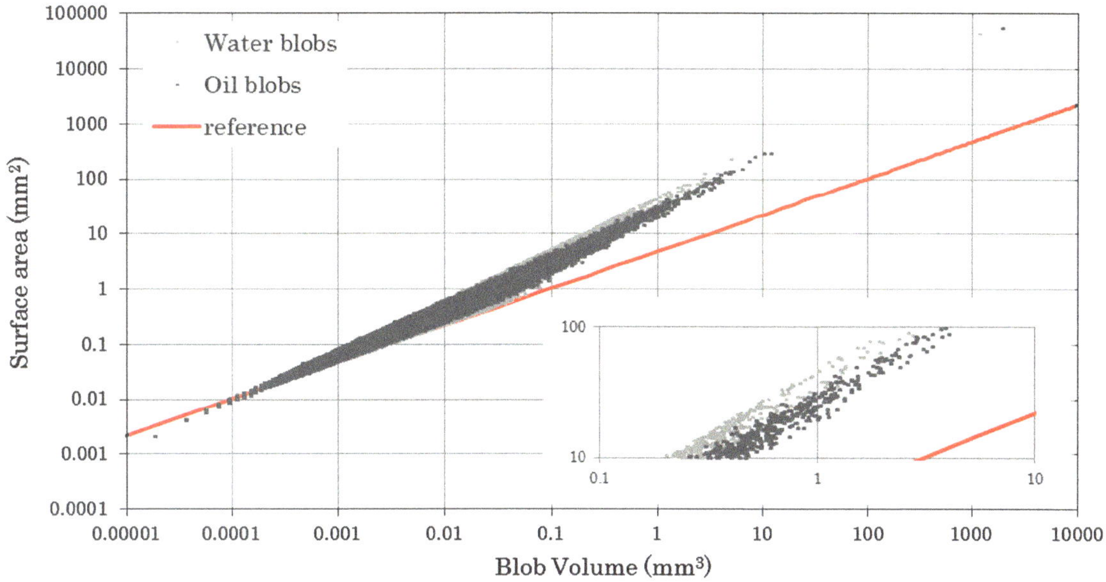

Fig. 7 Water blob surface area as function of blob volume after oil injection and oil blob surface area as a function of blob volume after water injection

Table 1 Average percent water saturation

Pore system	End of oil injection (%)	End of water injection
Mixed-wet	14.08	83.61 % (16.39 % oil)
Oil-wet	8.32	76.14 % (23.86 % oil)
Water-wet	–	82.28 % (17.72 % oil)

comparable to the mean pore volume. At volumes larger than 0.03 mm^3, the blobs no longer behave like spheres and start to extend from a pore through other interconnected pores. The porosity was found to be similar to that reported by Karpyn et al. (2010), given the grain size distribution and packing was almost the same in both samples. The largest blob volume, that of 10^3 mm^3, has the largest surface to volume ratio and was expected to have the highest extension to other interconnected pores. For the small blob volume, as these blobs were close to the reference line, their shape was likely to be almost spherical. The trend of surface area to blob volume for both oil and water was very similar, with a slight tendency of water blobs to reach a higher surface area for the same blob volume which can be attributed to stronger water wetting affinity to the solid surface than that of oil.

Percent water saturation

The average percent water saturation after the oil injection and the water injection was shown in Table 1. As noted earlier, the bead pack was saturated with water (brine) followed by the oil (kerosene) injection from the top until the amount of water collected at the outlet of the core was negligible, which meant that the water trapped inside the bead pack was at irreducible water saturation. The amount of trapped water was measured per slice and was averaged across the vertical direction. The average water saturation was found to be 14.08 % and the rest of the pore space was filled with oil (i.e., at 85.92 %).

In comparative analysis with findings reported by Landry et al. (2011) for acrylic bead packs, average oil saturation after the first imbibition (oil injection process) was reported to be 91.68 % such that only 8.32 % of the water was trapped. This behavior was expected, as acrylic beads are oil-wet; therefore, an interface force between the oil and the acrylic beads was stronger than that between the oil and glass. And given that oil is more likely to attach to acrylic beads, the result was higher percent oil saturation. This observation also corresponds with findings reported by Jadhunandan and Morrow (1995) on the relationship between wettability and irreducible water saturation in which water-wet rock has a water saturation of greater than 20–25 % whereas oil-wet rock commonly has a water saturation of less than 15 % of pore volume and frequently less than 10 %.

Moreover, McCaffery and Bennion (1974) found that irreducible water saturation tends to decrease with increasing oil wettability. The observation is based on the fact that in water-wet rocks, water covers the pore surfaces and thus exists as a continuous film; however, in oil-wet rocks in which the solid grains are mostly covered by oil, residual water is likely to behave like disconnected blobs in the center of the pore space Anderson (1987) and Dandekar (2006).

In the next stage, brine was injected into the core sample until the amount of oil collected at the outlet of the core was negligible, indicating that the oil phase was at the residual oil saturation. The average residual oil trapped was found to be 16.39 % and the rest of the pore space was filled with brine (i.e., at 83.61 %). Compared to related findings reported by Landry et al. (2011) for acrylic bead packs and by Karpyn et al. (2010) for glass bead packs, the average residual oil saturation after water injection was reported to be 23.86 and 17.72 %, respectively. In the case of a uniformly wet system (e.g., the water-wet system), the wetting fluid brine tends to form a continuous thin film over the solid surface, and to fill in small or medium pores while displacing the oil into the larger pores.

This results in the oil occupying the center of the larger pores such that continuous channels of oil are formed or oil becomes trapped as disconnected blobs. At slow flow rates, the water moving in a fairly uniform front, displaces oil by a piston-like mechanism (as the capillary pressure is greater than the threshold pressure), and by spontaneous imbibition (where capillary forces make the water phase spontaneously displace the oil from small or medium pores) Patzek (2001), Kumar et al. (2008), Wilkinson and Willemsen (1983) and Wardlaw (1982). As the water bypasses the oil, it swells (due to the decrease in capillary pressure) and breaks down the oil, forcing it into the large pores. This process is referred to as "snap-off" Lenormand et al. (1983), Oren et al. (1997) and Patzek (2001). For oil-wet systems, most of the oil in the large pores is recovered, whereas the remainder is trapped in the smaller pores due to capillary forces. And, as these pores are oil-wet, there is no driving force to cause the water to displace the oil from them.

Thus, the mixed-wet system significantly trapped smaller amount of residual oil compared to oil-wet system, which is consistent with Amott's finding that extreme wettability results in high residual oil, whereas weakly water-wet conditions or slightly oil-wet conditions result in low residual oil. It may be that the mixed wet system trapped less residual oil as a result of the oil-wet solid surface that allows the oil blobs that are close to each other to connect and form a continuous flow path. As such, an oil-wet solid surface prevents water from forming a continuous thin film, and prevents snap-off. Likewise, the oil trapped in the smaller pores is reduced as compared to an oil-wet system, because the small pores in a mixed-wettability system are water-filled Anderson (1987). According to Anderson (1987), as compared to the water-wet system, the mixed-wet system traps less oil because it traps less oil in the larger oil-wet pores. However, in this experiment, there was no significant difference between the percent oil trapping in mixed-wet (16.39 %) and water-wet systems (17.72 %). The residual oil was as small as that of the water-wet system or perhaps even smaller, suggesting that global phase trapping in porous media (residual saturation) is sensitive to the wetting characteristics of the solid matrix. Moreover, the saturation history of oil might affect the amount of residual oil trapped after the water injection, as the amount of oil saturation of the acrylic bead packs (91.68 %) was higher than that of mixed-bead packs (85.92 %), resulting in higher residual oil saturation.

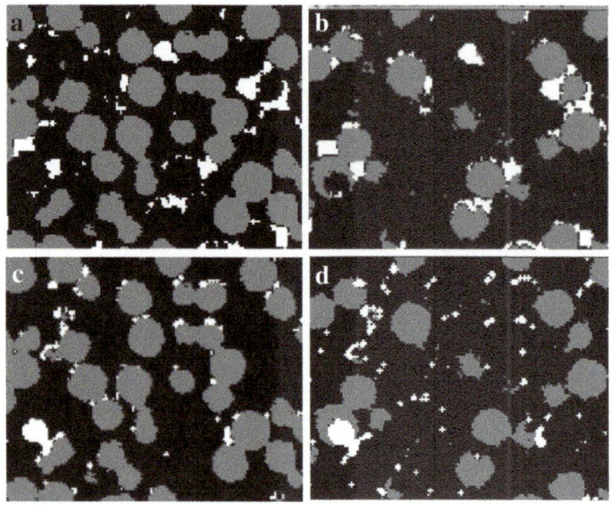

Fig. 8 Residual fluid structures (in *white*) and selected solid material (in *gray*) **a** oil and glass beads; **b** oil and plastic beads; **c** water and glass beads; **d** water and plastic beads

However, the change in residual oil saturation is small, whereas wettability ranges from one extreme to the other. Masalmeh (2003) also reported that in mixed-wet conditions, the small pore and grain contacts are preferentially water-wet but that the surfaces of the larger pores are oil-wet. As oil-wet paths were continuous through the larger pores, water could displace oil from the pores and very little oil would be held by capillary forces in small pores or at grain contacts. This resulted in low residual oil and high oil recovery, as reported by Salathiel (1973), and Anderson (1987).

Wetting affinity

To evaluate the wetting affinity of the glass and plastic beads, we examined the proximity of residual fluid structures to each solid surface (plastic and glass). Figure 8 displays sections of the mixed-bead pack relative to residual fluid structures. Figure 8a represented the residual oil structures and the glass beads whereas Fig. 8b represented the residual oil structures and the plastic beads. The results showed that the positions of the residual oil were closer to the plastic beads than to the glass beads such that we observed the plastic beads surrounded by the residual oil structures. In contrast to the residual oil, the positions of the residual water were closer to the glass beads than the plastic beads as shown in Fig. 8c, d, respectively, and we observed the glass bead surrounded by the residual water structures.

In addition, the preferential affinity of each fluid type to a particular solid surface was quantified by calculating the shortest distance between fluid and solid surfaces. The frequency distributions of the shortest distance were divided into 50 bins and plotted. The shortest measurable distance was 0.026 mm, given by the imaging resolution, which means that the pixel of the fluid blob is next to the pixel of the solid bead. In general, the shorter the distance between the fluid and the solid beads, the greater is the wetting affinity of the fluid to a particular solid; i.e., the more likely the fluid is to come into contact with and spread over or adhere to a solid surface. Results of this study are presented in Figs. 9 and 10. Figure 9 reveals a narrower distribution of oil-to-plastic distances than that of oil-to-glass. This is evidence that the plastic beads are preferentially oil-wet relative to the glass beads. A similar analysis of Fig. 10 confirms that glass beads are more water-wet than the plastic beads, and that the wetting affinity of the glass beads with regard to water was stronger than that of the plastic beads.

Conclusions

Immiscible fluid flow through a synthetic bead pack with mixed wettability was investigated. Three-dimensional imaging of this bead pack after cycles of oil and water injections allowed the characterization of the final saturation distribution and of the trapped fluid structures in the context of surface wettability. In addition, results were compared against previous experimental findings in a uniformly water-wet glass bead pack and oil-wet plastic bead pack. The average porosity of the mixed-wet system was 42.04 %, which is comparable to that of water-wet and oil-wet systems at 41.6 and 42.1 %, respectively, suggesting that the packing was mostly cubic.

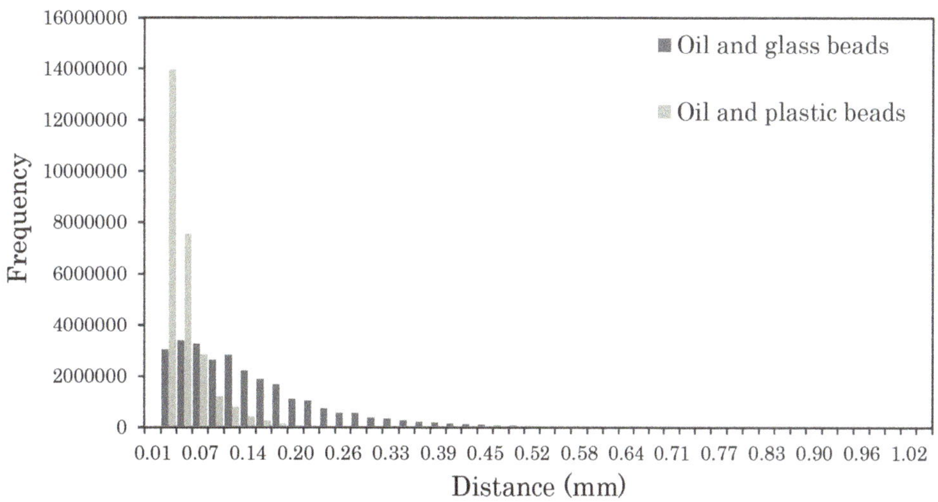

Fig. 9 Histogram of the shortest distance from the oil blobs to the plastic beads and the shortest distance from the oil blobs to the glass beads

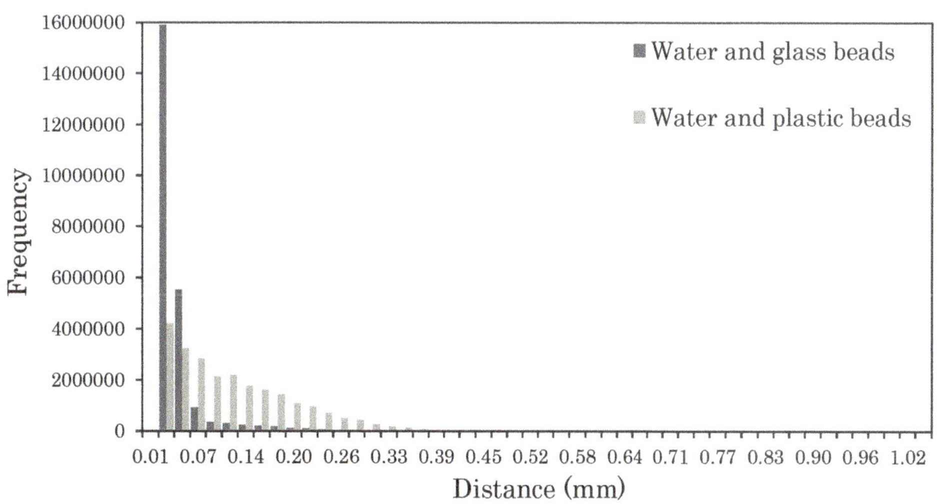

Fig. 10 Histogram of the shortest distance from the water blobs to the plastic beads and the shortest distance from the water blobs to the glass beads

The blob analyses presented indicate that the number of trapped water blobs at the end of kerosene injection was 2.4 times greater than that of the oil blobs at the end of water injection, for similar residual saturations. Moreover, the single largest oil blob found at residual oil saturation was larger than any trapped water blob, suggesting that water remained trapped at the crevices of grain-to-grain contacts and therefore a more dominant wetting phase in this mixed-wet system. In addition, most of the blobs fell in a size range between 0.0001 and 0.001 mm^3, making them 100 times smaller than those in the single wettability case.

The cumulative volume plot shows that oil blobs in range of 0.01–1 mm^3 and water blobs in range of 0.001–0.1 mm^3 account for an important portion of the total residual fluid trapped in the pore space. This result is consistent to the finding of Salathiel et al. (1973), whereby the smaller pores are occupied by water, whereas the oil preferentially wets the interconnected larger pores. The trend of surface area to blob volume for both oil and water was very similar, with a slight tendency of water blobs to reach a higher surface area for the same blob volume which can be attributed to stronger overall water wetting affinity to the solid surface than that of oil.

Moreover, the explicit imaging of plastic and glass beads was used to evaluate the wetting affinity of fluid to preferential solid surface by computing the shortest distance from each fluid surface (oil or water blobs) to a particular solid type (plastic or glass beads). The results confirm that oil blobs tend to adhere to plastic surfaces rather than to glass whereas water blobs tend to adhere to glass rather than to plastic. Results also suggest that glass beads have a stronger wetting affinity to water than that of plastic to oil, which is indicated by a narrower characteristic distance between water and glass surfaces.

The average percent saturation showed that irreducible water saturation tends to decrease with increasing oil wettability. In the case of the residual oil saturation, the mixed-wet sample had a trapped oil average of 16.39 %, whereas the residual oil in the water-wet system and the oil-wet system was found to be 17.72 and 23.86 %, respectively. Results from this investigation confirm that global phase trapping in porous media (residual saturation) is sensitive to the wetting characteristics of the solid matrix, but less intuitive is the evidence that the spatial distribution, characteristic size, and surface area of trapped fluid structures can vary drastically as a result of local wetting properties. These differences are then manifested in the magnitude and hysteretic attributes of capillary pressure curves that are so critical for the description of transport properties of natural permeable media.

References

Al-Futaisi A, Patzek TW (2003) Impact of wettability alteration on two-phase flow characteristics of sandstones: a quasi-static description. Water Resour Res 39(2):1042.

Al-Raoush RI, Willson CS (2005) A pore-scale investigation of a multiphase porous media system. J Contam Hydrol 77(1–2): 67–89.

Anderson WG (1987) Wettability literature survey-part 6: the effects of wettability on waterflooding. J Petrol Technol 39:1605–1622.

Chaouki J, Larachi F, Duduković MP (1997) Noninvasive tomographic and velocimetric monitoring of multiphase flows. Ind Eng Chem Res 36(11):4476–4503.

Culligan KA, Wildenschild D, Christensen BSB, Gray WG, Rivers ML (2006) Pore-scale characteristics of multiphase flow in porous media: a comparison of air-water and oil-water experiments. Adv Water Resour 29(2):227–238.

Dandekar AY (2006) Petroleum reservoir rock and fluid properties. FL: CRC/Taylor & Francis, Boca Raton

Delshad M, Najafabadi NF, Anderson GA, Pope GA, Sepehrnoori K (2006) Modeling wettability alteration in naturally fractured reservoirs. SPE/DOE Symp Improv Oil Recovery.

Denison C, Carlson WD, Ketcham RA (1997) Three-dimensional quantitative textural analysis of metamorphic rocks using high-resolution computed X-ray tomography: part I. Methods and techniques. J Metamorph Geol 15(1):29–44.

Donaldson EC, Thomas RD, Lorenz PB (1969) Wettability determination and its effect on recovery efficiency. Soc Petrol Eng J

Heiba AA, Davis HT, Scriven LE (1983) Effect of wettability on two-phase relative permeabilities and capillary pressures. In: SPE Annual Technical Conference and Exhibition, 5–8 October 1983. Society of Petroleum Engineers, San Francisco, California.

Jackson MD, Valvatne PH, Blunt MJ (2003) Prediction of wettability variation and its impact on flow using pore-to reservoir-scale simulations. J Petrol Sci Eng 39(34):231–246.

Jadhunandan PP, Morrow NR (1995) Effect of wettability on waterflood recovery for crude-oil brine rock systems. SPE Reserv Eng 10:40–46.

Jiang Z, Wu K, Couples G, van Dijke MIJ, Sorbie KS, Ma J (2007) Efficient extraction of networks from three-dimensional porous media. Water Resour Res 43(12):W12S03.

Karpyn ZT, Piri M, Singh G (2010) Experimental investigation of trapped oil clusters in a water-wet bead pack using X-ray microtomography. Water Resour Res 46(4):W04510.

Ketcham RA, Carlson WD (2001) Acquisition, optimization and interpretation of X-ray computed tomographic imagery: applications to the geosciences. Comput Geosci 27(4):381–400.

Kumar M, Senden TJ, Latham S, Sheppard AP, Knackstedt MA, SPE Australia National University, Cinar Y, SPE, Pinczewski WV (2008) Designing for mixed wettability. Paper presented at the SPE/DOE symposium on improved oil recovery

Landry CJ, Karpyn ZT, Piri M (2011) Pore-scale analysis of trapped immiscible fluid structures and fluid interfacial areas in oil-wet and water-wet bead packs. Geofluids 11(2):209–227.

Lebedeva EV, Fogden A (2011) Micro-CT and wettability analysis of oil recovery from sand packs and the effect of waterflood salinity and kaolinite. Energy Fuels 25(12):5683–5694.

Lenormand R, Zarcone C, Sarr A (1983) Mechanisms of the displacement of one fluid by another in a network of capillary ducts. J Fluid Mech 135:337–353

Masalmeh SK (2003) The effect of wettability heterogeneity on capillary pressure and relative permeability. J Petrol Sci Eng 39:399–408.

McCaffery FG, Bennion DW (1974) Effect of wettability on 2-phase relative permeabilities. J Can Petrol Technol 13:42–53.

Morrow NR, Mason G (2001) Recovery of oil by spontaneous imbibition. Curr Opin Colloid Interface Sci 6(4):321–337.

Oren PE, Bakke S, Arntzen OJ (1997) Extending predictive capabilities to network models. Paper presented at the SPE annual technical conference and exhibition, San Antonio, 5–8 Oct 1997

Otsu N (1979) A threshold selection method from gray-level histograms. IEEE Trans Syst Man Cybern 9(1):62–66 (citeulike-article-id:1116982)

Patzek TW (2001) Verification of a complete pore network simulator of drainage and imbibition. Soc Petrol Eng 6(2):144–156.

Porter ML, Wildenschild D (2010) Image analysis algorithms for estimating porous media multiphase flow variables from computed microtomography data: a validation. Study Comput Geosci 14(1):15–30.

Prodanović M, Lindquist WB, Seright RS (2007) 3D image-based characterization of fluid displacement in a Berea core. Adv Water Resour 30(2):214–226.

Radke CJ, Kovscek AR, Wong H (1992) A pore-level scenario for the development of mixed wettability in oil reservoirs. Paper presented at the SPE annual technical conference and exhibition, Washington, 4–7 Oct 1992

Raeesi B, Piri M (2009) The effects of wettability and trapping on relationships between interfacial area, capillary pressure and saturation in porous media: a pore-scale network modeling approach. J Hydrol 376(3–4):337–352.

Sahimi M (1995) Flow and transport in porous media and fractured rock: from classical methods to modern approaches. Wiley-VCH, Weinheim, Germany

Salathiel RA (1973) Oil recovery by surface film drainage in mixed-wettability rocks. J Petrol Technol 25:1216–1224.

Sezgin M, Sankur B (2004) Survey over image thresholding techniques and quantitative performance evaluation. J Electron Imaging 13(1):146–165

Speight JG (2009) Enhanced recovery methods for heavy oil and tar sands. Gulf Publishing Company, USA

Suicmez VS, Piri M, Blunt MJ (2008) Effects of wettability and pore-level displacement on hydrocarbon trapping. Adv Water Resour 31(3):503–512.

Tartakovsky AM, Meakin P (2006) Pore scale modeling of immiscible and miscible fluid flows using smoothed particle hydrodynamics. Adv Water Resour 29(10):1464–1478.

Tiab D, Donaldson EC (2004) Petrophysics: theory and practice of measuring reservoir rock and fluid transport properties. Gulf Professional Publishing, Boston, MA

van Dijke MIJ, Piri M (2007) Introduction to special section on modeling of pore-scale processes. Water Resour Res 43(12): W12S01.

Wardlaw NC (1982) The effects of geometry, wettability, viscosity and interfacial tension on trapping in single pore-throat pairs. J Can Petrol Technol 21:21–27.

Wilkinson D, Willemsen JF (1983) Invasion percolation: a new form of percolation theory. J Phys A: Math Gen 16:3365–3376.

Zhao X, Blunt MJ, Yao J (2010) Pore-scale modeling: effects of wettability on waterflood oil recovery. J Petrol Sci Eng 71(3–4): 169–178.

Oil displacement mechanisms of viscoelastic polymers in enhanced oil recovery (EOR): a review

Bing Wei · Laura Romero-Zerón · Denis Rodrigue

Abstract Polymer flooding has proved economically and technically successful in numerous enhanced oil recovery (EOR) projects, which can often increase oil recovery from 12 to 15 % of the original oil in place. When a reservoir is flooded with viscous polymer solution, the mobility ratio between the displacing fluid (i.e., water) and the displaced fluid (i.e., oil) becomes more favorable if compared to conventional water flooding. Therefore, the volumetric sweep efficiency and correspondingly the overall oil recovery are effectively improved. Currently, there is a widespread idea that polymer flooding is inefficient in improving the microscopic oil displacement (at pore scale). However, recent research based on laboratory studies and pilot field testing has proved otherwise. It seems that the viscoelastic properties of polymeric systems indeed improve the microscopic displacement efficiency of residual oil. This paper reviews and emphasizes the recovery mechanisms that have been proposed to explain oil displacement by polymer flooding within oil reservoirs. The aim of this review is to provide a synopsis of polymer flooding which is rapidly emerging as a popular and advantageous EOR process.

Keywords Enhanced oil recovery · EOR · Polymer flooding · Volumetric sweep efficiency · Microscopic displacement efficiency · Polymer viscoelasticity

B. Wei · L. Romero-Zerón
Department of Chemical Engineering, University of New Brunswick, Fredericton, NB E3B 5A3, Canada

D. Rodrigue
Chemical Engineering Department, Université Laval, Pavillon Adrien-Pouliot, 1065, Avenue de la Médecine, Local 3546, Quebec, QC G1V 0A6, Canada

Introduction

As early as the 1960s, polymer flooding has been suggested as an oil displacement process in enhanced oil recovery (EOR), with the main functionality of increasing the viscosity of the displacing fluid, which is commonly water (Sandiford 1964). This pioneer work conducted by Sandiford was then followed by wide research efforts with the purpose of recognizing the benefits of using non-Newtonian displacing fluids in oil recovery applications (Szabo 1975; Dominguez and Whillhite 1977; Gleasure 1990).

For most oil reservoirs, a large amount of oil is left behind in the formation trapped within the rock, even after extensive water flooding due to the unfavorable mobility ratio between the driving (displacing) water and the displaced oil. Once a preferential flow path is formed within the porous media, the driving water will flow directly from the injection well to the production well bypassing the oil bearing zones, which ultimately causes a relative low oil recovery particularly in heavy oil reservoirs.

To reach the bypassed oil zones, viscous polymer solution is usually injected into the reservoir as the driving fluid, strategy that is well-known as polymer flooding, by which the poor mobility ratio encountered in conventional water flooding is corrected. Consequently, polymer flooding increases the volumetric sweep efficiency of the flooded reservoir.

Currently, polymer flooding is considered as one of the most promising technologies in EOR process because of its technical and commercial feasibility. Particularly, the interest on polymer flooding applications worldwide has been stimulated by the outstanding results reported from the large-scale polymer flooding application in the Daqing oil field in China, with incremental oil productions of up to 300,000 barrels per day (Wang et al. 2001a).

In practice, two commercial polymers, hydrolyzed polyacrylamides (HPAM) and xanthan gums, are commonly used in oil field applications. HPAM is a water-soluble polyelectrolyte with negative charges on the polymer chains. Xanthan gums, which are polysaccharides, show excellent viscosifying ability, high tolerance to salinity, and temperature (Guo et al. 1999).

The well-established relationship between capillary number and oil recovery using Newtonian fluids, indicates that a substantial increase in oil recovery at the pore level (micro-scale) in rock formations can be obtained if the capillary number is increased by several thousand times. However, in the case of polymer flooding (non-Newtonian fluids), this number is normally increased less than 100 times (Stegemeier 1977; Wang et al. 2000). Consequently, it has been suggested for long time that polymer flooding can only improve the volumetric sweep efficiency with no effect whatsoever on the microscopic displacement efficiency (Du and Guan 2004). Nevertheless, the fact that the oil recovery factor was increased during polymer flooding by up to 13 % of the original oil in place in the Daqing oil field, made researchers to rethink if this significant incremental oil recovery could be explained just by the promotion of enhanced volumetric sweep efficiency, or could also be explained by the simultaneous increased efficiency of the microscopic-scale displacement. Improvements of the microscopic sweep efficiency by polymer flooding could be attributed to the distinctive flow characteristic of polymer solutions due to their viscoelastic properties.

Thus, this paper reviews the oil displacement mechanisms taking place during polymer flooding operations. Specifically, it aims to explain the individual contributions of the viscous component and the elastic component of the polymer injected on the overall incremental oil recovery.

Overall oil displacement efficiency

The overall oil displacement efficiency during oil recovery processes is defined by the product of the macroscopic displacement efficiency (macro-scale) and the microscopic displacement efficiency (micro-scale). Macroscopic displacement efficiency is a measure of the effectiveness of the displacing fluid(s) in contacting the oil zone volumetrically; while microscopic displacement efficiency refers to the effectiveness of the displacing fluid(s) in mobilizing oil trapped at the pore scale by capillary forces. Therefore, any displacement mechanism that can improve either the macro-scale or micro-scale oil sweep efficiency during polymer flooding is beneficial to increment oil production (Romero-Zerón 2012).

Fluid mobility control

The primary purpose of polymers in EOR processes is to control the mobility of the displacing phase (i.e., water). Mobility is defined as the ratio of the relative permeability of the fluid (i.e., water or oil) to the viscosity of the same fluid. The importance of controlling fluid mobility to increase macroscopic displacement efficiency has been well recognized by many researches (Pitts et al. 1995; Du and Guan 2004; Kotlar et al. 2007). The addition of polymer to the displacing phase (i.e., water) reduces its mobility by thickening the aqueous phase and significantly diminishing the formation of viscous fingerings and/or channels. A simultaneous mechanism taking place during polymer flooding is the reduction of the relative permeability to the displacing phase caused by the retention of polymer (adsorption and mechanical retention) within the porous media. Therefore, polymer flooding is very effective in improving the volumetric sweep efficiency.

Mobility ratio (M) is an important and useful parameter to quantify the mobility contrast between the displacing fluid (i.e., water) mobility and the displaced phase (i.e., oil) mobility. Thus, the mobility ratio for a waterflood is given by the following expression (Eq. 1) (Pitts et al. 1995):

$$M = {}^{\lambda_w}/_{\lambda_o} = \frac{k_{rw}/\mu_w}{k_{ro}/\mu_o} = \frac{k_{rw}\mu_o}{k_{ro}\mu_w} \quad (1)$$

where λ is the fluid mobility, k_r is the relative permeability, and μ is the fluid viscosity; the subscripts w and o denote water phase and oil phase, respectively.

The displacing fluid mobility should be equal or less than the total mobility of the displaced multiphase fluids (Sheng 2011; Lake 1989; Dyes et al. 1954). Figure 1 shows relative mobility as a function of water saturation. The plot indicates the individual mobility of water and oil, and the total mobility (λ_t), which is the summation of water and oil mobilities. The minimum total relative mobility, which corresponds to the minimum value of λ_t on the total mobility curve (λ_t at the corresponding S_w), is also indicated in Fig. 1.

In some cases, the total mobility of the fluids can not be calculated because the mobile fluid saturations are unknown; in these situations the minimum total mobility is often used to determine the target viscosity required for polymer flooding (Gogarty 1969; Gogarty et al. 1970).

The conventional concept of mobility ratio distinguishes "favorable" mobility conditions when $M \leq 1$ and "unfavorable" mobility conditions when $M > 1$. However, Sheng (2011) recently demonstrated that this existing concept of mobility ratio is invalid and proposed an updated definition of mobility ratio expressed as the ratio of the displacing fluid mobility to the oil mobility

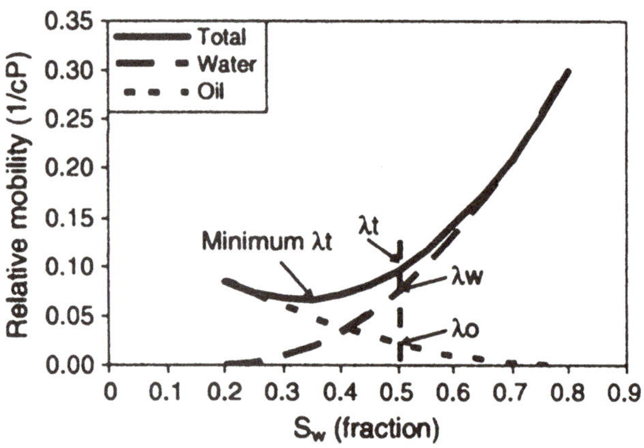

Fig. 1 Relative mobility as a function of saturation of water (source: Sheng 2011)

multiplied (corrected) by the normalized movable oil saturation ($\bar{S}_o$) as indicated by Eq. (2)

$$M = {}^{\lambda_w}/_{\lambda_o}\bar{S}_o = \frac{k_{rw}/\mu_w}{k_{ro}/\mu_o}\bar{S}_o = \frac{k_{rw}\mu_o}{k_{ro}\mu_w}\bar{S}_o. \quad (2)$$

This new definition of mobility ratio was demonstrated by numerical simulation validated using coreflooding data available for polymer flooding and alkali–surfactant–polymer. According to Sheng (2011), this new definition of mobility ratio should be used to determine the concentration of the mobility control agent (polymer) to be applied in the field.

Disproportionate permeability reduction (DPR)

Polymer solutions can reduce the mobility of the displacing phase by two mechanisms: by increasing the viscosity of the displacing phase and by causing disproportionate permeability reduction (DPR); i.e., polymer or gel considerably reduces water relative permeability (k_{rw}), while producing minimum reduction in the oil relative permeability (k_{ro}) (Schneider and Owens 1982; Taber and Martin 1983; Niu et al. 2006). The following mechanisms have been proposed to explain DPR: (1) segregation of flow pathways (water and oil); (2) shrinking/swelling of polymer depending on phase flow; (3) layer formation on pore wall by adsorbed polymer; and (4) wettability alteration (Mennella et al. 1998; Liang and Seright 1997; Nilsson et al. 1998; Barreau et al. 1997; Zaitoun and Kohler 1988). Among these mechanisms, polymer adsorption (layer formation) and segregation of flow pathways have been considered as the dominant mechanisms for DPR when polymers solutions (without cross-linker) are used.

Zheng et al. (2000) proposed an empirical model to correlate the reduction of relative permeability with polymer adsorption. They found that the adsorption of hydrolyzed polyacrylamides on both water-wet and mild oil-wet cores induces a selective reduction of the relative permeability to water with respect to the relatively permeability to oil. In the mild oil-wet core case, wettability change was also considered to be another mechanism for the reduction of the relative permeability to water. Córdova et al. (2002) observed selective permeability reduction due to adsorbed nonionic polyacrylamides on mica; however, wettability alteration was not detected in this study.

Al-Sharji et al. (2001) presented a visualization study showing the built-up of cationic polyacrylamide polymer layers on the grain crevices in water-wet models. Based on this work, the adsorption-entanglement mechanism or the dynamically formed layer of polymer was proposed to explain DPR. However, in the oil-wet case, DPR was not observed (Elmkies et al. 2001). This mechanism (adsorption-entanglement) is also applicable for the adsorption of anionic polyacrylamides, which are extensively used in oil fields (Grattoni et al. 2004; Ogunberu and Asghari 2005).

In addition, Denys et al. (2001) proposed a bridging adsorption mechanism for cationic polyacrylamides when polymer macromolecules are stretched under high rates. This mechanism was further investigated by Chauveteau et al. (2002).

Nilsson et al. (1998) and Stavland and Nilsson (2001) proposed another mechanism "the segregated flow path of oil and water at the pore level" to explain DPR during polymer flooding. Currently, it is normally considered that DPR occurs due to the combined action of several mechanisms. Nevertheless, few experimental works clearly demonstrated that polymer adsorption onto rock surface is the dominant mechanism for inducing DPR.

Polymer elasticity and/or flow resistance

Flow resistance is the third mechanism that can improve volumetric or macroscopic sweep efficiency during polymer flooding. Dehghanpour and Kuru (2010) investigated the effect of viscoelastic fluid rheology on the formation of an "internal cake" (frictional pressure drop). They observed that the fluid with higher elasticity exhibited significantly higher pressure drop during flow through porous media. This elastic effect that results in additional pressure drop could be enhanced by widening the molecular weight distribution without changing the shear viscosity. A similar trend was obtained by Urbissinova et al. (2010), who also evaluated the contribution of polymer's elasticity in EOR. Likewise, they observed that the polymer solution showing higher elasticity experienced higher flow resistance (pressure drop) to flow through porous media than the polymer with lower elasticity, even though their shear viscosities were identical, which rendered improved macro-scale sweep efficiency and oil recovery (Veerabhadrappa 2012). This mechanism may also improve the microscopic displacement efficiency in displacing residual oil immobilized in the core by capillary forces and rock configuration (Wang et al. 2000).

In the last decade, it has been proposed that polymer flooding can also increase the microscopic displacement efficiency by mobilizing and displacing residual oil saturation (trapped by capillary forces). This phenomenon has been attributed to the elasticity of the polymer solutions (Wang et al. 2000). This proposition has generated controversial arguments (Bakhitov et al. 1980; Schneider and Owens 1982; Pusch et al. 1987; Zaitoun and Kohler 1987, 1988).

Nonetheless, Wang et al. (2000) have studied this issue extensively by evaluating the effectiveness of polymer flooding in displacing "residual oil" after water flooding at different conditions as follows. (1) Residual oil in "dead ends"; (2) residual oil films on rock; (3) residual oil in pore throats trapped by capillary forces; 4) residual oil unswept in micro-scale heterogeneous portions of the porous media. In all cases, Wang et al. (2000) observed that residual oil was reduced after polymer flooding. Xia et al. (2008a) used a theoretical and an experimental approach to investigate the effect of the elastic characteristics of the driving fluid on the microscopic displacement efficiency. The results show that viscoelastic polymer solutions are efficient in displacing different types of residual oil without even increasing the pressure gradient. In contrast, Huh and Pope (2008) pointed out that a tertiary polymer flooding did not reduce the water flooding residual oil saturation in an homogeneous, water-wet Berea sandstone core and in Antolini core with small-scale heterogeneity, but a secondary polymer flooding did displace oil below the residual oil saturation. Hou et al. (2009) observed the increase of microscopic displacement efficiency in a polymer flooding using an industrial CT system and argued that the essence of this phenomenon is that polymer flooding increases the oil/water mobility ratio and diverts the flow profile of the displacing fluid that causes the redistribution of oil saturation. The combination of the effects of diverted waterways and polymer viscoelasticity mobilizes and displaces the residual oil. Cheng et al. (2010) analyzed the distribution of residual oil in the Daqing oil field after polymer flooding. They determined that the thickness of the volume of the reservoir contacted by polymer was increased by 21.4 % and that oil saturation was reduced to 11.9 % as compared to water flooding. The residual oil saturation after polymer flooding in the Daqing oil field is mainly controlled by capillary forces. Chen et al. (2011) modeled polymer flooding from the Daqing oil field. In this work, the mechanism of polymer elasticity was introduced. The field production performance of polymer flooding was matched, and the results suggested that high concentration polymer flooding cannot only enhance volumetric sweep efficiency, but also increased microscopic displacement efficiency. Meybodi et al. (2011) used an image processing technique to analyze and compare microscopic and macroscopic displacement behaviors during polymer flooding in oil-wet and water-wet porous media. They concluded that a favorable microscopic displacement could enhance the macroscopic sweep efficiency in strongly water-wet medium, but in a strongly oil-wet medium, there are more factors that influence the microscopic displacement, front stability, and oil recovery.

Mechanisms enhancing microscopic displacement efficiency

The proposed mechanisms to explain the enhancement of the microscopic displacement efficiency during polymer flooding are briefly reviewed in the subsequent paragraphs.

Pulling effect mechanism

Wang et al. (2007, 2011) indicated that if a fluid with elastic properties flows over dead ends, normal stresses between oil and polymer solution are generated in addition to the shear stresses resulting from the long molecular chains. Thus, polymer imposes a larger force on oil droplets and pulls them out of dead ends. The amount of residual oil pulled out from dead ends is proportional to the elasticity of the driving fluid; these observations are presented in Fig. 2.

The viscoelastic polymer (HPAM) pushes the fluid ahead and pulls the fluids besides and behind, while the non-elastic fluids (water and glycerin) as presented in Fig. 2 are capable of pushing the fluids ahead but cannot "pull" out oil from the dead end.

Likewise, Luo et al. (2006) compared the effect of polymer elasticity on microscopic oil displacement efficiency. Figure 3 displays the pictures of the experiments carried out in a dead end pore (glass-etched model), in which displacing fluids having different viscoelastic properties were injected over the dead end to displace the trapped oil. The ratio of the elastic modulus (G') to the viscous modulus (G'') is useful to represent the elasticity of a fluid. In this experiment, the dead end pore model was flooded with solutions with increasing elasticity and/or increasing G'/G'' values (0, 0.92, 1.75, and 2.72). The residual oil saturation after water (i.e., non-elastic fluid) injection over the dead end is shown in Fig. 3a, and the saturation of residual oil after water flooding followed by polymer flooding is shown in Fig. 3b, c. It can be seen that

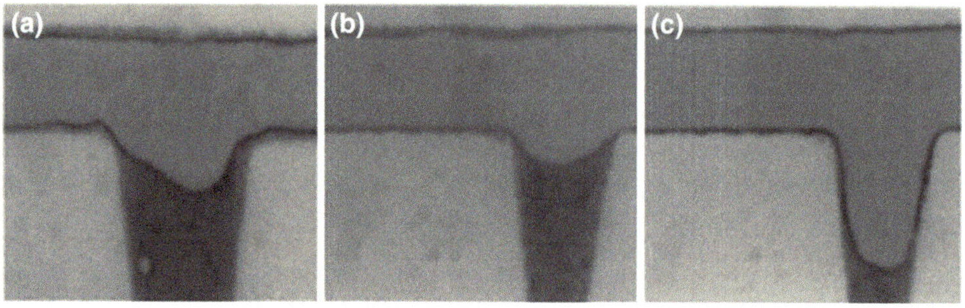

Fig. 2 Residual oil (*darker color*) in "dead ends" after **a** water, **b** glycerin, and **c** HPAM floods (source: Wang 2001)

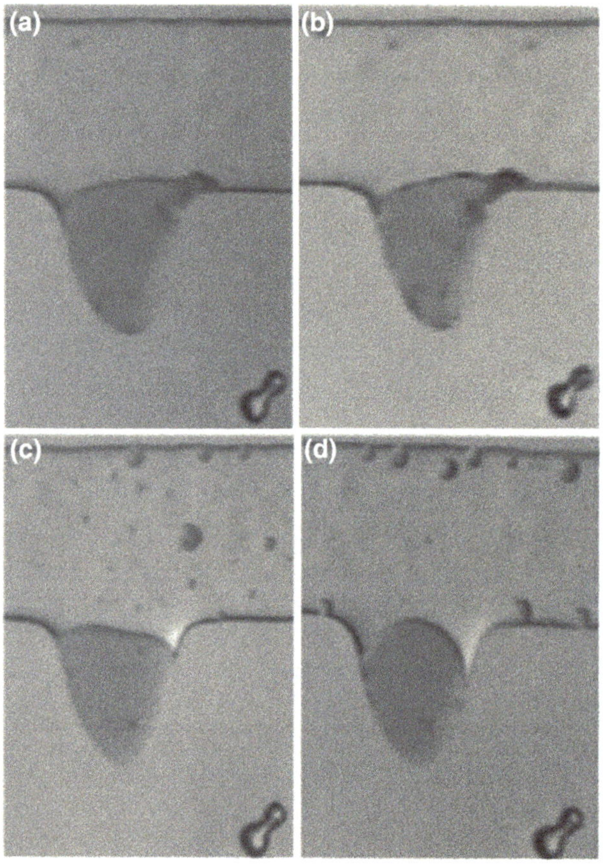

Fig. 3 Distribution of residual oil by water and polymer solutions with different viscoelastic properties: **a** $G'/G'' = 0$, $E_D = 0.0$; **b** $G'/G'' = 0.92$, $E_D = 0.0$; **c** $G'/G'' = 1.75$, $E_D = 0.13$; **d** $G'/G'' = 2.72$, $E_D = 0.18$ (source: Luo et al. 2006)

an increase in the elasticity of the displacing fluid decreases the saturation of the oil trapped in the dead end pore.

This effect has been further investigated through numerical simulation of polymer flooding over dead ends (Yin et al. 2006, 2012; Xia et al. 2008a, b; Zhang and Yue 2008). The pulling effect also works for residual oil trapped in a configuration that have both ends open to flow. This type of residual oil is immovable mainly because of capillary forces. In this case, there is at least a 50 % chance that the capillary force retaining the oil droplet can be reduced by the pulling effect (Wang et al. 2000).

Stripping mechanism

In oil-wet porous media, residual oil is attached to the rock surface in the form of a continuous oil film. Wang et al. (2000) reported the velocity profile of Newtonian and non-Newtonian fluids in capillary tubes and observed that the velocity gradient near the capillary wall for elastic fluids is considerably greater than for Newtonian fluids. Therefore, a stronger force is produced during flow of polymer solutions as compared to water, enhancing stripping oil films off the rock surface which eventually leads to a promotion of oil displacement efficiency (Jiang et al. 2008; Xia et al. 2004, 2008b; Wang 2001; Wang et al. 2001c; Sedagha et al. 2013). The alteration in wettability from oil-wet to more water-wet due to this oil stripping effect would also increase oil recovery.

Oil thread or column flow mechanism

The third possible mechanism bearing on the elastic property of polymers is oil thread flow. Elastic polymer solutions are able to stabilize oil threads due to the normal stress. As Wang et al. (2001a, c) and Luo et al. (2006) observed, oil is pulled by the polymer solution into oil columns and then forms oil threads as they aggregate with residual oil downstream. The normal force acting on the convex surface of the oil thread is supposed to be larger than the normal force acting on the concave surface. Thus, the essential function of the normal stress is to prevent oil threads from deforming, which potentially improves the displacement efficiency (Sedagha et al. 2013). Moreover, this force contrast increases with the Deborah number (N_{Deh}), which is a dimensionless number commonly employed to describe the elasticity of polymer solutions in porous media. In regards to this mechanism, Huh and Pope (2008) indicated the possibility that the interfacial tension between oil and polymer solution could destabilize the long oil column and break it into oil droplets. However, the elasticity of polymer solutions could resist this interfacial deformation. Thereby, the oil column could be either drained to a thinner cross-section causing it to break it up into smaller ganglia, which results in lower residual oil saturation, or the thinned oil column could be broken into longer-length oil ganglia that have more chance to be mobilized.

Shear thickening effect

Theoretically, the rheological behavior of polymer solutions as a function of shear rate (flow velocity gradient) in porous media could be divided into three types: Newtonian, shear thinning, and shear thickening. When polymer molecules flow through a series of pore bodies and pore throats, the flow field undergoes elongation and contraction. Shear thickening occurs only when polymer solutions flow through porous media and the flow velocity is too high that the polymer molecules do not have sufficient relaxation time to stretch and recoil to adjust to the flow velocity and the elastic chains cause a high apparent viscosity (shear thickening). This behavior can help driving fluid to rapidly displace the mobile but hard to displace oil or to displace the bypassed oil in small-scale heterogeneities more effectively (Delshad et al. 2008). Therefore, the shear thickening effect is considered as one of the possible mechanisms for the enhanced microscopic displacement efficiency during polymer flooding. A simple analysis elaborating on this point has been made by Jones (1980). Moreover, the high apparent viscosity is also beneficial for improvement of macro-scale sweep efficiency (Seright et al. 2010).

Although, shear thickening behavior is an important polymer flow regime in polymer flooding, it is still a challenge to accurately describe it. The reason is that this behavior cannot be quantified through bulk rheological measurements even at comparable shear rates, as is the case for the shear thinning regime (Green and Willhite 1998). To quantify this flow behavior, several models or equations have been proposed (Hirasaki and Pope 1974; Masuda et al. 1992; Chen et al. 1998). Recently, more reliable and precise predictions of the shear thickening regime in terms of the apparent viscosity (pressure drop) dependence on the effective shear rate (flow velocity) have been developed. Garrocuh and Gharbi (2006) proposed an empirical viscoelastic model analogous to Darcy's law to study the flow behavior of HPAM and Xanthan gum in porous media. This model accounts for polymer elasticity, viscosity, and porous media structure (porosity and permeability). This study shows that the Deborah number may not be an adequate parameter to characterize viscoelastic flow in porous media. Instead, a dimensional number called "viscoelasticity number" (N_v) seems to be more adequate because it clearly distinguishes viscous flow from viscoelastic flow. This model may allow predicting the relationship between pressure drop and flow velocity in the shear thickening regime.

Delshad et al. (2008) established another model for HPAM polymer solutions. One notable distinction of this model when compared with previous models is that the shear thickening viscosity does not increase indefinitely with the Deborah number and a plateau value is reached, which seems more reasonable. Delshad et al. (2008) also attempted to combine equations for the shear thinning regime and the shear thickening region to obtain a viscoelastic model applicable to the entire shear rate range. This model accounts for both rheological behaviors and was successfully used to history-matched data taken from published laboratory reports. Kim et al. (2010) used this viscoelastic model to develop a database for HPAM polymer solutions for a wide range of polymer concentration, salinity, and hardness, and temperature making it more practical to quantify the apparent viscosity of the polymer solution at reservoir conditions. Recently, further improvement of the parameters of this model was presented by Sharma et al. (2011).

A constitutive equation was first developed by Zhang et al. (2011) to represent the viscoelastic behavior of HPAM solutions in porous media. This equation indicates that the increase in viscosity due to elasticity is proportional to the relaxation time, particle diameter, porosity, and it is inversely proportional to the Darcy velocity and the square root of tortuosity. This constitutive equation gives satisfactory agreement between calculated and experimental values, which validates its applicability for quantifying shear thickening behavior in porous media.

Cheng and Cao (2010) divided the flow performance of polymer solutions through porous media into three stages: entrance constriction, pore passage, and extrusion. Therefore, pressure drop could be built individually combining viscous drop and elastic drop, which allows establishing a constitutive model of viscous-elastic fluid in porous media that considered the polymer's flow characteristics and the porous media's features. This new model was tested against experimental data obtained from coreflooding, and accurately characterized the polymer solution's rheological behavior in porous media including the shear thinning and the shear thickening regimes.

Likewise, Stavland et al. (2010) and Norris (2011) presented a comparable but simpler rheological model that also covered the shear thinning and shear thickening viscosity regions. Unexpectedly, this model showed that the apparent viscosity decreased beyond the shear thickening regime with increasing shear rate during coreflooding tests. This phenomenon could be interpreted by the mechanical degradation of the polymer molecular structure caused by chain rupture. Thereby, an improved rheological model multiplied by a degradation term was proposed that can be effectively used to predict viscosity change in porous media.

Conclusions

The primary objective of this paper was to review the oil displacement mechanisms of polymer flooding aiming to provide an up to date overview of this topic. Based on the discussion presented, the following conclusions can be made:

1. At the macro-scale, polymer flooding improves the sweep efficiency by reducing the mobility ratio between the displacing phase and the displaced phase through increasing water viscosity and selectively decreasing the relative permeability to water (DPR effect).
2. DPR effect due to polymer flooding is strongly dependent on the porous media characteristics. In strongly oil-wet porous media, DPR is much less pronounced as compared to water-wet porous media. Polymer adsorption onto rock surface seems to be the dominant mechanism for inducing DPR.
3. At the micro-scale, polymer flooding improves displacement efficiency due to the elasticity of the polymer solution.
4. The pulling effect, oil stripping, oil thread and/or column flow, and polymer shear thickening mechanisms have been proposed to explain the enhanced microscopic displacement efficiency taking place during polymer flooding. However, more research is required to verify these mechanisms or to explore other possible explanations.

Acknowledgments The authors would like to acknowledge the financial support provided by the Canadian Foundation for Innovation (CFI), the Natural Sciences and Engineering Research Council of Canada (NSERC), and the New Brunswick/Québec Cooperation in Advanced Education.

References

Al-Sharji HH, Grattoni CA, Dawe RA et al (2001) Disproportionate permeability reduction due to polymer adsorption entanglement. In: SPE 68972 presented at European formation damage conference, The Hague, 21–22 May 2001

Bakhitov GG, Ogandzhanyants VG, Polishchuk AM (1980) Experimental investigation into the influence of polymer additives in water on the relative permeabilities of porous media. Fluid Dyn 15(4):611–615.

Barreau P, Bertin H, Lasseux D, Glénat P, Zaitoun A (1997) Water control in producing wells: influence of an adsorbed-polymer layer on relative permeabilities and capillary pressure. SPE Reserv Eng 12(4):234–239

Chauveteau G, Denys K, Zaitoun A (2002) New insight on polymer adsorption under high flow rates. In: SPE 75183 presented at SPE/DOE improved oil recovery symposium, Tulsa, 13–17 April 2002

Chen TL, Tang JX, Peng KZ, Xie F (1998) HPAM rheology in subsurface flow. In: Gang QL (ed) Chemical flooding symposium–research results during the eighth five-year period (1991–1995), vol I. Petroleum Industry Press, Beijing, pp 112–119

Chen G, Han PH, Shao ZB et al (2011) History matching method for high concentration viscoelasticity polymer flood pilot in Daqing oilfield. In: SPE 144538 presented at enhanced oil recovery conference, Kuala Lumpur, 19–21 July 2011

Cheng LS, Cao RY (2010) Constitutive model of viscous-elastic polymer solution in porous media. Pet Sci Technol 28:1170–1177.

Cheng J C, Wei J G, Song K P et al (2010) Study on Remaining Oil Distribution After Polymer Injection. In: SPE 133808 presented at Annual Technical Conference and Exhibition, Florence, 19–22 Sept 2010

Córdova M, Mogollón JL, Molero H et al (2002) Sorbed polyacrylamides: selective permeability parameters using surface techniques. In: SPE 75210 presented at SPE/DOE improved oil recovery symposium, Tulsa, 13–17 April 2002

Dehghanpour H, Kuru E (2010) A new look at the viscoelastic fluid flow in porous media: a possible mechanism of internal cake formation and formation-damage control. In: SPE 121640 presented at SPE international symposium on oilfield chemistry. The Woodlands, 20–22 April 2010

Delshad M, Kim DH, Magbagbeola OA et al (2008) Mechanistic interpretation and utilization of viscoelastic behavior of polymer solutions for improved polymer-flood efficiency. In: SPE 113620 presented at SPE/DOE symposium on improved oil recovery, Tulsa, 19–23 April 2008

Denys KM, Fichen C, Zaitoun A (2001) Bridging adsorption of cationic polyacrylamides in porous media. In: SPE 64984 presented at SPE international symposium on oilfield chemistry, Houston, 13–16 Feb 2001

Dominguez JG, Whillhite GP (1977) Retention and flow characteristics of polymer solution in porous media. SPE J 17:112–121.

Du Y, Guan L (2004) Field-scale polymer flooding: lessons learnt and experiences gained during past 40 years. In: SPE 91787 presented at SPE international petroleum conference, Mexico, 7–9 Nov 2004

Dyes AB, Caudle BH, Erickson RA (1954) Oil production after breakthrough—as influenced by mobility ratio. J Pet Technol 6:27–32.

Elmkies PH, Bertin H, Lasseux D et al (2001) Further investigations on two-phase flow property modification by polymers: wettability effect. In: SPE 64986 presented at SPE international symposium on oilfield chemistry, Houston, 13–16 Feb 2001

Garrocuh AA, Gharbi RB (2006) A novel model for viscoelastic fluid flow in porous media. In: SPE 102015 presented at SPE annual technical conference and exhibition, Texas, 24–27 Sept 2006

Gleasure RW (1990) An experimental study of non-newtonian polymer rheology effects on oil recovery and injectivity. SPE Reserv Eng 5:481–486.

Gogarty WB (1969) Process for the simultaneous displacement of petroleum and water in formations. US Patent No. 3,443,635, 13 May 1969

Gogarty WB, Meabon HP, Milton HW (1970) Mobility control design for miscible-type waterfloods using micellar solutions. J Pet Technol 22:141–147.

Grattoni CA, Luckham PF, Jing XD et al (2004) Polymer as relative permeability modifiers: adsorption and the dynamic formation of thick polyacrylamide layers. J Pet Sci Eng 45:233–245.

Green DW, Willhite GP (1998) Enhanced oil recovery. Soc Pet Eng, Dallas

Guo XH, Li DW, Tian J, Liu YZ (1999) Pilot test of xanthan gum flooding in Shengli oilfield. In: SPE 57294 presented at SPE Asia Pacific improved oil recovery conference, Kuala Lumpur, 25–26 Oct 1999

Hirasaki GJ, Pope GA (1974) Analysis of factors influencing mobility and adsorption in the flow of polymer solution through porous media. SPE J 14:337–346.

Hou J, Li ZQ, Zhang SK et al (2009) Computerized tomography study of the microscopic flow mechanism of polymer flooding. Transp Porous Media 79:407–418.

Huh C, Pope GA (2008) Residual oil saturation from polymer floods: laboratory measurements and theoretical interpretation. In: SPE 113417 presented at SPE/DOE symposium on improved oil recovery, Oklahoma, 20–23 April 2008

Jiang HF, Wu WX, Wang DM et al (2008) The effect of elasticity on displacement efficiency in the lab and results of high-concentration polymer flooding in the field. In: SPE 115315 presented at SPE annual technical conference and exhibition, Colorado, 21–24 Sept 2008

Jones WM (1980) Polymer additives in reservoir flooding for oil recovery: shear thinning or shear thickening. J Phys D Appl Phys 13:L87–L88.

Kim DH, Lee SJ, Ahn CH et al (2010) Development of a viscoelastic property database for EOR polymers. In: SPE 129971 presented at SPE improved oil recovery symposium, Oklahoma, 24–28 April 2010

Kotlar HK, Selle O, Torsaeter O (2007) Enhanced oil recovery by comb flow: polymer floods revitalized. In: SPE 106421 presented at international symposium on oilfield chemistry, Houston, 28 Feb–2 March 2007

Lake LW (1989) Enhanced oil recovery. Prentice-Hall, USA

Liang JT, Seright RS (1997) Further investigations of why gels reduce water permeability more than oil permeability. SPE Prod Facil 12:225–230.

Luo JH, Liu YZ, Zhu P (2006) Polymer solution properties and displacement mechanisms. In: Shen PP, Liu YZ, Liu HR (eds) Enhanced oil recovery–polymer flooding. Petroleum Industry Press, Beijing, pp 1–72

Masuda Y, Tang K, Miyazawa M, Tanaka S (1992) 1D simulation of polymer flooding including the viscoelastic effect of polymer solution. SPE Reserv Eng 7:247–252.

Mennella A, Chiappa L, Bryant SL, Burrafato G (1998) Pore-scale mechanism for selective permeability reduction by polymer injection. In: SPE 39634 presented at SPE/DOE improved oil recovery symposium, Oklahoma, 19–22 April 1998

Meybodi HE, Kharrat R, Wang XQ (2011) Study of microscopic and macroscopic displacement behaviors of polymer solution in water-wet and oil-wet media. Transp Porous Media 89:97–120.

Nilsson S, Stavland A, Jonsbraten HC (1998) Mechanism study of disproportionate permeability reduction. In: SPE 39635 presented at SPE/DOE improved oil recovery symposium, Oklahoma, 19–22 April 1998

Niu JG, Chen P, Shao ZB, Wang DM, Sun G, Li Y (2006) Research and development of polymer enhanced oil recovery. In: Cao HQ (ed) Research and development of enhanced oil recovery in Daqing. Petroleum Industry Press, Beijing, pp 227–325

Norris UL (2011) Core-scale simulation of polymer flow through porous media. MS thesis. University of Stavanger, Stavanger

Ogunberu AL, Asghari L (2005) Water permeability reduction under flow-induced polymer adsorption. J Can Pet Technol 44:56–61.

Pitts MJ, Campbell TA, Surkalo H et al (1995) Polymer flood of the Rapdan pool. SPE Reserv Eng 10:183–186.

Pusch G, Lotsch T, Muller T (1987) Investigation of the oil displacing efficiency of suitable polymer products in porous media, aspects of recovery mechanisms during polymer flooding. DGMK-report, 295–296, German. Society for petroleum science coal chemistry, Hamburg, FRG

Romero-Zerón L (2012) Advances in enhanced oil recovery. In: Romero-Zerón L (ed) Introduction to enhanced oil recovery (EOR) processes and bioremediation of oil contaminated sites. InTech, Rijeka. ISBN 978-953-51-0629-6

Sandiford BB (1964) Laboratory and field studies of water floods using polymer solutions to increase oil recoveries. J Pet Technol 16:917–922.

Schneider FN, Owens WW (1982) Steady state measurements of relative permeability for polymer/oil systems. SPE J 22:79–86.

Sedagha MH, Ghazanfari MH, Parvazdavani M, Moshedi S (2013) Experimental investigation of micro/macroscopic efficiency of polymer flooding in fractured heavy oil five spot systems. ASME J Energy Resour Technol 135:1–9.

Seright RS, Fan TG, Wavrik K (2010) New insights into polymer rheology in porous media. In: SPE 129200 presented at SPE improved oil recovery symposium, Oklahoma, 24–28 April 2010

Sharma A, Delshad M, Huh C et al (2011) A practical method to calculate polymer viscosity accurately in numerical reservoir simulators. In: SPE 147239 presented at SPE annual technical conference and exhibition, Colorado, 30 Oct–2 Nov 2011

Sheng JJ (2011) Modern chemical enhanced oil recovery: theory and practice. Gulf Professional Publishing, Elsevier, USA

Stavland A, Jonsbraten HC et al (2010) Polymer flooding—flow properties in porous media versus rheological parameters. In: SPE 131103 presented at SPE EUROPEC/EAGE annual conference and exhibition, Spain, 14–17 June 2010

Stavland A, Nilsson S (2001) Segregated flow is the governing mechanism of disproportionate permeability reduction in water and gas shutoff. In: SPE 71510 presented at SPE annual technical conference and exhibition, Louisiana, 30 Sept–3 Oct 2001

Stegemeier GL (1977) Mechanisms of entrapment and mobilization of oil in porous media. In: Shah DO, Schechter RS (eds) Improved oil recovery by surfactant and polymer flooding. Academic Press, New York, pp 55–91

Szabo MT (1975) Laboratory investigation of factors influencing polymer flood performance. SPE J 15:338–346.

Taber JJ, Martin FD (1983) Technical screening guides for the enhanced recovery of oil. In: SPE 12069 presented at SPE annual technical conference and exhibition, California, 5–8 Oct 1983

Urbissinova TS, Trivedi J, Kuru E (2010) Effect of elasticity during viscoelastic polymer flooding: a possible mechanism of increasing the sweep efficiency. J Can Pet Technol 49:49–56.

Veerabhadrappa SK (2012) Study of effects of polymer elasticity on enhanced oil recovery by core flooding and visualization experiments. MS thesis. University of Alberta, Alberta

Wang DM (2001) Development of new tertiary recovery theories and technologies to sustain Daqing oil production. PGODD 20:1–7

Wang DM, Cheng JC, Yang QY et al (2000) Viscous-elastic polymer can increase microscale displacement efficiency in cores. In: SPE 63227 presented at SPE annual technical conference and exhibition, Texas, 1–4 Oct 2000

Wang DM, Xia HF, Liu ZC, Yang QY (2001a) Study of the mechanism of polymer solution with viscoelastic behavior increasing microscopic oil displacement efficiency and the forming of steady "oil thread" flow channels. In: SPE 68723 presented at SPE Asia Pacific oil and gas conference and exhibition, Indonesia, 17–19 April 2001

Wang KL, Liao GZ, Yang ZY, Li JL (2001b) Experimental study of the effect of displacing fluid viscosity in polymer and ASP flooding. Oilfield Chem 18:354–357

Wang DM, Cheng JC, Xia HF et al (2001d) Viscous-elastic fluids can mobilize oil remaining after water-flood by force parallel to the oil-water interface. In: SPE 72123 presented at SPE Asia Pacific improved oil recovery conference, Kuala Lumpur, 6–9 Oct 2001

Wang DM, Wang G, Wu W et al (2007) The influence of viscoelasticity on displacement efficiency–from micro to macro scale. In: SPE 109016 presented at SPE annual technical conference and exhibition, California, 11–14 Nov 2007

Wang DM, Wang G, Xia HF (2011) Large scale high visco-elastic fluid flooding in the field achieves high recoveries. In: SPE 144294 presented at SPE enhanced oil recovery conference, Kuala Lumpur, 19–21 July 2011

Xia HF, Wang DM, Wu JZ (2004) Elasticity of HPAM solutions increases displacement efficiency under mixed wettability conditions. In: SPE 88456 presented at SPE Asia Pacific oil and gas conference and exhibition, Australia, 18–20 Oct 2004

Xia HF, Wang DM, Wang G, Ma WG, Liu J (2008a) Mechanism of the effect of micro-forces on residual oil in chemical flooding. In: SPE 115315 presented at SPE/DOE improved oil recovery symposium, Oklahoma, 19–23 April 2008

Xia H, Wang D, Wang G et al (2008b) Effect of polymer solution viscoelasticity on residual oil. Pet Sci Technol 26:398–412.

Yin HJ, Wang DM, Zhang H (2006) Study on flow behaviors of viscoelastic polymer solution in micropore with dead end. In: SPE 101950 presented at SPE annual technical conference and exhibition, Texas, 24–27 Sept 2006

Yin HJ, Wang DM, Zhong SY et al (2012) Flow characteristics of viscoelastic polymer solution in micro-pores. In: SPE 154640 presented at SPE EOR conference at oil and gas West Asia, Oman, 16–18 April 2012

Zaitoun A, Kohler N (1987) The role of adsorption in polymer propagation through reservoir rocks. In: SPE 16274 presented at SPE international symposium on oilfield chemistry, Texas, 4–6 Feb 1987

Zaitoun A, Kohler N (1988) Two phase flow through porous media: effect of adsorbed polymer layer. In: SPE 18085 presented at SPE annual technical conference, Texas, 2–5 Oct 1988

Zhang LJ, Yue XA (2008) Displacement of polymer solution on residual oil trapped in dead ends. J Cent South Univ Technol 15:84–87.

Zhang Z, Li JC, Zhou JF (2011) Microscopic roles of "viscoelasticity" in HPMA polymer flooding for EOR. Transp Porous Media 86:199–214.

Zheng CG, Gall BL, Gao HW et al (2000) Effect of polymer adsorption and flow behavior on two-phase flow in porous media. SPE Reserv Eval Eng 3:216–223.

5

Estimation of the potential of an anaerobic thermophilic oil-degrading bacterium as a candidate for MEOR

Isty Adhitya Purwasena · Yuichi Sugai · Kyuro Sasaki

Abstract We investigated the decrease in oil viscosity caused by the biodegradation of crude oil by a bacterium (AR80) isolated from an oil reservoir, and estimated the potential for this bacterium for use in microbial enhanced oil recovery (MEOR). AR80 degraded long-chain *n*-alkanes preferentially, and anaerobically increased the ratio between the short-chain and long-chain *n*-alkane concentrations in the crude oil. The long-chain *n*-alkane metabolism by AR80 decreased the oil viscosity. AR80 grew well in a reservoir of brine supplemented with yeast extract, and decreased the oil viscosity to approximately 60 % of its original value. Adding a small amount of yeast extract (0.05 g/L) was necessary to stimulate the AR80 activity. MEOR can, therefore, be achieved using AR80 without incurring excessive costs. AR80 can grow at temperatures up to 80 °C and grows well at between 50 and 70 °C. AR80 can grow at a salinity of up to 90 g/L and grows well at a salinity of <30 g/L. The AR80 activity was not affected very much by high pressures (such as 6.0 MPa). Core flooding experiments were performed by injecting AR80 (in brine supplemented with yeast extract) into Berea sandstone cores. Gas chromatography analysis of the effluent oil suggested that long-chain *n*-alkanes in the residual oil were preferentially degraded by the AR80 in the porous rocks. The core flooding experiments showed that the AR80 activity in the porous rocks caused the oil recovery to be enhanced, so AR80 could be a suitable candidate for MEOR.

Keywords Microbial enhanced oil recovery · Oil-degrading bacteria · Oil viscosity · Yeast extract · Reservoir conditions · Core flooding

Introduction

Microbial enhanced oil recovery (MEOR) is well known as both an economically feasible and an environmentally friendly enhanced oil recovery technique that uses biotechnology or microbial biological activity. MEOR generally involves injecting effective microorganisms and nutrients into oil reservoirs where they are then incubated. Decreasing the viscosity of crude oil through biodegradation and/or bioproducts, such as surfactants and polymers, leads to improved oil recovery.

Molasses, which is produced as a byproduct of sugar refining, has been used in many MEOR field trials as a nutrient for the microorganisms used (Bass and Lappin-Scott 1997). Molasses is cheap and contains nutrients, such as sucrose, minerals, organic compounds, and vitamins (Hanaa et al. 2011). However, the cost of molasses has increased progressively over the last few decades because of its extensive use as a raw material for bioethanol production. For example, Chinese molasses rose in price from US$40/ton in 2000 to US$200/ton in 2007 (Fujiwara et al. 2008). The cost of transporting molasses is high if there are no sugar refinery plants near the oilfield being treated. A sterilization facility is necessary so that molasses, which often contains various bacteria, can be sterilized before use in MEOR. The activity of the injected microorganism(s) may

I. A. Purwasena
School of Life Science and Technology, Bandung Institute of Technology, Ganesha 10, Bandung 40132, West Java, Indonesia

Y. Sugai · K. Sasaki
Department of Earth Resources Engineering, Faculty of Engineering, Kyushu University, 744 Motooka, Nishi-ku, Fukuoka 819-0395, Japan

be hindered by the growth of unwanted bacteria in an oil reservoir because molasses is rich in nutrients that can stimulate the growth of a wide range of bacteria. It is, therefore, predicted that using molasses during MEOR will become unattractive in the near future.

Using microorganisms that specifically consume crude oil as their sole carbon source in MEOR processes could be a way of avoiding the need to add molasses. These microorganisms can break long-chain hydrocarbons in crude oil into shorter chains, resulting in a decrease in the oil viscosity and improved oil mobility. Lazar et al. (1999) suggested that microorganisms could be used to decrease the proportion of long-chain *n*-alkanes present in biotechnological processes that involve waxy crude oil, to improve the recovery of the oil. Yijiang et al. (1999) described a successful MEOR field trial using oil-degrading bacteria without the injection of any nutrients. They concluded that the oil recovery was enhanced because of a decrease in the oil viscosity caused by the injected bacteria. However, a number of unknown bacterial strains had been injected into the reservoir in that study, and it is unclear which bacteria had caused the decrease in viscosity, so the reproducibility of the effects was left as an open question. Later, single species of hydrocarbon-degrading bacteria, including *Bacillus* spp. (Wankui et al. 2006), *Brevibacillus* spp. (Wang et al. 2008), *Pseudomonas* spp. (Sadeghazad and Ghaemi 2003), and *Acinetobacter* spp. (Kotlar et al. 2007), were used in MEOR. However, all of these species are aerobic and/or facultative anaerobic bacteria, and need oxygen to degrade the hydrocarbons. To achieve this, an aerated nutrient solution was, therefore, injected into the oil reservoirs in the MEOR field trials using aerobic bacteria (Youssef et al. 2009). Using this method, the growth of the bacteria was limited to near the injection wells because the dissolved oxygen was entirely consumed there. This resulted in the oil viscosity only decreasing near the injection wells, so the enhancement in the oil recovery was not particularly great. Using anaerobic hydrocarbon-degrading bacteria in MEOR would, therefore, be a way of improving the MEOR performance.

Dong-Hui et al. (2008) described a novel anaerobic paraffin-degrading *Gordonia* sp. bacterium and suggested that this bacterium had the potential for being used in MEOR. However, the enhanced oil recovery potential of the bacterium was not demonstrated in their core flooding experiments. This bacterium may not be practically used for MEOR because it can only live at below 40 °C, which is lower than the temperature in many oil reservoirs. There have been few reports of thermophilic anaerobic hydrocarbon-degrading bacteria that could be practically used for MEOR so far. We are investigating the use of thermophilic anaerobic hydrocarbon-degrading bacteria for MEOR. We have successfully isolated an anaerobic bacterium, and identified it as a strain belonging to the *Petrotoga* sp., from an oil reservoir in Japan (Isty et al. 2014). Although six species of *Petrotoga* have been isolated from several oilfields so far (Miranda-Tello et al. 2007), none of them has been found to degrade hydrocarbons or be candidates for use in MEOR. Our isolate (named AR80) can grow at $\geq$50 °C in a culture medium containing crude oil as the sole carbon source. This species may, therefore, anaerobically degrade hydrocarbons and decrease the viscosity of oil at high temperatures. If this is the case it could be a candidate for use in MEOR.

The purpose of the study presented here was to demonstrate the potential use of AR80 for MEOR using incubation experiments and core flooding experiments, including typical reservoir conditions. In particular, we investigated the influences of several incubation conditions, both controllable and uncontrollable during MEOR, on the AR80 activity. The types of substrates and their concentrations (which are controllable conditions), and reservoir conditions, such as temperature, salinity, and pressure (which are uncontrollable conditions), were studied. The results of these investigations allow us to propose an effective method for using AR80 in MEOR and to describe the reservoir conditions under which AR80 can be used.

Experimental

Bacterial strain and culture medium preparation

An AR80 cell is rod-shaped, 0.25–0.75 μm wide, and 2.5–7.0 μm long, and it has a characteristic outer sheath-like structure and polar flagella, as shown in Fig. 1. The motility of AR80 cells can be expected to allow them to penetrate the micropores in reservoir rocks.

Pure culture experiments were performed using AR80 to estimate its ability to decrease oil viscosity. A culture medium consisting of natural brine and crude oil was used. The natural brine and the crude oil had both been collected from an oil-producing well in the Yabase oilfield, Akita, Japan. Table 1 shows the brine composition used in the study. The salinity of the brine was approximately 1.0 % and the API gravity of the crude oil was approximately 33°.

Filter-sterilized brine (50 mL) was put into autoclaved serum bottles (100 mL), and autoclaved crude oil (1.0 mL) was added to act as the sole carbon source. Each bottle was sealed with a butyl rubber cap and an aluminum crimp. The headspace was replaced with 10 % CO_2 in N_2 in all of the experiments because AR80 had been isolated from natural brine that had been incubated in the presence of CO_2 (Isty

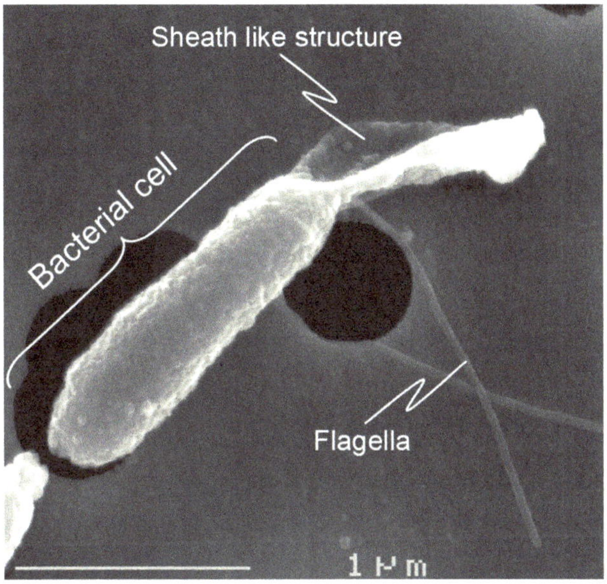

Fig. 1 Scanning electron microscopy image of the AR80 bacterium

Table 1 Compositions of the brine used for the culture medium

Component	Concentration (mg/l)
Na^+	3,075
K^+	30
Ca^{2+}	5
Mg^{2+}	3
Cl^-	2,500
I^-	1.5
NH_4^+	15
HBO_2^{2-}	250
T-Fe	2
HCO_3^-	4,000
Lactic acid	<0.1
Formic acid	2.2
Acetic acid	4.3
Propionic acid	<0.1

et al. 2014). An AR80 cell suspension (5 mL) was inoculated into each bottle containing culture medium and the bottles were incubated at 50 °C. The initial AR80 concentration in the culture medium in each bottle was 1.0×10^6 cells/mL. Several bottles with identical conditions were prepared for each experiment and an un-inoculated control was also prepared and incubated for each experiment.

Investigating the influence of controllable conditions on the AR80 activity

The controllable condition assessed was the nitrogen source available to the AR80. The abilities of yeast extract, urea, and ammonium nitrate to act as nitrogen sources were assessed, and they were added to the medium at concentrations of 0.100, 0.025, and 0.030 g/L, respectively, to give the same nitrogen concentration in each test medium. AR80 was also incubated in the culture medium without a nitrogen source as a control. The optimum concentration of the most suitable nitrogen source was assessed by performing incubation experiments with culture media containing different concentrations of that nitrogen source.

Investigating the influence of uncontrollable conditions on the AR80 activity

The influence of temperature on the AR80 activity was assessed. AR80 was incubated in the culture medium (containing the optimum concentration of the chosen nitrogen source) at 50, 60, 70, and 80 °C. The influence of salinity on the AR80 activity was assessed in incubation experiments using the medium containing the optimum concentration of the chosen nitrogen source with 10, 30, 60, and 90 g/L of NaCl present. The influence of pressure on the AR80 activity was assessed in incubation experiments using the medium containing the optimum concentration of the chosen nitrogen source in high-pressure cylinders at 6.0 MPa and 60 °C. The pressure was increased by injecting 10 % CO_2 into the cylinders at 0.2 MPa and then injecting pure N_2 into the cylinders to reach 6.0 MPa.

Analytical methods

Culture solution (100 μL) was extracted from each bottle every few days during the incubation experiments, and the growth of AR80 was determined by direct cell counting in a 0.02-mm deep bacteria chamber (Sunlead Glass Corp., Koshigaya, Japan) under phase contrast microscopy (BH-2; OLYMPUS Corp., Tokyo, Japan).

The crude oil was extracted into benzene (5 mL) that was placed into each bottle every few days during the incubation experiments. The benzene was completely evaporated from the extracted solution at 50 °C. The viscosity of the extracted crude oil was measured using a rotational viscometer (Programmable Rheometer DV-III; Brookfield Eng., Middleboro, MA, USA) at 50 °C. The crude oil was also analyzed by gas chromatography (GC) with a flame ionization detector (FID) (PU 4600; Philips, Eindhoven, the Netherlands) and the ratio of lighter to heavier *n*-alkanes components in the crude oil was estimated. A DB-1 chromatography column (Agilent J&W, Santa Clara, CA, USA) was used for the analysis, and the carrier gas was nitrogen, used at a flow rate of 2.5 mL/min. The oven temperature was increased from 60 to 280 °C at a rate of 10 °C/min and kept at the final temperature for 27 min. The extracted crude oil was diluted with n-hexane and injected into the column using splitless injection mode. Each hydrocarbon component separated was detected by FID. Pure n-alkanes with carbon numbers of 10–32 were used as standards. The relative abundance of each n-alkane in the crude oil was calculated using Eq. 1.

$$R_i = \frac{A_i}{\sum A_i}, \tag{1}$$

where R_i is the relative abundance of an n-alkane with a carbon number of i and A_i is the peak area of that n-alkane. The data presented are the means of the results from triplicate culture bottle tests and the analyses were repeated at least twice.

Core flooding experiment procedure

Duplicate core flooding experiments were performed with microbial treatment (T1 and T2) and without microbial treatment (NT1 and NT2). A schematic of the experimental setup used for the core flooding experiments is shown in Fig. 2. The properties of the Berea sandstone cores that were used in the study are shown in Table 2. The cores were 70.0 mm long and 38.1 mm wide. The absolute permeability of each core was 300 millidarcy. The cores were sterilized at 180 °C for 6 h. Each core was then vacuum saturated with filter-sterilized brine to determine its porosity and water wettability. The porosity of each core was found to be approximately 20.5 %. The brine-saturated cores were installed into Hassler-type core holders and 1.0 MPa radial confining pressure was introduced using pure nitrogen. Then, 3.5 pore volume (PV) of culture medium containing 1.0 × 10^6 cells/mL AR80 was injected into cores T1 and T2 anaerobically, at a constant pressure of 0.02 MPa, to achieve a high concentration of bacteria spread through the rock pores. Similarly, 3.5 PV of sterile culture medium was injected into cores of NT1 and NT2 at the same injection pressure. The culture medium was natural brine supplemented with the nitrogen source chosen from the results of the incubation experiments. The culture medium was injected into the cores under pressure provided by applying 10 % CO_2 in nitrogen. Crude oil that had been sterilized (in an autoclave) was then injected into the cores at a similar injection pressure until no more brine flowed out. Approximately 6.0 mL of water effluent was collected from each core during the crude oil injection process, so the original oil in place (OOIP) volume and the initial oil saturation were approximately 6.0 mL and 36.6 %, respectively.

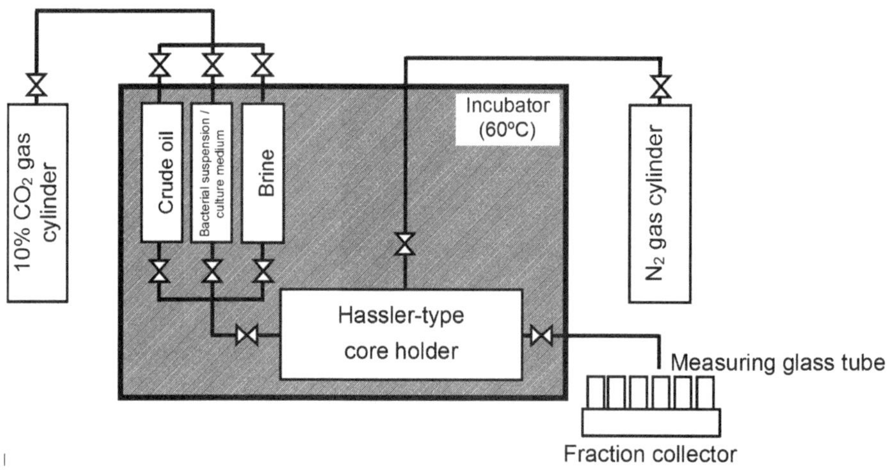

Fig. 2 Schematic image of the experimental setup for the core flooding experiments

Table 2 Properties of the Berea sandstone cores before the treatments

	Experimental number			
	T1	T2	NT1	NT2
Treatment	Treated with AR80		Non-microbial treatment (Control)	
Core length (mm)	70.0			
Core diameter (mm)	3.81			
Pore volume (mL)	16.48	16.31	16.26	16.55
Porosity (%)	20.65	20.44	20.37	20.73
Absolute permeability (mD)	300			
Oil viscosity at 60 °C (cP)	7.0			
Original oil in place (mL)	6.37	6.10	5.78	5.75
Initial oil saturation (%)	38.65	37.40	35.55	34.75

The culture medium containing 1.0×10^6 cells/mL of AR80 was again injected into cores T1 and T2 at the same injection pressure until no more oil was observed in the effluent. Similarly, the sterile culture medium was injected into cores NT1 and NT2. The effluent was collected into glass measuring tubes at specific time intervals, and the amount of effluent oil and water was determined volumetrically. At this stage of the experiment, the crude oil in the water effluent was called 'primary oil recovery'. All of the core holders (for both the bacterial treatments and the controls) were then closed and incubated at 60 °C for 2 weeks, then 3.5 PV of filter-sterilized brine was injected into each core at the same injection pressure. The effluent was collected into glass measuring tubes at specific time intervals. The crude oil obtained during this stage was called 'secondary oil recovery'. The bacterial cell numbers in the effluent collected during the secondary recovery were counted by phase contrast microscopy. The primary oil recovery and secondary oil recovery crude oil samples were analyzed by GC-FID.

Results

Influence of controllable conditions on the AR80 activity

Type of nitrogen source

Figure 3 shows the AR80 growth curves in the culture media supplemented with yeast extract, ammonium nitrate, and urea. AR80 grew well in all of the culture media, but particularly well in the culture medium supplemented with yeast extract. The AR80 growth rates in the culture media supplemented with yeast extract, ammonium nitrate, and urea were 0.154, 0.149, and 0.055 h^{-1}, respectively. Figure 4 shows the decreases in oil viscosity in the culture media supplemented with yeast extract, ammonium nitrate, and urea. The highest decrease in oil viscosity was found in the culture medium supplemented with yeast extract. The oil viscosity decreased from 26.6 to 16.6 cP after 2 weeks of incubation with AR80 and yeast extract, whereas it only decreased to 17.2 and 19.5 cP with ammonium nitrate and urea, respectively. Figure 5 shows a chromatogram of crude oil that had been treated with AR80 in the medium supplemented with yeast extract for 2 weeks. For comparison, a chromatogram of the crude oil in a control sample is also shown. Almost all of the *n*-alkanes in the treated crude oil were degraded slightly, suggesting that AR80 simultaneously degraded a broad range of *n*-alkanes in the crude oil. Figure 6 shows the *n*-alkane composition ratio in the crude oil that had been treated for 2 weeks under the four different test conditions. The C26–C30 fraction was lower

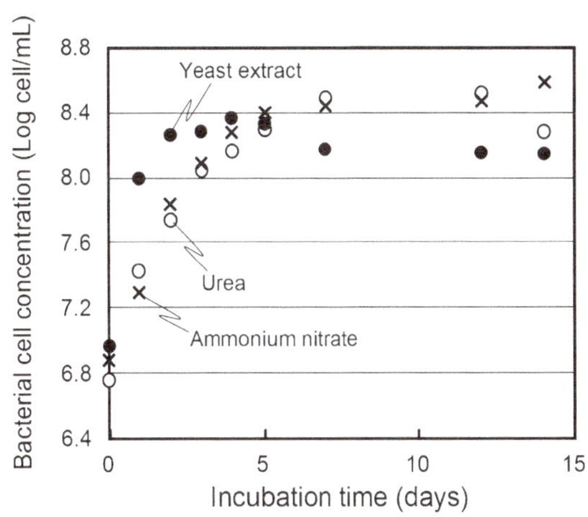

Fig. 3 AR80 growth curves with different nitrogen sources

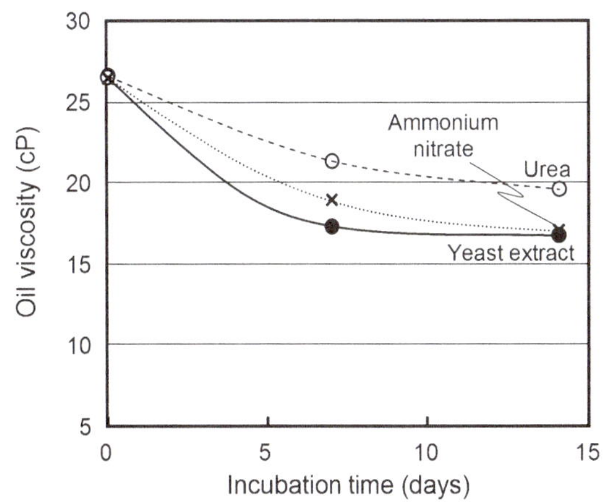

Fig. 4 Viscosity of oil treated with AR80 with different nitrogen sources

in all of the treated samples than in the controls. The C16–C20 fraction was also lower in the treatments supplemented with yeast extract and urea than in the controls. In contrast, the short-chain *n*-alkane (such as the C10–C15) fractions were higher in the treatments supplemented with yeast extract and urea than in the controls.

These results show that yeast extract is the most suitable nitrogen source for stimulating AR80 activity, so yeast extract was used as the nitrogen source in the subsequent experiments.

Yeast extract concentration

AR80 was incubated in culture media containing 0.05, 0.10, 0.20, and 0.50 g/L of yeast extract, and the AR80 growth curves are shown in Fig. 7. The AR80 growth rate increased as the yeast extract concentration increased. The growth rate was 0.92 h^{-1} at the highest yeast extract concentration, and 0.81, 0.76, and 0.71 h^{-1} at yeast extract concentrations of 0.20, 0.10, and 0.05 g/L, respectively.

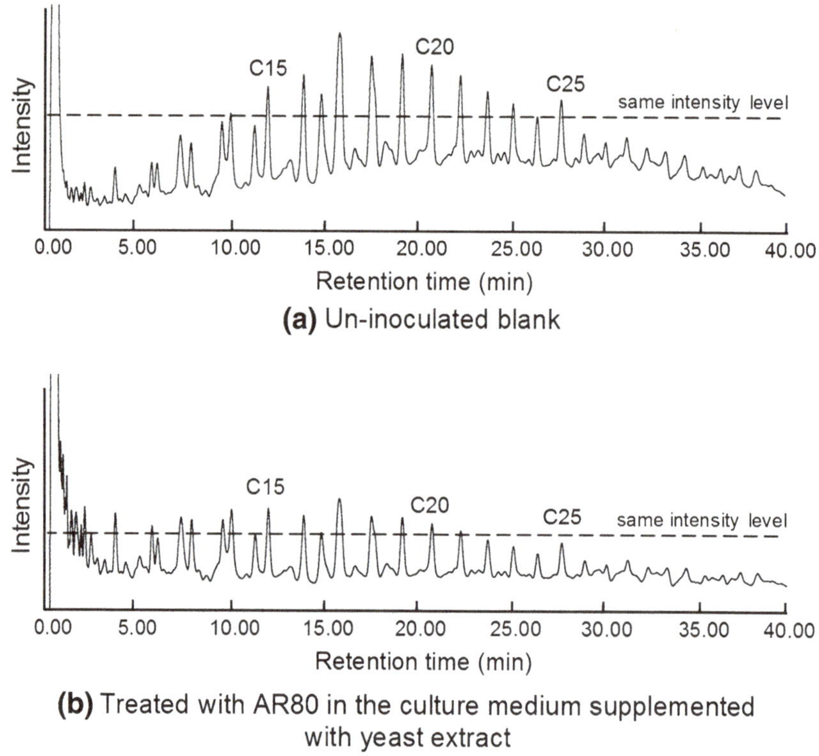

Fig. 5 Chromatograms of crude oil treated with **a** un-inoculated blank and **b** AR80 in the culture medium supplemented with yeast extract

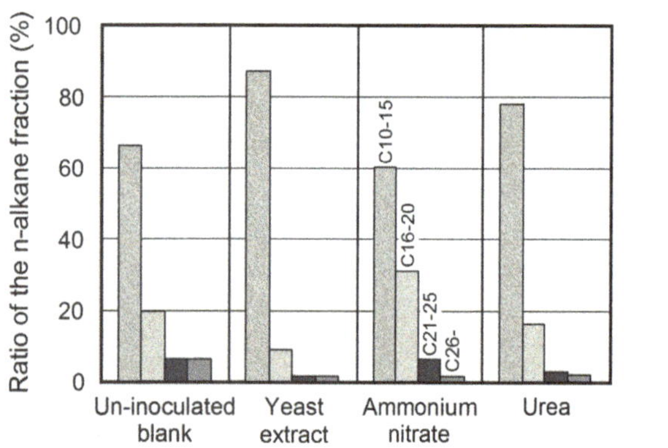

Fig. 6 Comparison of the *n*-alkane fractions in the oil treated using different nitrogen sources

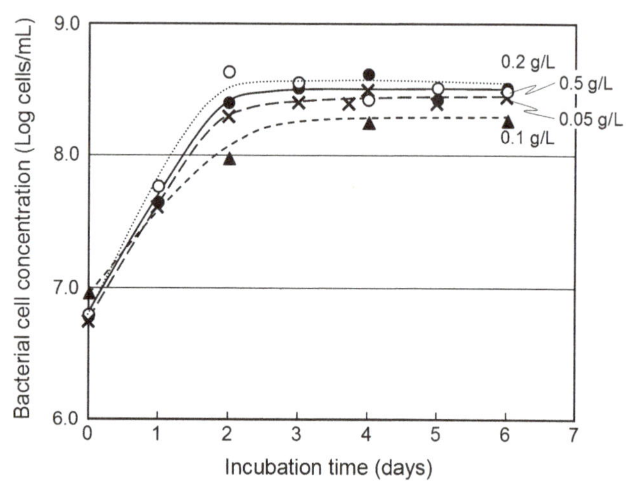

Fig. 7 Growth curves for AR80 incubated with different yeast extract concentrations

Although the growth rate was different for the different treatments, the bacterial cell number increased during the incubation period, to 2.0–3.0 × 10^8 cells/mL, in all of the treatments. The decrease in the oil viscosity was also similar for each treatment (data not shown). In the subsequent experiments, 0.05 g/L of yeast extract was used.

Influence of uncontrollable conditions on the AR80 activity

Temperature

Growth curves for AR80 incubated at different temperatures are shown in Fig. 8. The AR80 growth rate decreased with increasing temperature. The growth rates were 0.66, 0.48, 0.25, and 0.15 h^{-1} at 50, 60, 70, and 80 °C, respectively.

The viscosity of the crude oil increased in the control samples that had been incubated at above 60 °C, and this was caused by the light components in the crude oil evaporating at such high temperatures. Therefore, we estimated the decrease in the oil viscosity as the relative viscosity, that is, relative to the viscosity of the oil in the corresponding control sample. Figure 9 shows the relative viscosities of the oil after it had been treated with AR80 at

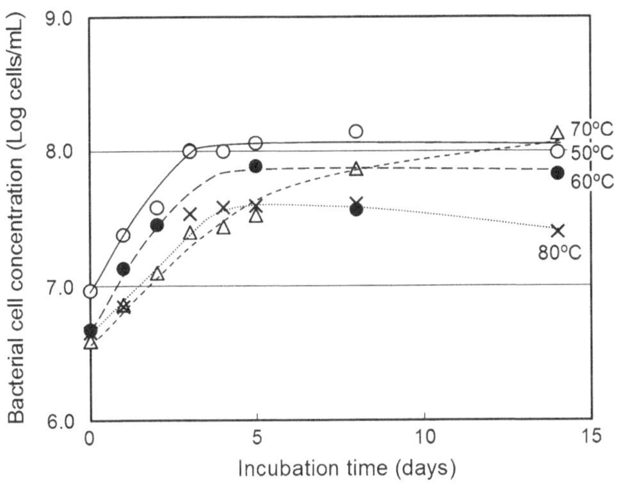

Fig. 8 Growth curves for AR80 incubated at different temperatures

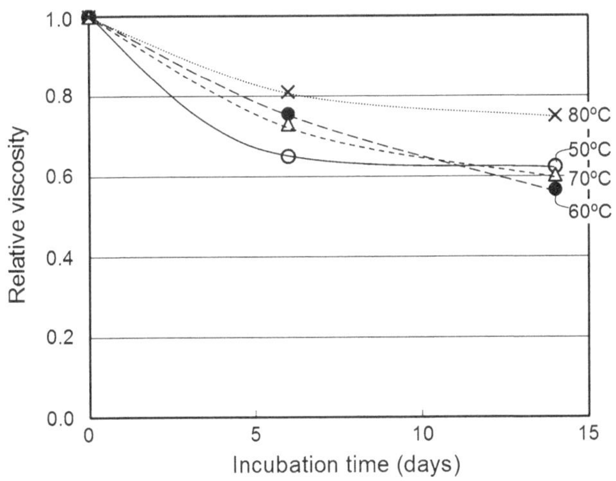

Fig. 9 Viscosity of oil treated with AR80 at different temperatures

50, 60, 70, and 80 °C. The relative viscosity decreased to 0.62, 0.57, 0.60, and 0.75 after 2 weeks of incubation with AR80 at 50, 60, 70, and 80 °C, respectively.

Salinity

Figure 10 shows the AR80 growth curves at different salinities. AR80 was able to grow at each salinity that was tested, but its growth rate and maximum number of cells decreased as the salinity increased. The specific growth rate of AR80 was 0.66 h^{-1} at a salinity of 10 g/L, which was original salinity of the brine, and 0.49, 0.31, and 0.22 h^{-1} at salinities of 30, 60, and 90 g/L, respectively. The bacterial cell numbers reached 2.0×10^8, 1.5×10^8, 5.0×10^7, and 2.5×10^7 cells/mL at salinities of 10, 30, 60, and 90 g/L, respectively. The oil viscosity could not be measured accurately in this experiment because the oil formed a mousse with the highly saline media during the incubation period. The decrease in oil viscosity can be expected to have become smaller as the salinity increased based on the AR80 growth results at each salinity condition described above.

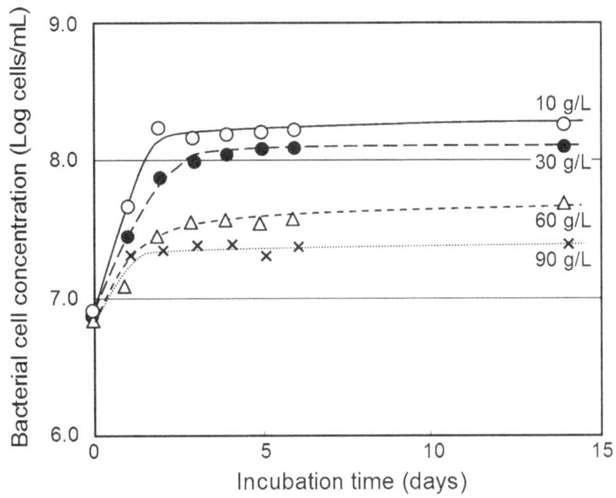

Fig. 10 Growth curves for AR80 incubated at different salinities

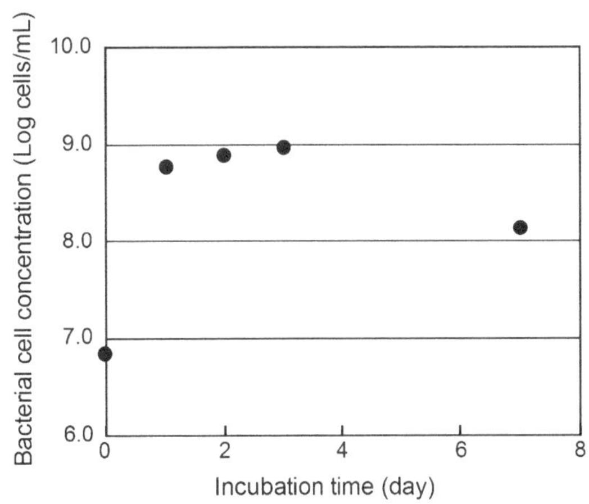

Fig. 11 Growth curve for AR80 incubated at 6.0 MPa

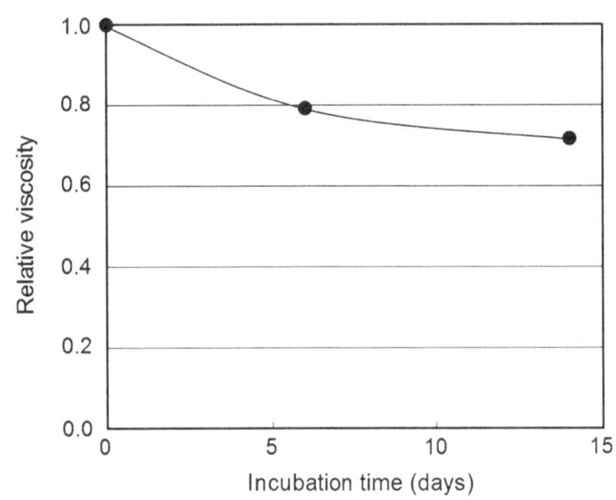

Fig. 12 Viscosity of oil treated with AR80 at 6.0 MPa

Pressure

Figure 11 shows the growth curve of AR80 incubated at 6.0 MPa and 60 °C. The AR80 had a specific growth rate of 0.65 h^{-1} and the number of cells increased to 1.0×10^9 cells/mL, which was the highest number of cells achieved in this study. Figure 12 shows the changes in the relative viscosity of the oil during the incubation period. The relative viscosity decreased to 0.65 after a 2-week incubation period.

Core flooding experiments

AR80 started to be detected in the effluent brine once 0.3 PV of the microbial suspension had been injected. The number of AR80 cells in the effluent became relatively constant, at around 6.5×10^5 cells/mL (35 % less than the original bacterial suspension concentration), once 0.7 PV of the microbial suspension had been injected.

Figure 13 shows the cumulative oil recovery from each experiment. There was little difference in the cumulative oil recovery between the experiments, with 49.9, 41.6, 39.6, and 41.7 % of the OOIP being recovered in the T1, T2, NT1, and NT2 experiments, respectively, as primary oil recovery. The injected bacterial suspension broke through the core later in experiment T1 than in the other experiments because the flow velocity was lower in experiment T1 than in the other experiments. Therefore, the primary oil recovery was higher in experiment T1 than in the other experiments. A negligible amount of crude oil was drained from the core in the early stages of secondary oil recovery. The drainage of the remaining oil started to be clearly detected after 0.2–0.4 PV of brine had been injected in both of the microbial treatment experiment cores, while the remaining oil was not found to drain in either of the control experiments. Drainage of the remaining oil was observed until 1.5 PV of brine had been injected in both of the microbial treatment experiments. The final oil recovery in experiments T1 and T2 reached 51.4 and 42.8 %, respectively, i.e., 1.5 and 1.2 % of the OOIP had been drained from the cores because of the microbial treatments (Table 3).

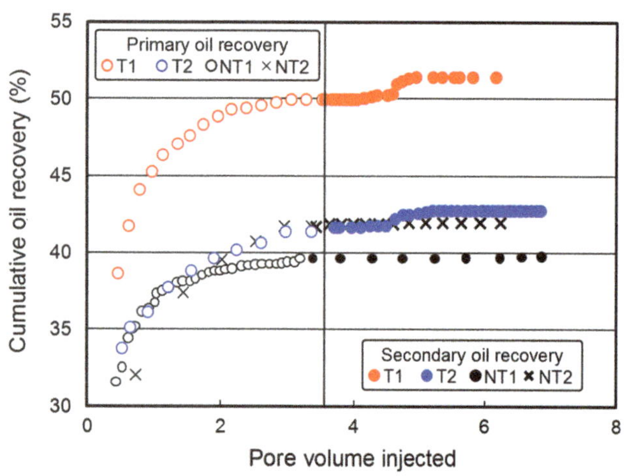

Fig. 13 Cumulative oil recovery in each core flooding experiment

Table 3 Results of the core flooding experiments

	Experimental number			
	T1	T2	NT1	NT2
Pore volume of injected culture medium in the primary oil recovery	3.90	3.74	3.18	3.35
Primary oil recovery (%)	49.92	41.64	39.62	41.70
Residual oil saturation after the primary oil recovery (%)	19.35	21.83	21.46	20.26
Pore volume of injected culture medium	3.0 (with AR80)		3.0 (without AR80)	
Shut-in time (days)	14			
Pore volume of injected brine in the secondary oil recovery	2.60	3.16	4.08	2.99
Secondary oil recovery vs. OOIP (%)	1.46	1.21	0.29	0.33
Secondary oil recovery vs. residual oil (%)	2.92	2.90	0.49	0.57

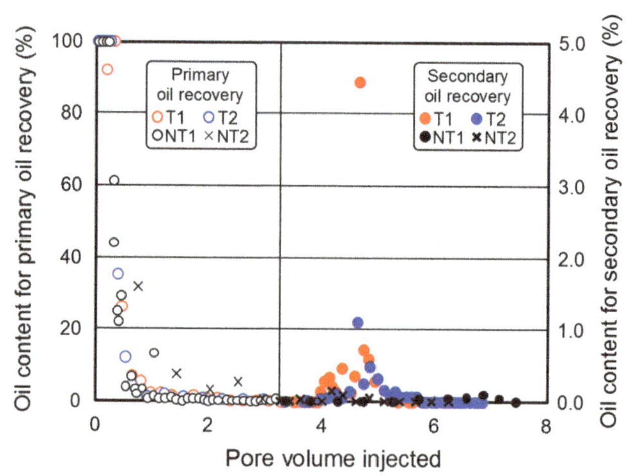

Fig. 14 Oil content in the effluent from each core flooding experiment

The changes in the oil content in the effluent from each experiment are shown in Fig. 14. The oil content decreased sharply once 0.2–0.4 PV of brine had been injected, and oil drainage was not observed after 3.0 PV of brine had been injected during the primary oil recovery. The oil content started to increase once 0.2–0.4 PV of brine had been injected in both of the microbial treatment experiments, as described above, during secondary oil recovery. Maximum

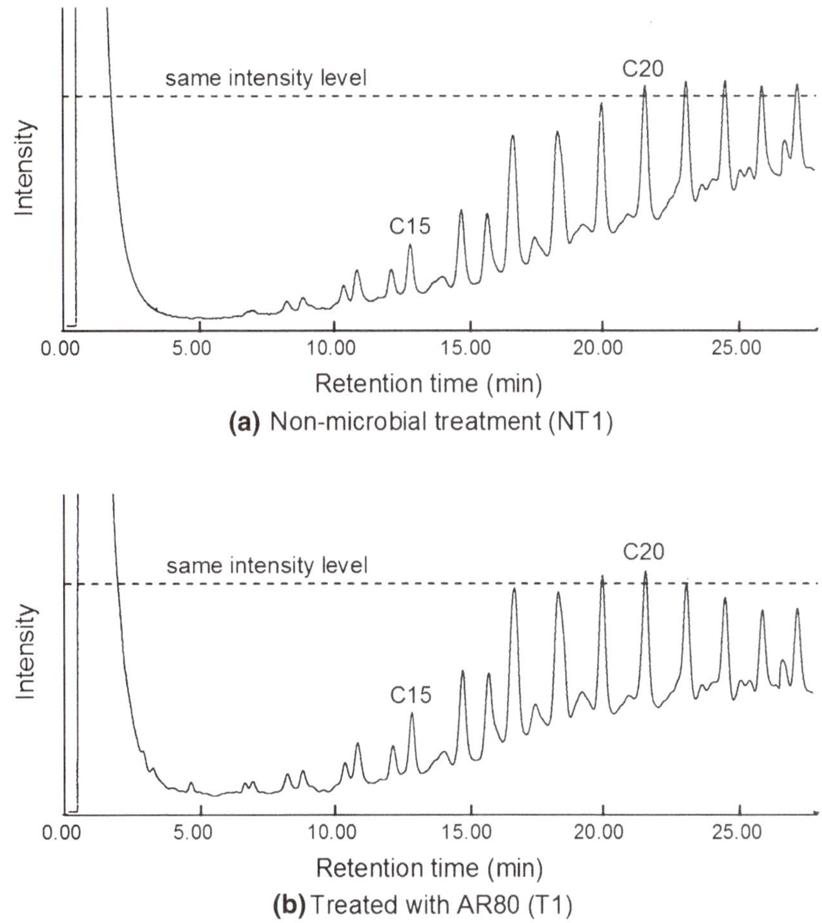

Fig. 15 Chromatograms of crude oil treated with **a** non-microbial control (NT1) and **b** AR80 (T1)

oil contents of 4.4 and 1.1 % were observed once 1.1 PV of brine had been injected in experiments T1 and T2, respectively. Drainage of the remaining oil finished once 1.5–2.0 PV of brine had been injected.

The number of AR80 cells in the effluent that flowed from the microbial-treated cores was observed when between 0.2 and 0.3 PV of brine had been injected. The numbers of AR80 cells in the T1 and T2 effluents were 1.4×10^7 and 1.8×10^7 cells/mL, respectively. Fluorescence microscopy revealed that AR80 cells remained in the cores after the secondary oil recovery (data not shown). The effluent oil chromatograms for experiments T1 and NT1 are shown in Fig. 15. The C20 and lower *n*-alkane fractions were found at higher proportions in the cores treated with AR80 than in the controls, while the C21 and higher *n*-alkane fractions were found at lower proportions in the cores treated with AR80 than in the controls. Although the viscosity of the effluent oil could not be measured because enough oil to make the measurements could not be recovered in the experiments, a decrease in the oil viscosity in the cores could be expected based on the GC analysis results.

Discussion

Mechanisms involved in AR80 causing the oil viscosity to decrease

AR80 can degrade long-chain *n*-alkanes preferentially, increasing the ratio between the short-chain and long-chain *n*-alkane concentrations in crude oil. Metabolism by AR80 caused the decrease in oil viscosity described above. Although a large number of microorganisms, with a wide range of physiological properties, have been isolated from oil reservoirs, only a very small number of anaerobic thermophilic oil-degrading bacteria have been reported. Rueter et al. (1994) identified only one anaerobic thermophilic oil-degrading bacterium that was potentially capable of inhabiting deep reservoirs. That was a sulfate-reducing bacterium that generated hydrogen sulfide when it degraded hydrocarbons, so it should not be used in MEOR to avoid reservoir souring. AR80 can degrade crude oil without generating hydrogen sulfide, so it is a candidate for use in MEOR.

Supplementary nutrition for AR80 activity

The highest AR80 growth rates, and the highest numbers of cells, were found in the culture medium supplemented with yeast extract. Similarly, the greatest decrease in oil viscosity was found in the culture medium supplemented with yeast extract. Yeast extract contains many amino acids, which bacteria can take up easily, so the AR80 activity was stimulated in the culture medium supplemented with yeast extract. The yeast extract concentration had little or no effect on the maximum number of AR80 cells or the decrease in oil viscosity. From the standpoint of costs for MEOR operations, 0.05 g/L yeast extract is necessary and is sufficient to stimulate AR80 activity. Yeast extract for food industry use is not very expensive, at US$1–10/ton, although the cost of high-quality yeast extract for research use (used in this study) is more than US$ 300/kg. If AR80 activity can also be stimulated by yeast extract intended for food industry use, MEOR using AR80 could be performed without impairing its economic advantage.

Screening criteria for MEOR using AR80

AR80 grew at all of the temperatures tested, although the growth rate decreased as the temperature increased. Although the AR80 growth rate was lower at 70 °C than at 50 or 60 °C, the number of bacterial cells reached the same level at 70 °C as at 50 or 60 °C after 2 weeks of incubation. As a result, the decrease in the relative viscosity was almost the same for the oil samples treated at 50, 60, and 70 °C after 2 weeks of incubation. The relative viscosity was also decreased even at 80 °C, although it was higher at that temperature than at the other temperatures. The most suitable reservoir temperature for MEOR using AR80 is, therefore, 50–70 °C, but AR80 can be used in reservoirs with temperatures below 80 °C. Although AR80 grew even at a high salinity (e.g., 90 g/L), the growth rate became lower as the salinity increased. The most suitable reservoir salinity for MEOR using AR80 is <30 g/L, but AR80 can be used in reservoirs with a salinity below 90 g/L. Although the AR80 growth improved, the decrease in oil viscosity became smaller at higher pressures, such as 6.0 MPa. This result agrees with that reported by Atlas (1981), who showed that 94 % of hexadecane was used by bacteria within 8 weeks at 1 bar but that it took 40 weeks to achieve a similar degradation level by the same bacteria at 500 bars. It appears that the rate of hydrocarbon degradation by AR80 is lower at high pressures than at ordinary pressures. However, the relative viscosity of the oil decreased to 0.72 after a 2-week incubation period at high pressure, so the AR80 activity will not be seriously affected by higher reservoir pressures.

Mechanisms involved in oil recovery during MEOR using AR80

Some of the injected AR80 remained in the porous rocks during the injection of the bacterial suspension, which might be caused by their adsorption onto the rock surface and/or their clogging of the pore space. However, adsorption and clogging were assumed not to affect the injectivity of AR80 because the effluent rate remained relatively constant throughout the injection (data not shown). AR80 can be used in MEOR from the viewpoint of its ability to flow through porous rocks.

More than 10^7 cells/mL of AR80 were found in the effluent sampled during secondary oil recovery, suggesting that AR80 increased more than tenfold in the cores during the incubation period. Assuming that some of the AR80 remained in the cores, AR80 could be expected to increase to higher cell numbers within the rock than were observed in the effluent brine. These results suggest that AR80 can grow actively in porous rocks.

The bacterial cell concentration was 35 % lower in the effluent brine than it was in the original bacterial suspension because of the adsorption of bacterial cells onto the sandstone surface, as described above. There is also a possibility that some components of the culture medium adsorbed onto the sandstone surface. For these reasons, we assumed that there were bacterial concentration and/or nutrient concentration gradients in the cores after the primary oil recovery stage, and that their concentrations would have been higher in the upstream parts of the cores than in the downstream parts of the cores. As a result, the remaining oil in the upstream parts of the cores would have been actively degraded by AR80, and its viscosity would have become lower in the upstream than in the downstream parts of the cores. This would have led to a delay in finding enhanced oil recoveries during the secondary oil recovery stage (in Figs. 13, 14) relative to the average decrease in the oil viscosity in the core.

A decrease in the viscosity of oil treated with AR80 in the cores can be expected because an increase in the ratio between the short-chain *n*-alkane and long-chain *n*-alkane concentrations was found in the effluent oil, as described above. The decrease in viscosity would have improved the oil–water mobility ratio and the water sweep efficiency during secondary oil recovery. The brine injected during secondary oil recovery would, therefore, have flowed into regions that had not been flooded by brine during primary oil recovery. The remaining oil in those regions would have been displaced by the brine and an oil bank would have formed ahead of the slug of brine injected during secondary oil recovery. A high oil content fluid drained from the core when the oil bank reached the core outlet, as shown in Fig. 14.

The production of effective bioproducts, such as surfactants, gases, or polymers, by AR80 was not observed in the incubation experiments using the culture medium, so the additional oil recovery appears to have solely been caused by the decrease in viscosity of the remaining oil by AR80.

Conclusions

We assessed the potential for using an anaerobic thermophilic bacterium named AR80, which we isolated from an oilfield, as a candidate for MEOR. We assessed the ability of AR80 to anaerobically degrade hydrocarbons in crude oil and decrease the crude oil viscosity in a number of incubation experiments. We assessed, from the MEOR standpoint, the influence of controllable conditions, such as the type of nutrient present and uncontrollable conditions, such as the temperature, salinity, and pressure, on the AR80 activity. We also performed core flooding experiments to evaluate the enhancement of oil recovery during MEOR that could be achieved using AR80. Our main findings are described below.

1. AR80 was characterized as an anaerobic hydrocarbon-degrading microorganism, which could preferentially degrade long-chain *n*-alkanes in crude oil and improve the ratio between the short-chain and long-chain *n*-alkane concentrations in the crude oil. The viscosity of the crude oil was decreased by this selective degradation.
2. Yeast extract is the most suitable nitrogen source for AR80 activity. The oil viscosity was decreased to 60 % of its original value after incubation with AR80 in the culture medium supplemented with yeast extract. From the standpoint of the cost of MEOR, 0.05 g/L is a necessary and sufficient yeast extract concentration for stimulating AR80 activity.
3. The most suitable temperature for AR80 activity is 50–70 °C, but AR80 can be used in reservoirs with temperatures below 80 °C.
4. A salinity of below 30 g/L is suitable for AR80 activity, but AR80 can be used in reservoirs with salinities below 90 g/L.
5. The AR80 activity is not be seriously affected by the reservoir pressure.
6. AR80 can be injected into and grow in porous rocks. AR80 can degrade long-chain *n*-alkanes preferentially and improve the ratio between the short-chain and long-chain *n*-alkane concentrations in crude oil in porous rocks.
7. AR80 activity enhanced the oil recovery from porous rocks in the core flooding experiments, so AR80 could be a good candidate for use in MEOR.

Acknowledgments This work was supported by JSPS KAKENHI Grant Number 21760678 and Kyushu University G-COE program Novel Carbon Resource Science. We gratefully acknowledge the INPEX Corporation in Japan for their support and cooperation in providing the brine sample. We are also grateful to Dr. Kazuhiro Fujiwara for his technical advice. We appreciate technical support for our anaerobic incubation experiments provided by the Research Support Center, Graduate School of Medical Sciences, Kyushu University.

References

Atlas RM (1981) Microbial degradation of petroleum hydrocarbons: an environmental perspective. Microbiol Rev 45:180–209

Bass C, Lappin-Scott H (1997) The bad guys and the good guys in petroleum microbiology. Oilfield Rev 9:17–25

Dong-Hui H, Jian-Qun L, Xin S, Jian-Qiang L, Yu-Jie S, Yin-Bo Q (2008) Isolation, identification, and performance studies of a novel paraffin-degrading bacterium of *Gordonia amicalis* LH3. Biotechnol Bioproc Eng 13:61–68

Fujiwara K, Sugai Y, Chui J, Enomoto H (2008) Recent advances and prospects in microbial enhanced oil recovery (MEOR)—movement towards practical EOR technology. J Jpn Assoc Pet Technol 73:296–306

Hanaa AS, Sanket J, Yahya AW, Saif AB, Abdulkadir E, Ali AB (2011) Microbial biotechnology for enhancing oil recovery: current developments and future prospects. Biotechnol Bioinf Bioeng 1:147–158

Isty AP, Sugai Y, Sasaki K (2014) Utilization of natural reservoir brine in enrichment culture medium: an alternative approach for isolation of anaerobic bacteria from an oil reservoir. Pet Sci Technol (to be published)

Kotlar HK, Wetzel A, Thorne-Holst M, Zotchev S, Ellingensen T (2007) Wax control by biocatalytic degradation in high-paraffinic crude oils. SPE paper 106420, presented at the SPE international symposium on oilfield chemistry, Houston

Lazar I, Voicu A, Nicolescu C, Mucenica D, Dobrota S, Petrisor IG, Stefanescu M, Sandulescu L (1999) The use of naturally occurring selectively isolated bacteria for inhibiting paraffin deposition. J Pet Sci Eng 22:161–169

Miranda-Tello E, Fardeau M-L, Joulian C, Magot M, Thomas P, Tholozan J-L, Olliver B (2007) *Petrotoga halophila* sp. nov., a thermophilic, moderately halophilic, fermentative bacterium isolated from an offshore oil well in Congo. Int J Syst Evol Microbiol 57:40–44

Rueter P, Rabus R, Wilkes H, Aeckersberg F, Rainey FA, Jannasch HW, Widdel F (1994) Anaerobic oxidation of hydrocarbons in crude-oil by new types of sulfate-reducing bacteria. Nature 372:455–458

Sadeghazad A, Ghaemi N (2003) Microbial prevention of wax precipitation in crude oil by biodegradation mechanism. SPE paper 80529, presented at the SPE asia pacific oil and gas conference and exhibition, Jakarta

Wang J, Ma T, Zhao L, Lv L, Li G, Zhang H, Zhao B, Liang F, Liu R (2008) Monitoring exogenous and indigenous bacteria by PCR-DGGE technology during the process of microbial enhanced oil recovery. J Ind Microbiol Biotechnol 35:619–628

Wankui G, Chengfang S, Zhenyu Y, Zhaowei H, Rui J, Ying W, Jiyuan Z, Guoghao S (2006) Microbe-enhanced oil recovery technology obtains huge success in low-permeability reservoirs in Daqing oilfield. SPE paper 104281, presented at the SPE eastern regional meeting, Canton

Yijiang Z, Zhengshun X, Ping J, Weihong H (1999) Microbial EOR laboratory sudies and application results in Daqing oilfield. SPE paper 54332, presented at the SPE asia pacific oil and gas conference and exhibition, Jakarta

Youssef N, Elshahed MS, McInerney MJ (2009) Microbial process in oil fields: culprits, problems and opportunities. In: Allen IL, Sima S, Geoffrey MG (eds) Advances in applied microbiology. Academic press, Burlington, pp 141–251

6

Seismic-attribute identification of brittle and TOC-rich zones within the Eagle Ford Shale, Dimmit County, South Texas

Osareni Ogiesoba · Ursula Hammes

Abstract Analysis of 3D poststack seismic attributes can be used to identify areas of high exploration potential within shale resource plays. We integrated seismic attributes and acoustic impedance (AI) with wireline logs to determine total organic carbon (TOC) distribution within the Eagle Ford Shale in South Texas. We computed TOC from wireline logs using the Δ Log R method and then used seismic attributes to predict TOC and deep-resistivity log distribution, and identify brittle zones within the seismic survey. Our results show that high-TOC and high-resistivity zones are laterally more continuous in the south part of the survey. In the north, continuity of these properties is broken by NE–SW-trending faults having throws ranging from about 10 to 100 ft (3–30 m). High resistivity occurs in high-quality-factor (Q) attribute zones. Although the relationship is nonlinear, resistivity and TOC increase as Q increases. That is, both properties increase with increasing bed resistance suggesting increasing carbonate. Two high-resistivity zones, an upper resistive bed and a lower resistive bed, are identified within the Eagle Ford Shale. Additionally, because a strong positive linear relationship exists between AI and Q, Q can be used to identify brittle zones. Compared to other attributes used in identification of brittle zones, Q is faster and cheaper to compute from 3D poststack seismic data. Therefore, Q could serve as a quicker, alternate method of identifying brittle zones within the Eagle Ford Shale.

Keywords Inversion · TOC · Resistivity · Brittleness · Faults · Attributes

O. Ogiesoba · U. Hammes
Bureau of Economic Geology, Jackson School of Geosciences, The University of Texas at Austin, Austin, USA

Introduction

The Cretaceous Eagle Ford Shale in South Texas is one of the most prolific shale resource plays in the United States. Covering an area of more than 11,000 mi^2 ($\sim$28,160 km^2) (Treadgold et al. 2011), it stretches from the East Texas basin, which is bordered by the Sabine uplift in the NE, across the San Marcos arch, to the Maverick Basin in the SW (Hentz and Ruppel 2011) (Fig. 1a). The study area is located in the SW corner of the productive trend (Fig. 1a). Since the Eagle Ford Shale was recognized as a major gas shale play in 2008, exploration there has been geared toward delineating fault-related fracture zones, as well as stratigraphic layers having abundant minor faults and brittle zones, to find the most productive intervals (Kuich 1989; Treadgold et al. 2011). Advances in horizontal-well construction and hydraulic stimulation have made it possible to generate fractures in rocks that would enable fluid migration into the well bore, thereby transforming shale-gas resources into economic reserves (Rickman et al. 2008; Sondergeld et al. 2010). The success or failure of hydraulic stimulation, however, depends on our ability to accurately identify and delineate brittle and ductile zones (Laubach et al. 2009) within the shale-gas interval. Brittle zones are those rock layers characterized by high velocity—that is, high Young's modulus—whereas ductile zones are layers characterized by low velocity—that is, low Young's modulus (Koesoemadinata et al. 2011; Sena et al. 2011). In this regard, seismic technology has been employed successfully through amplitude-versus-offset (AVO) inversion, which computes elastic properties—the $\lambda\mu$ and $\mu\rho$ attributes (where λ and μ are Lamé's constants and ρ is density) used to determine brittleness of the rock (Goodway et al. 1998; Koesoemadinata et al. 2011; Sena

(a)

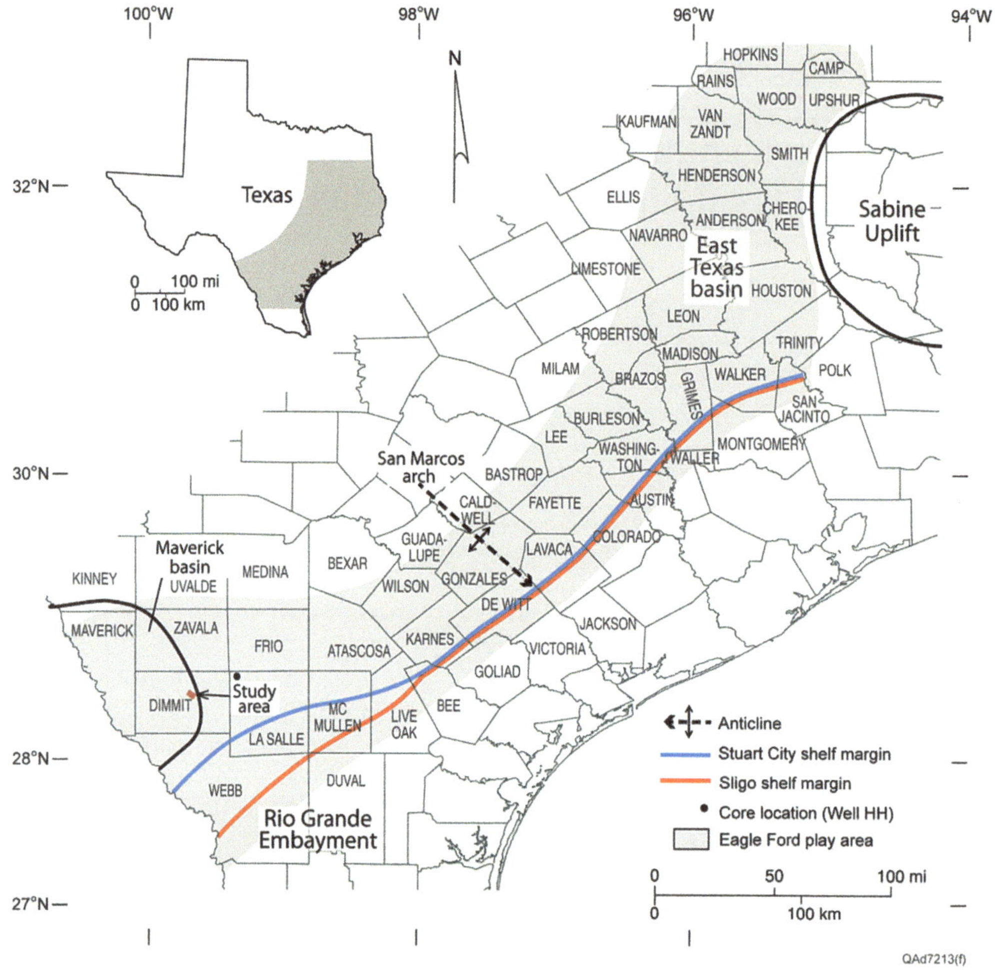
100°W
98°W
96°W
94°W
32°N
30°N
28°N
27°N
N
Texas
0 100 mi
0 100 km
HOPKINS
RAINS
CAMP
WOOD
UPSHUR
KAUFMAN
VAN ZANDT
SMITH
HENDERSON
ELLIS
ANDERSON
CHEROKEE
NAVARRO
Sabine Uplift
East Texas basin
LIMESTONE
HOUSTON
LEON
ROBERTSON
TRINITY
MADISON
MILAM
BRAZOS
GRIMES
WALKER
POLK
SAN JACINTO
BURLESON
LEE
WASHINGTON
MONTGOMERY
WALLER
San Marcos arch
BASTROP
CALDWELL
FAYETTE
AUSTIN
GUADALUPE
LAVACA
COLORADO
BEXAR
WILSON
GONZALES
DE WITT
Maverick basin
KINNEY
UVALDE
MEDINA
JACKSON
MAVERICK
ZAVALA
FRIO
ATASCOSA
KARNES
VICTORIA
GOLIAD
DIMMIT
Study area
MC MULLEN
LIVE OAK
BEE
LA SALLE
WEBB
DUVAL
Rio Grande Embayment
Anticline
Stuart City shelf margin
Sligo shelf margin
Core location (Well HH)
Eagle Ford play area
0 50 100 mi
0 100 km
QAd7213(f)

(b)

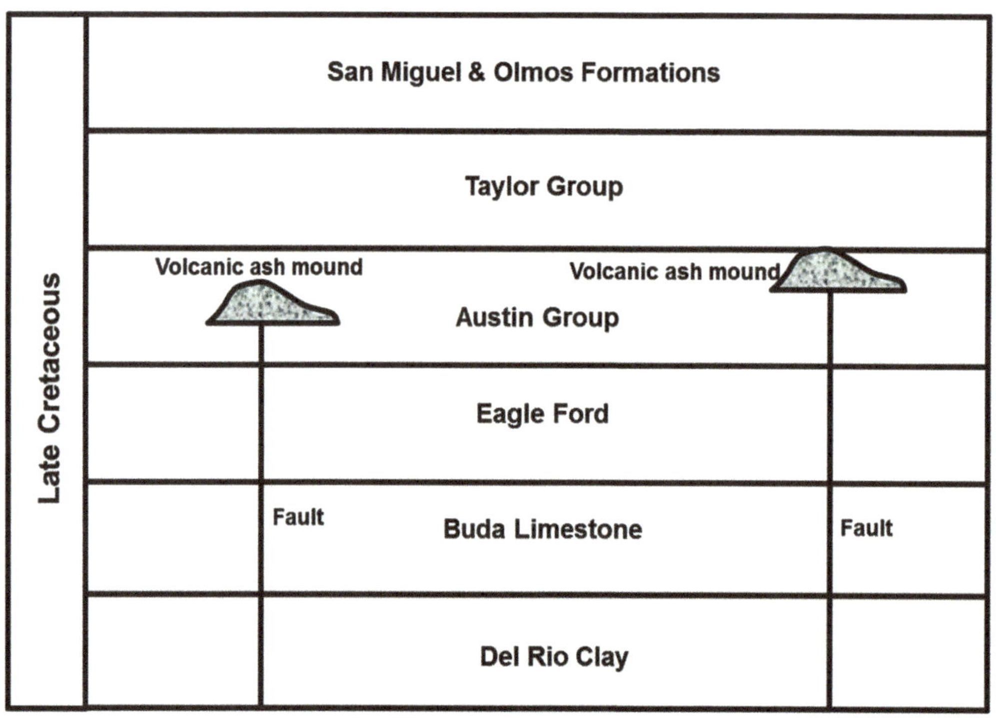
Late Cretaceous
San Miguel & Olmos Formations
Taylor Group
Volcanic ash mound
Volcanic ash mound
Austin Group
Eagle Ford
Fault
Buda Limestone
Fault
Del Rio Clay

◀**Fig. 1** **a** Map showing areal extent of Eagle Ford Shale play in South Texas and location of study area (*red box*). Figure modified from Hentz and Ruppel (2011). **b** Schematic diagram showing stratigraphic succession in South Texas during Late Cretaceous

et al. 2011; Treadgold et al. 2011). This method requires a prestack data set that meets special AVO requirements.

Because quality and prolificacy of shale resource plays depend on the richness and maturity of the shale, areas having high total organic carbon (TOC) typically yield the best production (e.g., Hill et al. 2007; Hammes et al. 2011). Matching high-TOC-bearing zones with fracture clusters might provide a clue to increased productivity in these areas. Therefore, it is necessary not only to delineate fracture zones but also to determine TOC distribution within the survey area and to identify those fracture zones that are located within high-TOC zones. Three methods can be used to detect TOC concentration within a rock interval: (1) measurement of TOC from core taken from the well bore, (2) computation of TOC from wireline logs using the Δ Log *R* method (Passey et al. 1990), and (3) calculation of TOC and lithology using the Multi Min method (Eastwood and Hammes 2010). The Δ Log *R* method (where *R* is resistivity) yields results of log-calculated TOC in reasonable agreement with limited core measurements (Fig. 2). For example, the maximum deviations in TOC values between the two methods occur at depths of 6,870, 6,990, and 7,300 ft (2,094, 2,130, and 2,224 m), being ~2, ~1, and 1.5 %, respectively. At these depths, core-measured TOC values are ~1, ~3.4, and ~2 %, whereas the corresponding log-calculated TOC values are ~3, 2.4, and 0.5 %, respectively. Apart from these differences, the difference between core-measured and log-calculated TOC values at other depths is either 0 or < 1 % (Fig. 2). Although these methods can provide TOC values, their

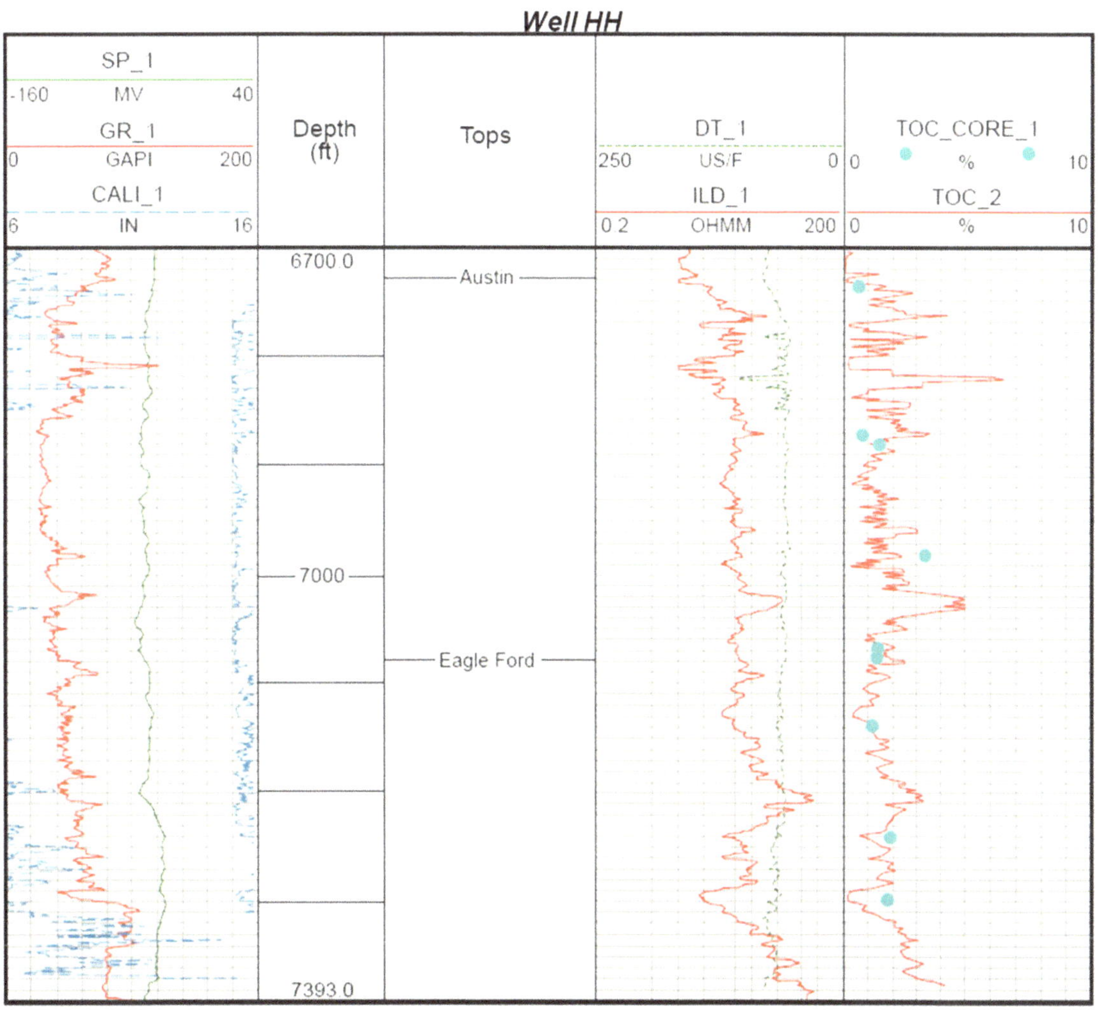

Fig. 2 Comparison between total organic carbon (TOC) computed using Δ Log *R* method and measurement from core in well HH (last track in figure). TOC measurements from core = *blue*, TOC log derived from Δ Log *R* method = *red*. Note reasonable agreement between methods. Maximum difference between two methods ~2 % at depth of 6,870 ft but ≤1.5 % at all other depths (see text for details). First track shows gamma-ray (*red*), spontaneous-potential (*green*), and caliper (*blue*) logs; second track shows sonic (*green*) and resistivity (deep induction) (*red*) logs

application is limited to well bore locations. In order to analyze TOC away from the well bore and characterize TOC throughout the entire seismic 3D volume, alternate methods are necessary.

In the subsurface, fractures are not easily identified. However, it is possible to identify the rock intervals in which fractures are most likely to occur. These intervals are composed of brittle rock characterized by relatively higher Young's modulus, higher velocity, higher AI, and higher Q than the surrounding, more ductile rock. At this juncture, 3D seismic method seems to be the best technology to employ because these physical properties of the rock can be derived at many hundreds of closely spaced locations [e.g., 82 ft (25-m) spacing] within a 3D survey area (Hampson et al. 2001; Ogiesoba 2010). This method not only offers interpreters a way of viewing lateral distribution of the rock physical property but also enables the volumetric computation of that physical property. In this paper, we adopt an integrated approach by combining results from petrophysical analysis with seismic attributes to determine brittle zones, TOC, and resistivity distributions within a 3D survey located in Dimmit County, South Texas.

Geological background and project objectives

The Eagle Ford Shale of South Texas, late Cenomanian to Turonian in age, is underlain by the lower Cenomanian Buda Limestone and overlain by the Coniacian to Santonian Austin Chalk (Fig. 1b). The Buda Limestone acts as a distinctive seismic marker bed throughout the areal extent of the Eagle Ford Shale, the Eagle Ford Shale ranging from less than 40 to 400 ft (~12–122 m) in thickness. Thickness correlates with onlap and truncation of unconformities on the top and bottom of the Eagle Ford across the depositional area from the Maverick Basin in the south to the East Texas basin in the north (Donovan and Staerker 2010; Fig. 1a). In the study area, thickness is fairly uniform at about 310 ft (~95 m). The Eagle Ford was deposited during a worldwide second-order transgression, and deposition of organic-rich marine shales above the Buda Limestone formed the organic-rich lower Eagle Ford Shale (Dawson 2000; Hentz and Ruppel 2011; Treadgold et al. 2011). A major condensed section occurs within the Eagle Ford Shale, separating the upper member of the Eagle Ford that composes the highstand sequence (Dawson 2000; Donovan and Staerker 2010). Whereas the lower Eagle Ford unit is characterized by dark, laminated shales, the upper, regressive unit consists of thinly intercalated shales, limestones, and carbonaceous, quartzose siltstones (Dawson 2000). Smaller scale cycles are recorded by intercalated, thin, carbonate-rich layers containing marine organic-rich shales. The Eagle Ford, bound by sequence boundaries at its base and top, comprises one complete depositional sequence, Eagle Ford strata being unconformably overlain by the Austin Chalk (Grabowski 1981). Overlying the Austin Chalk is the Taylor Group, which is composed primarily of shale; San Miguel and Olmos sands overlie the Taylor Group.

Deposition of the Upper Cretaceous formations was accompanied by volcanic eruptions, during which volcaniclastic rocks were interlayered with formations such as the Cenomanian Del Rio Clay (Baldwin and Adams 1971; Ewing and Caran 1982). Ewing and Caran (1982) found that the magma came through NE–SW-oriented, strike-slip, vertical to subvertical regional faults. The period of magmatism continued throughout deposition of the Austin Chalk and was at its peak during Austin-Taylor deposition, particularly in the Maverick Basin in the SW (Spencer 1969; Ewing and Caran 1982). As eruptions continued, the magma built volcanic islands that created platforms for the deposition of shallow-water carbonates having good porosity (Luttrell 1977). Because of intense seawater diagenesis, the magma cones were altered into permeable and porous palagonite. Overlying the Austin Chalk is the Taylor Group, which is primarily composed of shale rich in organic matter from the sea transgressing yet again (Ewing and Caran 1982). Finally, with the recession of the sea, the San Miguel and Olmos Formations were deposited over the Taylor Group.

Two main conventional hydrocarbon plays lie within these stratigraphic units: (1) a structural play within the San Miguel and Olmos Formations and (2) a stratigraphic play within the porous palagonite and porous shallow-water carbonate (Austin Chalk) (Lewis 1977; Ewing and Caran 1982). The Eagle Ford Shale was recently recognized as a major unconventional resource play having gas, condensate, and oil potential. The shale interval is actively being exploited, and 3,000 wells were drilled between 2009 and 2012 (IHS Energy Information 2012).

A typical seismic line within the survey showing several features is displayed along transect A–A' (Fig. 3a). Shown in this figure are two sets of NE–SW trending faults: (1) the almost-45°-dip-angle faults (black) and (2) the subvertical to vertical faults (red) cutting the top Del Rio. In some cases, the subvertical faults (red) cut into the Austin Chalk, intersecting the black-colored faults. In the coherency transect (Fig. 3b), several faults can be seen within the Taylor Group that did not penetrate the Austin Chalk below. Numerous small faults, as well as two major faults, can be seen in the interval between the top Austin Chalk and top Edwards (Fig. 3b). Whereas weak faults are located mostly in the SE, major faults, particularly two that cut through the Edwards into the Austin Chalk, are located in

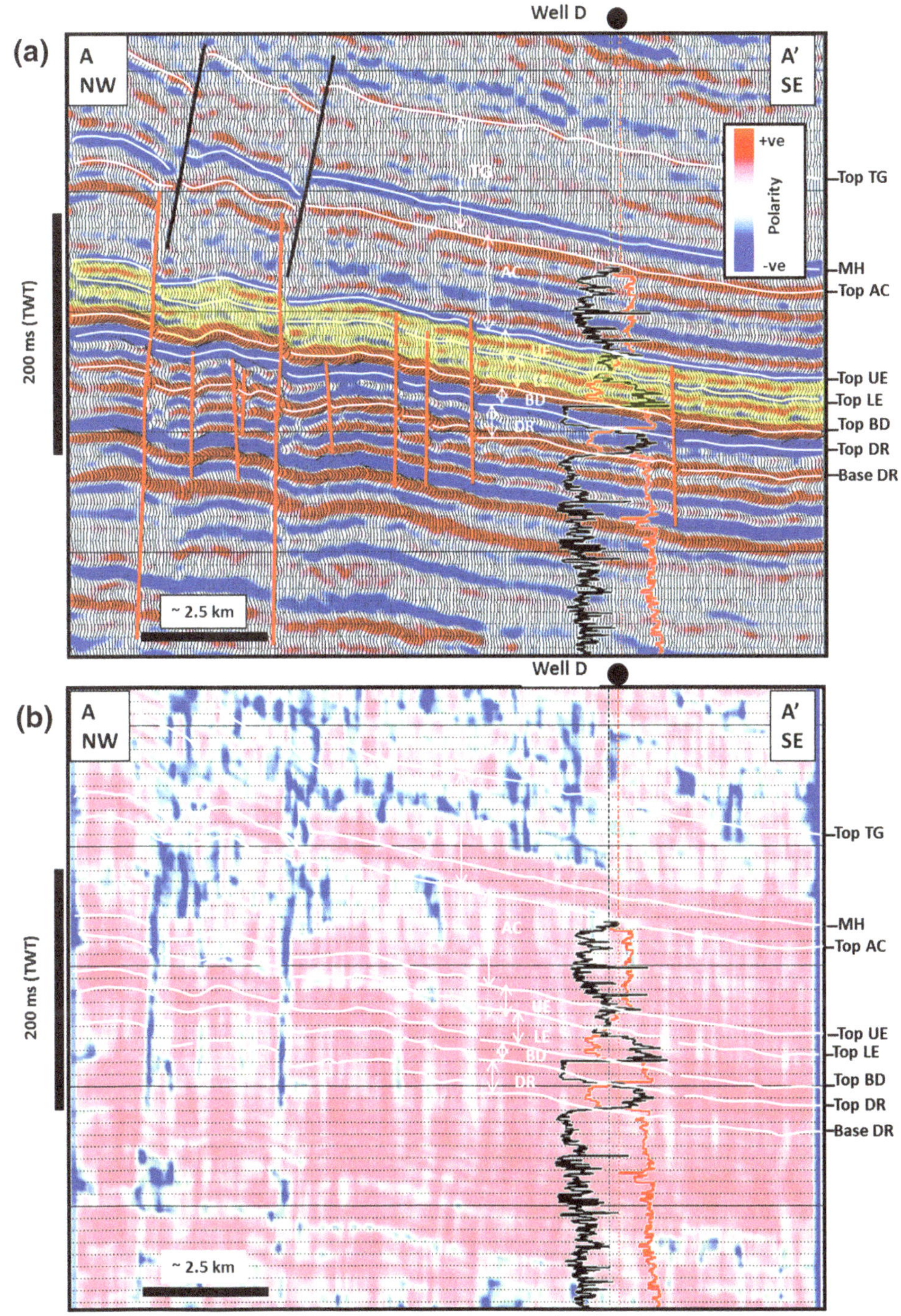

Fig. 3 **a** Typical seismic section A–A' through Well D, showing interpreted stratigraphic surfaces and faults. Note vertical to subvertical faults (*red*) having predominantly SE throw direction and faults (*black*) having NW throw direction. **b** Corresponding coherency section showing two prominent major faults that cut the Del Rio into the Austin Chalk and other small faults. *Light brown* in coherency section indicates areas of high coherence; *dark-blue* vertical and subvertical stripes are faults. Weak faults having small displacements are represented by *white* vertical-to-subvertical stripes. *AC* Austin Chalk, *TG* Taylor Group, *UE* upper Eagle Ford, *MH* mapped horizon, *LE* lower Eagle Ford, *BD* Buda, *DR* Del Rio. Area in *yellow* is entire Eagle Ford interval. *Red log curve* sonic; *black log curve* gamma ray. On sonic curve, deflections to right = increasing interval velocity; deflections to left = decreasing interval velocity. On gamma-ray curve, deflections to right = increasing gamma ray; deflections to left = decreasing gamma ray. *TWT* two-way time. Definitions of these abbreviations, as well as area in *yellow*, apply to all other figures

the NW. Note the downthrown direction of these two sets of faults. Whereas the black-colored faults are downthrown to the NW, the red-colored faults are downthrown to the SE (Fig. 3a). The Eagle Ford Shale interval (yellow zone) is about 310 ft (~95 m) thick. Given the foregoing geologic scenario, our objective was to use the existing well database in conjunction with seismic attributes to identify the high-resistivity and TOC-rich zones within the Austin Chalk and Eagle Ford that could be used to augment hydrocarbon exploration.

Methodology and database

Our database consists of a 3D seismic survey covering a 32-mi^2 area (~82 km^2). The stacking bin size is 34 × 34 m, and the sampling interval during acquisition was 2 ms. The well database is composed of six wells (Fig. 4a). Shown on the base map are three seismic traverses, A–A', B–B', and C–C'. The method of study is an integrated workflow in which we first conducted a petrophysical analysis to determine TOC. A well-to-seismic tie

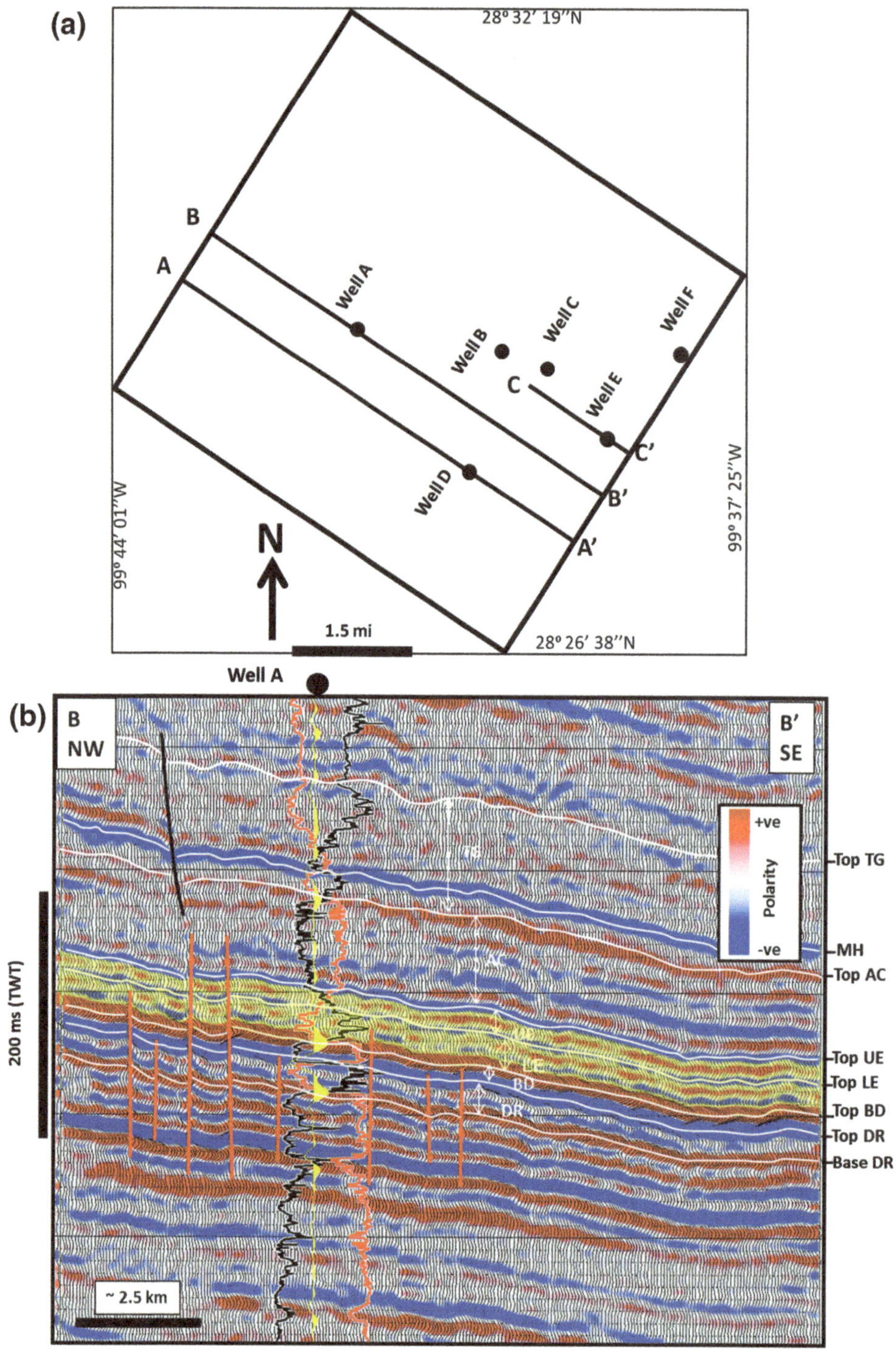

Fig. 4 **a** Base map of study area displaying three seismic traverses and well locations discussed in paper. **b** Seismic section B–B' showing well seismic tie at Well A. Note dominant fault-throw direction of red faults is NW, in contrast to those in section A–A'. *Yellow curve* synthetic seismogram

was established through a synthetic seismogram (Fig. 4b) to identify key stratigraphic surfaces that we later interpreted to generate horizon maps. Note the strong amplitude reflection associated with the top Buda (base Eagle Ford) (Fig. 4b). This strong character enabled identification of the base Eagle Ford throughout the survey. In addition, note that in seismic transect B–B', the dominant fault-throw direction of the faults at base Eagle Ford Shale (red) is NW, whereas the fault-throw direction of faults within the Austin Chalk (black) is SE. Fault-throw directions in this transect are in contrast with those seen along transect A–A' (Fig. 3b), where the faults at base Eagle Ford Shale (red) are downthrown to the SE and the faults within the Austin Chalk (black) are downthrown to the NW. Furthermore, the thickness of the Eagle Ford Shale interval (yellow zone) at Well A is approximately the same as that at Well D—about 310 ft ($\sim$95 m). Generated horizons maps, together with necessary log properties, were used in acoustic-impedance

(AI) inversion. Results from AI, horizons maps, and seismic attributes computed from 3D poststack data were input into an industry-standard multiattribute analysis algorithm to predict gamma ray (lithology), TOC, and resistivity (deep induction log) using the multiattribute algorithm. The procedure involved has been described by Hampson et al. (2001), Ogiesoba (2010) and Ogiesoba and Eastwood (2013). As such, it will not be repeated in this paper. It is sufficient to say that the choice of the optimal number of attributes used in the prediction process is determined by the algorithm from the plot of mean squared error between the predicted and actual log property versus the number of attributes. The number of attributes where the error convincingly stops decreasing is taken as the optimal number of attributes. This plot is demonstrated under the result section in this paper.

Because quality factor Q featured prominently in the prediction of TOC and resistivity, we closely examined the use of Q in seismic interpretation. By definition, Q is a transmissive attribute, similar to instantaneous and interval velocities; it indicates relative absorptive characteristics of bedrocks (Johnston et al. 1979; Toksoz et al. 1979; Johnston and Toksoz 1981). The attribute has a strong relationship with porosity, permeability, and fractures (Johnston et al. 1979; Robinson and Treitel 2008). High-porosity rocks are more absorptive and are generally characterized by low Q, whereas low-porosity (more compacted, high-velocity) rocks are less absorptive and are characterized by high Q (Hamilton 1972a, b; Toksoz et al. 1979; Johnston and Toksoz 1981). Fault zones, having high porosity and permeability, are characterized by low Q. Because high-velocity rocks have higher Q values, Q also indicates the relative Young's modulus of the rock. That is, high Q suggests high Young's modulus, and low Q suggests low Young's modulus (Hamilton 1972b; Johnston et al. 1979; Johnston and Toksoz 1981). The Q attribute was computed according to the following expression (Barnes 1992):

$$Q(t) = \frac{\pi f(t)}{\frac{d}{d_t}(\log_e A(t))} \tag{1}$$

using a 50-m sliding window. In Eq. 1, $A(t)$ is instantaneous amplitude (amplitude envelope), and $f(t)$ is instantaneous frequency. Further simplification shows the denominator on the right-hand side of Eq. 1 to be the derivative of the amplitude envelope divided by the envelope, suggesting the energy-decay rate (Barnes 1992, 1993).

Seismic attribute rock property prediction—conceptual review

Before resistivity and TOC prediction using seismic attributes and well log data are discussed, pertinent rock physical properties pertaining to resistivity and P-wave velocity need to be examined. As documented in the literature (e.g., Grant and West 1965), the electrical conductivity exhibited by a rock depends on its permeability and porosity and on the conductivity of the fluids that it contains, rather than on its mineralogical composition. Grant and West (1965) discussed a relationship between resistivity and porosity and saturation for sedimentary rocks (sandstones and carbonates), given as.

$$\text{Resistivity } \rho = IF\rho_w, \tag{2}$$

where I is resistivity index, F is formation factor, and ρ_w is resistivity of water in the formation. Formation factor F depends on porosity and is given as

$$F = a\varphi^{-m}, \tag{3}$$

where φ is porosity of the rock and a and m are positive constants. From Eqs. 2, 3, resistivity can be seen to be higher for denser and more compact rocks, such as dense limestone. That is, rocks having high AI (high velocity) have high resistivity. However, if such compact rock contains pockets of clay having formation water, overall resistivity is decreased. Note that fluid content is emphasized.

Furthermore, Faust (1953) provided a relation linking seismic-wave (P-wave) velocity to resistivity of rocks, given as

$$V = CL^{1/6}Z^{1/6}, \tag{4}$$

where $C = 2 \times 10^3$; L is lithology, a quantity proportional to formation resistivity; and Z = depth of rock. Although this relationship was derived more than five decades years ago, it still sheds light on the possibility of predicting resistivity using seismic attributes. In Eq. 4, note that resistivity of rock increases as velocity increases (i.e., with increasing compactness). If the porous zone in a carbonate rock contains pockets of clay rich in TOC, or if the porous zone is filled with hydrocarbon, resistivity within that zone is increased because TOC or hydrocarbon does not conduct electricity. Because velocity can be derived as seismic waves propagate, waves traveling through a porous medium filled with hydrocarbon or TOC-rich clay will have slower velocity. In this case, the low velocity indicates increased porosity. As such, longer transit times are recorded in such a medium, and instantaneous frequencies are lower than the frequencies of the surrounding rocks (Ogiesoba and Eastwood 2013). This correlation suggests that seismic attributes, such as instantaneous frequencies, dominant frequencies, and other frequency-related and amplitude-related attributes (such as envelope, quality factor, etc.), can be employed to predict resistivity. Herein are the physical principles upon which resistivity and TOC prediction, by way of seismic attributes, are based.

Results

Petrophysical analysis

The intent of the petrophysical study was to derive TOC log data at well locations that would be used in the mutiattribute prediction process (Fig. 5). The first track contains gamma-ray (GR), caliper (CALI), and spontaneous potential (SP) logs; the well bore is generally in gauge; and the GR response is generally 80–150 API. The second track contains the resistivity log (ILD) and sonic log (DT). Values for these logs range from about 4.3 Ωm in the Austin Chalk to more than 200 Ωm in places in the Eagle Ford Shale. This track also displays the separation of DT and ILD used in the manner of Δ Log *R* analysis (Passey et al. 1990); ILD is made to overlie DT, where low ILD values are seen. The logs overlie one another in the uppermost Eagle Ford and at the top of the lower Eagle Ford. The gray shading between DT and ILD indicates kerogen and/or hydrocarbons. This pair of logs was used to compute TOC because sonic logs are generally available for wells in this study. The TOC log that was computed using the Δ Log *R* method by employing the Passey et al. (1990, 2010) equation is displayed in the third track (in brown; Fig. 5), and limits range from 0 to 10 %. Calculated TOC values are in reasonable agreement with measured TOC values from a core taken from a nearby well (Fig. 2).

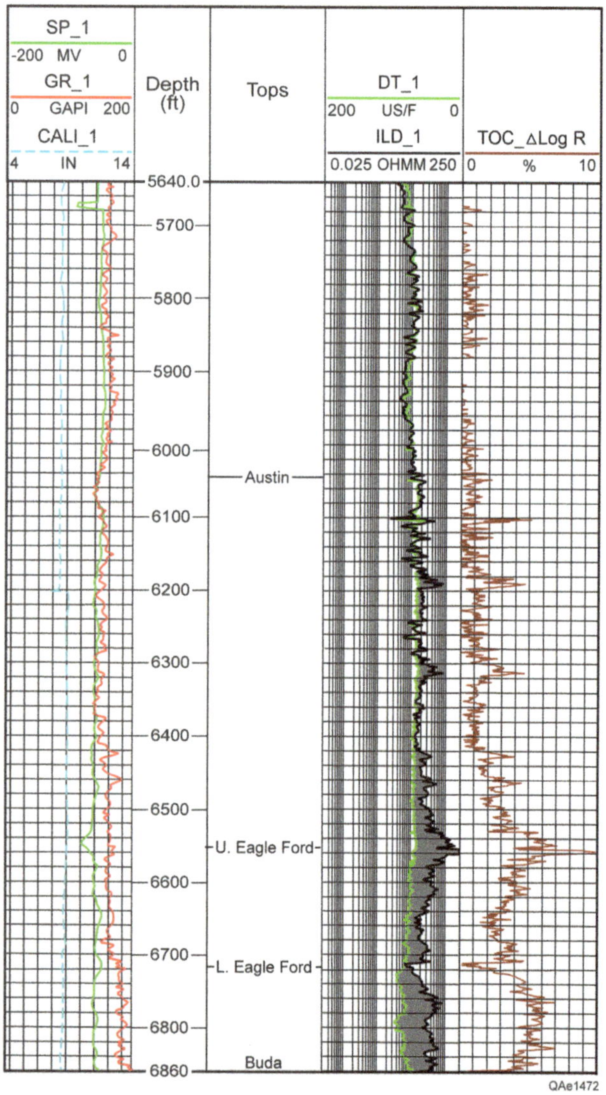

Fig. 5 Petrophysical analysis results showing derived total organic carbon (TOC) log (last track). All other tracks discussed in text. *L. Eagle Ford* lower Eagle Ford, *U. Eagle Ford* upper Eagle Ford

Seismic inversion—acoustic impedance

Results of AI inversion are now examined (Fig. 6). Along transect B–B' (Fig. 6a), AI results show that some variations in AI values, in both temporal and lateral directions, occur within the Austin Chalk, upper Eagle Ford, and lower Eagle Ford intervals. Note that the lowest AI values are in green and the highest are in magenta (see color bar). Cross plots between inverted and original AI (Fig. 6b) show a high degree of correlation, with a correlation coefficient of 0.99. Note the wide variation in AI values. For example, between a time of 1,100 and 1,167 ms (67 ms interval), the AI spans from about 28,000–45,000 ft/s g/cm^3 (difference = 17,000 ft/s g/cm^3). This relatively wide range in AI values within a relatively short interval of time shows that although the duration may be short, the interval can be composed of rocks having different facies. Close inspection of the lower Eagle Ford (Fig. 6a) shows that it can be subdivided into three units, and some stratigraphic interpretations can be deduced on the basis of the AI values. The lowermost unit at this location is ~9 m thick; it is a highstand-transgressive marine shale (Treadgold et al. 2011; Slatt et al. 2012) having a low AI value that sits atop the Buda. This shale is followed by a layer of ~8 m thick having higher AI values; it is probably richer in calcareous material. This second interval was most likely deposited during sea-level highstand (Grabowski 1981). Over this unit, a third layer ~9 m thick composed of shale and having low AI values and rich organic matter was deposited when the sea transgressed the previously deposited Eagle Ford layers. Deposition of carbonate-rich rocks commenced again when the sea-level became highstand once more; however, the carbonate sediments intertongued with some shales owing to the interchange of highstand and transgressive processes. This unit, which has relatively higher AI values, constitutes the upper Eagle Ford (Fig. 6a). The unit grades into more carbonate-rich rock which was also deposited during a highstand of sea-level—the Austin Chalk having high AI values (magenta, Fig. 6a). Finally, the sea transgressed yet again, leaving shale-rich

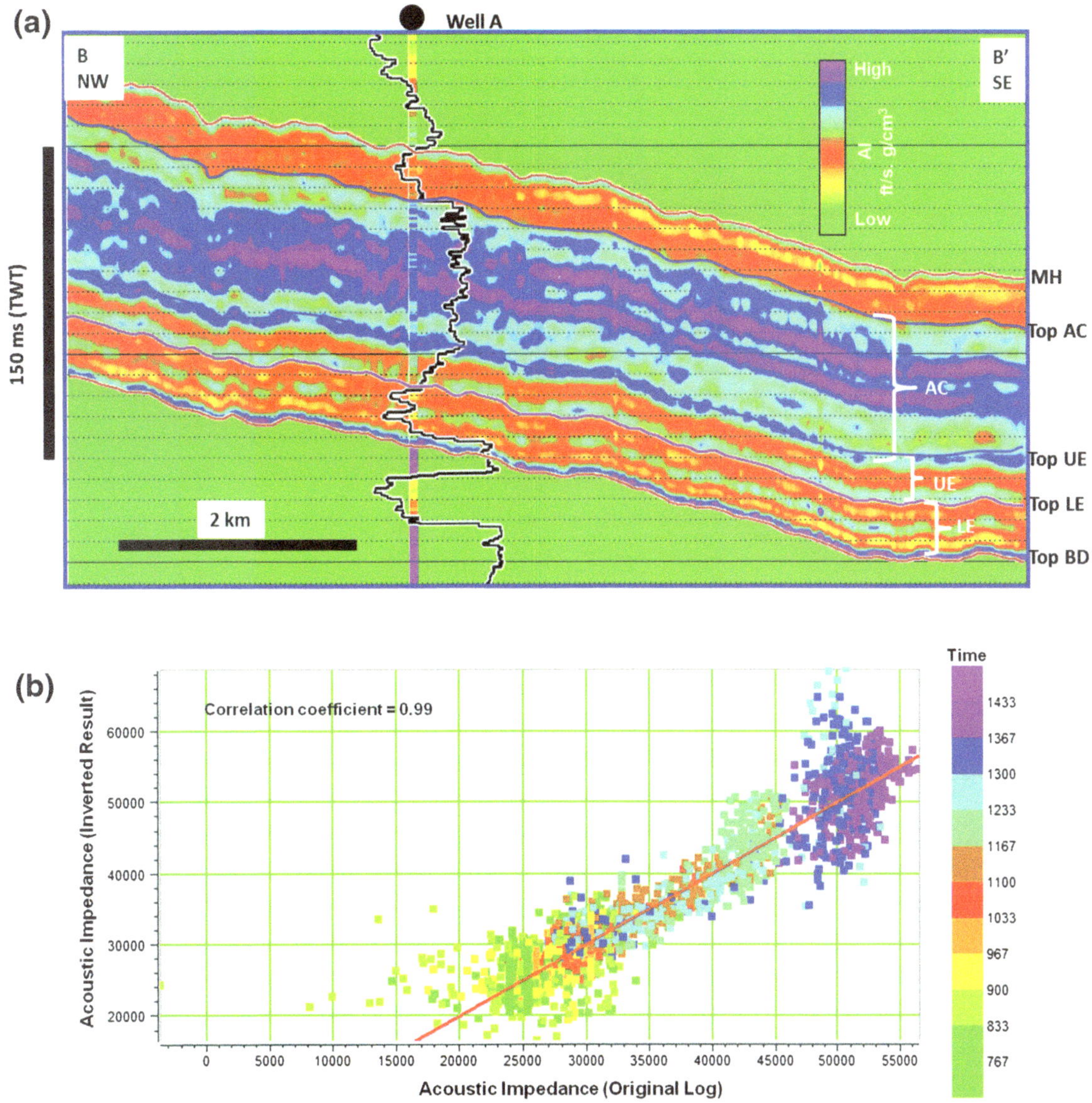

Fig. 6 Results from acoustic impedance (AI) inversion showing **a** vertical and lateral variations in AI along transect B–B'. **b** Cross plot of inverted versus original AI also showing vertical variations in AI (see text for discussion)

sediments having low AI (Grabowski 1981; Ewing and Caran 1982).

Prediction of gamma ray (lithology), TOC, and resistivity (deep induction log)

Using seismic attributes derived from the poststack data set, together with AI as an external attribute, we performed multiattribue analysis so as to predict the gamma ray. Because the gamma-ray log is used as the lithology log, predicting the gamma-ray log is considered equivalent to predicting lithology. From an array of computed poststack seismic attributes, the multiattribue analysis algorithm was used to select relevant attributes through a method of validation and testing (Hampson et al. 2001; Ogiesoba 2010). Our intent was to obtain the distribution of shale-rich zones within the Eagle Ford interval throughout the survey area. Note that in the vertical distribution of gamma ray (lithology) (Fig. 7a), the lower Eagle Ford has the highest concentration of gamma ray. Because this interval on the AI transect corresponds to the lowest AI values, we conclude that it must be rich in organic materials. Gamma ray gradually decreases within the upper Eagle Ford and finally grades into low gamma ray in the Austin Chalk. Capping the Austin Chalk is an interval of moderately high gamma ray. We found the point at which the error stops decreasing to be at attribute #3, suggesting that three attributes were sufficient to predict gamma ray having a correlation coefficient of 0.95 (Fig. 7b, c). Although the error decreased slightly at attribute #4 and then starts to increase again at

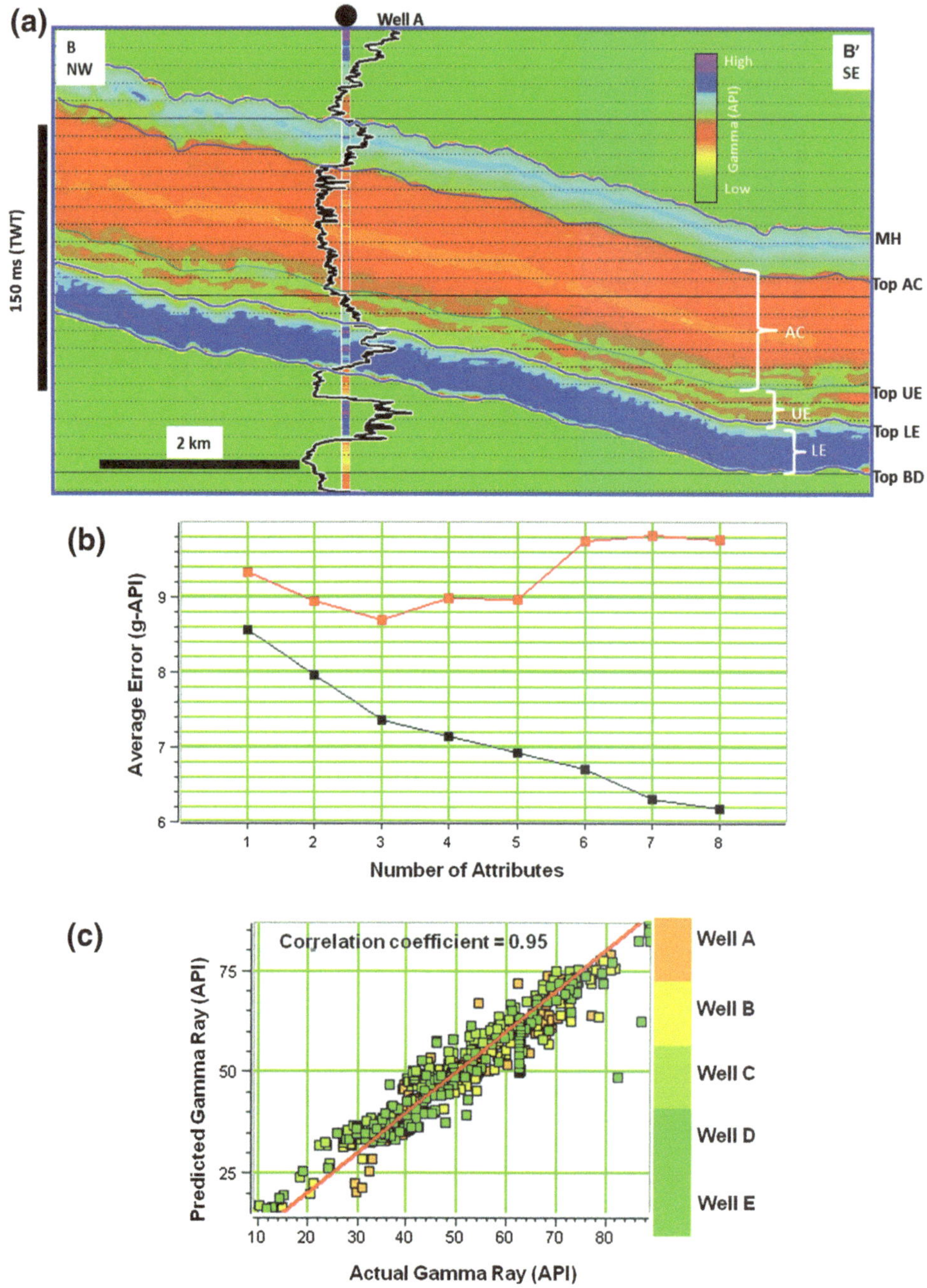

Fig. 7 Gamma-ray prediction showing **a** gamma-ray distribution along transect B–B'. **b** Plot of average error versus number of attributes to determine optimal number of attributes to use in prediction processes. Note *red curve* is mean squared error between predicted and actual log property when a well is hidden and remaining wells together with number of attributes are used to predict log property; whereas *black curve* is mean squared error that is obtained when all wells together with number of attributes are used to predict log property. For example, if there are five wells, *red curve* is obtained when four wells are used at a time. That is, at every attribute combination, prediction is performed five times, each time hiding one well and previously hidden well is brought back into the prediction process; while *black curve* is obtained by using all five wells every time during the prediction process. For detained explanation, refer to Hampson et al. (2001) and Ogiesoba (2010). **c** Cross plot of actual gamma ray versus predicted gamma ray showing high correlation coefficient of 0.95

#5, the difference between the error and attribute #5 is insignificant. Therefore, we used three attributes in the prediction.

To invert for resistivity (deep induction log, ILD) and TOC, we input gamma-ray and AI volumes, serving as external attributes, into the multiattribue analysis package, together with other poststack-derived seismic attributes. Relevant attributes to predict TOC and ILD were also selected by the multiattribute analysis algorithm. Lateral variation of resistivity is shown along transect B–B' (Fig. 8a), and a combination of four attributes is clearly optimal for prediction of ILD (Fig. 8b) having a correlation coefficient of 0.98 (Fig. 8c). Although the error starts to decrease again at attribute #5, the difference between attribute #4 and #5 is huge and far more significant compared to that at attribute #5. Therefore, instead of using all the attributes, which would have led to overprediction, we chose four attributes for prediction of resistivity. The vertical distribution of TOC can be seen along transect B–B' (Fig. 9a). Additionally, a combination of four attributes was also sufficient to predict this property with a high correlation coefficient of 0.98 (Fig. 9b, c). Note the absence of difference in average error between the fourth and fifth attributes (Fig. 9b). Because the point at which the

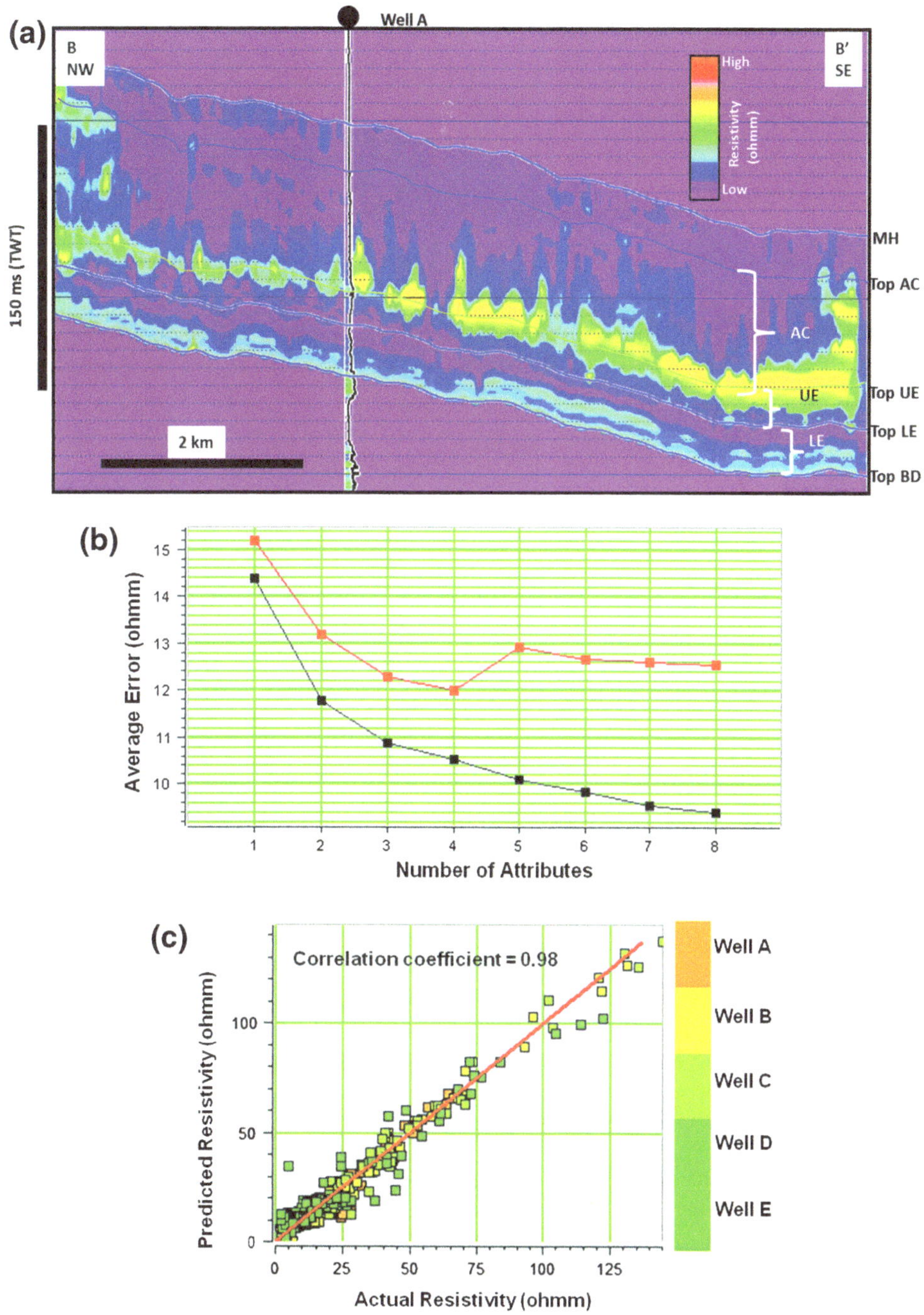

Fig. 8 Results from resistivity (deep induction, ILD) inversion showing **a** ILD distribution along transect B–B' and **b** plot of average error versus number of attributes. Optimal number of attributes to predict ILD is four. Note big difference in average error between first and second attributes compared with small differences associated with remaining two attributes. Note that *black curve* indicates error using all wells, whereas *red curve* shows error plot when one well is removed. **c** Cross plot of predicted versus original ILD showing high degree of correlation

average error stops decreasing occurs at the fourth attribute, four attributes must be used to predict TOC.

Note the similarity in vertical distribution of ILD and TOC (Figs. 8, 9, respectively). Note also the almost continuous nature of ILD and TOC within the upper and lower Eagle Ford intervals south of Well A. The ILD and TOC maps extracted at 2 m below the top Eagle Ford horizon (Fig. 10a, b) show the lateral distribution of TOC and ILD, respectively. On these maps, a maximum concentration of both properties lies to the south and southwest of the survey area. In the south half of the survey, ILD and TOC distribution appears to be more continuous than it is in the north. In addition, note the similarity between the lateral distribution of ILD and TOC. High-ILD zones correspond to high-TOC zones. Areas of low TOC and low ILD (light-green zones in both cases) are major fault zones, as confirmed by the superposition of some fault traces from the coherency map (Fig. 10c) on top of the ILD and TOC maps. Coherency attributes were also extracted at about 12 m below the top lower Eagle Ford (Fig. 11a). As can be seen on this map, fault trends are mostly NE–SW, and numerous small faults are present. Although some of these

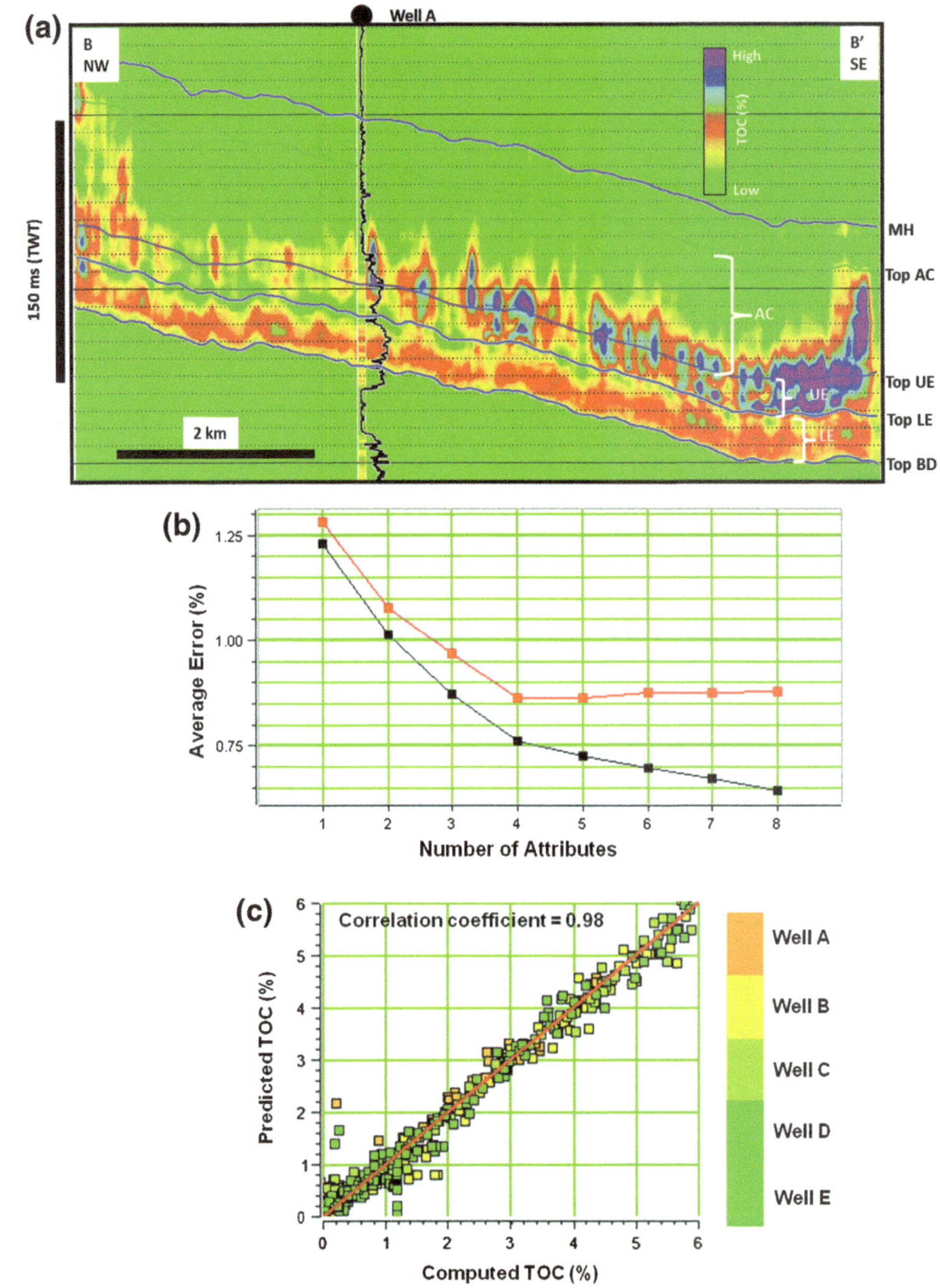

Fig. 9 Results from total organic carbon (TOC) prediction showing **a** TOC distribution along transect B–B' and **b** plot of average error versus number of attributes; note absence of difference in error between fourth and fifth attributes. Therefore we conclude that optimal number of attributes is four. Note that *black curve* indicates the error using all wells, whereas *red curve* shows the error plot when one well is removed. **c** Cross plot of derived TOC and predicted TOC showing high correlation coefficient of 0.98

small faults are seen in the north, they are concentrated mostly in the south part of the survey. Superposition of major fault traces on top of the TOC map at this horizon (Fig. 11b) shows that the faults tend to occur in areas of low TOC (green areas). Note the area of high TOC concentration within the black dotted line to the south.

The optimal number of attributes in each multiattribue analysis process is tabulated vertically in order of importance (Table 1). The first most significant attribute to predict gamma ray (lithology) is the inverse of the acoustic impedance (1/AI). The other two attributes are narrow-band frequency filters, which can be thought of as one in that both are low-frequency filters, suggesting the dominance of low-frequency zones within the interval of investigation. The low frequencies within the interval could be caused by the presence of faults, fractures, hydrocarbon zones, and some shale-rich zones. Essentially two attributes are therefore used to predict lithology—low-frequency attribute and 1/AI. Of these, 1/AI is the primary attribute, with a correlation coefficient of 0.82, higher than the coefficients of the other two attributes (Table 1). Similarly, the four attributes used to predict resistivity (ILD) can be grouped into (1) amplitude-related and (2) frequency-related attributes. Frequency-related attributes

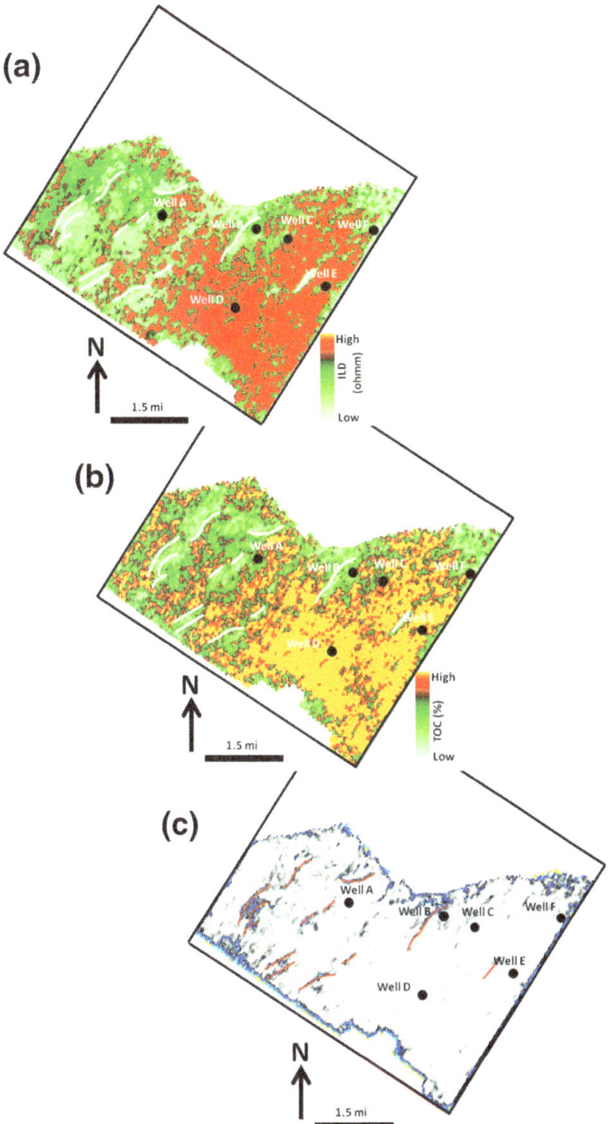

Fig. 10 Inversion results showing lateral-distribution maps of predicted properties extracted along horizon phantomed 2 m below top upper Eagle Ford: **a** resistivity (ILD) map and **b** total organic carbon (TOC) map. **c** Coherency map also generated along same horizon. Fault traces, *red* in **c**, superimposed on ILD and TOC maps; fault traces shown by *white lines* in both **a** and **b**

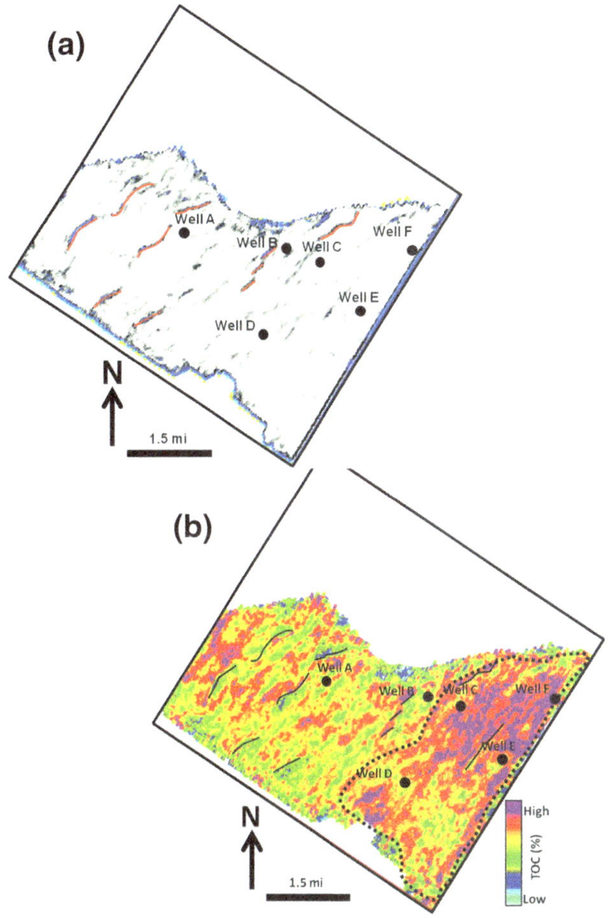

Fig. 11 Major fault traces extracted along horizon phantomed 12 m below top lower Eagle Ford: **a** coherency map; note fault traces in *red*. **b** Total organic carbon (TOC) map at same horizon. *Black lines* fault traces from coherency map. Note concentration of high TOC values (*black dotted* outline) in south half of survey area. Fault traces in TOC map occur in low-TOC zones (*light-green* and *blue* zones)

(average frequency and filter 15-20-25-30) are both low-frequency attributes, again suggesting the low-frequency composition of the zone of investigation as seen in lithology prediction. The quality attribute is both amplitude and frequency related. Either way, the square of the instantaneous quality factor (Q^2) is the most significant attribute to predict resistivity. It has a correlation coefficient of 0.73, whereas the other attributes have lower correlations (Table 1). That Q^2 plays a major role can be seen from the plot of average error versus number of attributes (Fig. 8c), in which the difference between Q^2 and the second-most-significant attribute is larger (2.0) than the small differences (1.0 and 0.2) between the other attributes. Like resistivity, four attributes were found to be optimal in the prediction of TOC. Apart from gamma ray^2, the other three attributes are amplitude related; as such, they can be grouped into (1) amplitude-related attributes (Q^2, derivative of seismic trace, and amplitude-weighted cosine of phase) and (2) gamma ray^2. It is important to note that gamma ray^2 was chosen by the algorithm because it gave the best linear relationship between gamma ray and TOC (see further explanation under the Discussion section). Although gamma ray was not the primary attribute for predicting TOC, the relationship between lithology (gamma ray) and TOC has been discussed by several authors (e.g., Sondergeld et al. 2010). These authors noted that high-gamma-ray intensity is recorded on wireline logs in zones having anomalously high uranium content in organic matter (TOC). However, high gamma ray response does not necessarily suggest high TOC; for example, in this study, where gamma ray is highest, the corresponding TOC is low (Fig. 5). Because associated AI value is low, we conclude that identified high-gamma-ray zone is shale. Of the amplitude-related attributes, Q^2 was also found to be the most significant attribute, with a correlation coefficient of 0.64, higher than correlation coefficients of the other three attributes (Table 1).

Discussion

In this section we focus on the instantaneous quality factor (Q), resistivity (ILD), and acoustic impedance (AI). When the pore spaces within a resistant rock are filled with fluids, calculated Q within the pores are low, whether the fluids are hydrocarbons or brine. However, the corresponding resistivity is high if the fluids are hydrocarbons, but low if the fluid is brine. Note that the resistivity in this case is not caused by the matrix but by the fluid; however, for the matrix, which constitutes the brittle part of the rock, resistivity is high and Q is also high. Note that low Q is an indication of porosity (Ogiesoba 2010), and the degree of porosity determines the values of Q. That is, Q decreases as porosity increases (Ogiesoba and Eastwood 2013).

In the case of entirely organic-rich shale, resistivity is high, but Q is low; in this situation, there is no brittle zone. However, shales are not stratigraphically or spatially homogeneous (Slatt and Abousleiman 2011). They contain some fine-grained sandstone and sometimes carbonate materials that can be more resistant than the encasing shales. In this case, the more resistant rock materials would have relatively higher Q than would the encasing rock. This is the case with the Eagle Ford Shale. The shale contains about 45–64 % carbonate, 13 % quartz, and 27 % clay minerals (Hildred et al. 2011; Sondhi 2011; Slatt et al. 2012). Based on the mineralogical composition, it becomes obvious why the Q attribute can be an effective discriminating tool in identifying brittle zones within the Eagle Ford Shale. The carbonate intervals are more resistive and thus, having higher Q than the encasing shale, they act as the brittle zones. A comparison of Q and resistivity transects (Fig. 12a, b; transect C–C') shows a good correlation between Q and resistivity.

During the prediction process, the square of the quality factor, Q^2, was used to predict the deep induction log (ILD). The choice of Q^2 is based on the following explanation. Because the prediction algorithm makes use of the linear relationship between log property and attributes, if the cross plot shows nonlinearity, the algorithm tests for linear relationships using nonlinear transforms. For example, if the cross plot between a target log and an attribute resembles a curve, the algorithm will test different nonlinear transforms such as log, square, square root, etc., to see whether a linear relationship is created by any of the transforms. From these tests, the transform that creates the best linear relationship is chosen. In our case, Q^2 was found to provide the best linear relationship between ILD and Q (Fig. 12c), suggesting that the relationship between ILD and Q is nonlinear. We therefore used Q^2 in the prediction process instead of Q. The interval having the highest resistivity (i.e., highest Q values) is the upper Eagle Ford (Fig. 12).

The reason Q^2 is the primary attribute for predicting TOC can be found in the equation used in TOC computation (Passey et al. 1990):

$$\mathrm{TOC(wt\%)} = \Delta\ \mathrm{Log}\ R\ 10^{(2.297 - 0.1688\ \mathrm{LOM})}, \quad (5)$$

where Δ Log R is the separation between the sonic log (DT) and resistivity (ILD) (Fig. 5). This separation in the context of Δ Log R implies that the scale of the resistivity log is a 4-decade logarithmic scale and that of the sonic log is a 200 μs/ft difference. The separation is then digitally measured and input into the algorithm to compute Δ Log R. LOM (level of organic maturity) is a fixed value; LOM ranges from 0 to 20 (Passey et al. 2010). In this study, LOM was assigned the value of 9 because the maturity of the Eagle Ford source rock in the study area places it in the oil window according to the calculated R_o value ($R_o = 0.7$). R_o is the source maturity measured from vitrinite reflectance; a R_o value from 0.5 to 0.8 indicates oil-prone source rock; Δ Log R was computed from Eq. 1 of Passey et al. (1990). In Eq. 5, the only

Table 1 Selected optimal number of attributes in each prediction process arranged in order of importance with the corresponding correlation coefficients

Gamma ray (Lithology)	Resistivity (ILD)	TOC
1/(acoustic impedance) correlation coefficient = 0.82	(Quality factor)^2 correlation coefficient = 0.73	(Quality factor)^2 correlation coefficient = 0.64
Filter 15/20—25/30 correlation coefficient = 0.08	Average frequency correlation coefficient = 0.47	Gamma ray^2 correlation coefficient = 0.47
Filter 5/10—15/20 correlation coefficient = 0.07	Derivative correlation coefficient = 0.42	Derivative instantaneous amplitude correlation coefficient = 0.31
	Filter 15/20—25/30 correlation coefficient = 0.19	Amplitude-weighted cosine phase correlation coefficient = 0.09

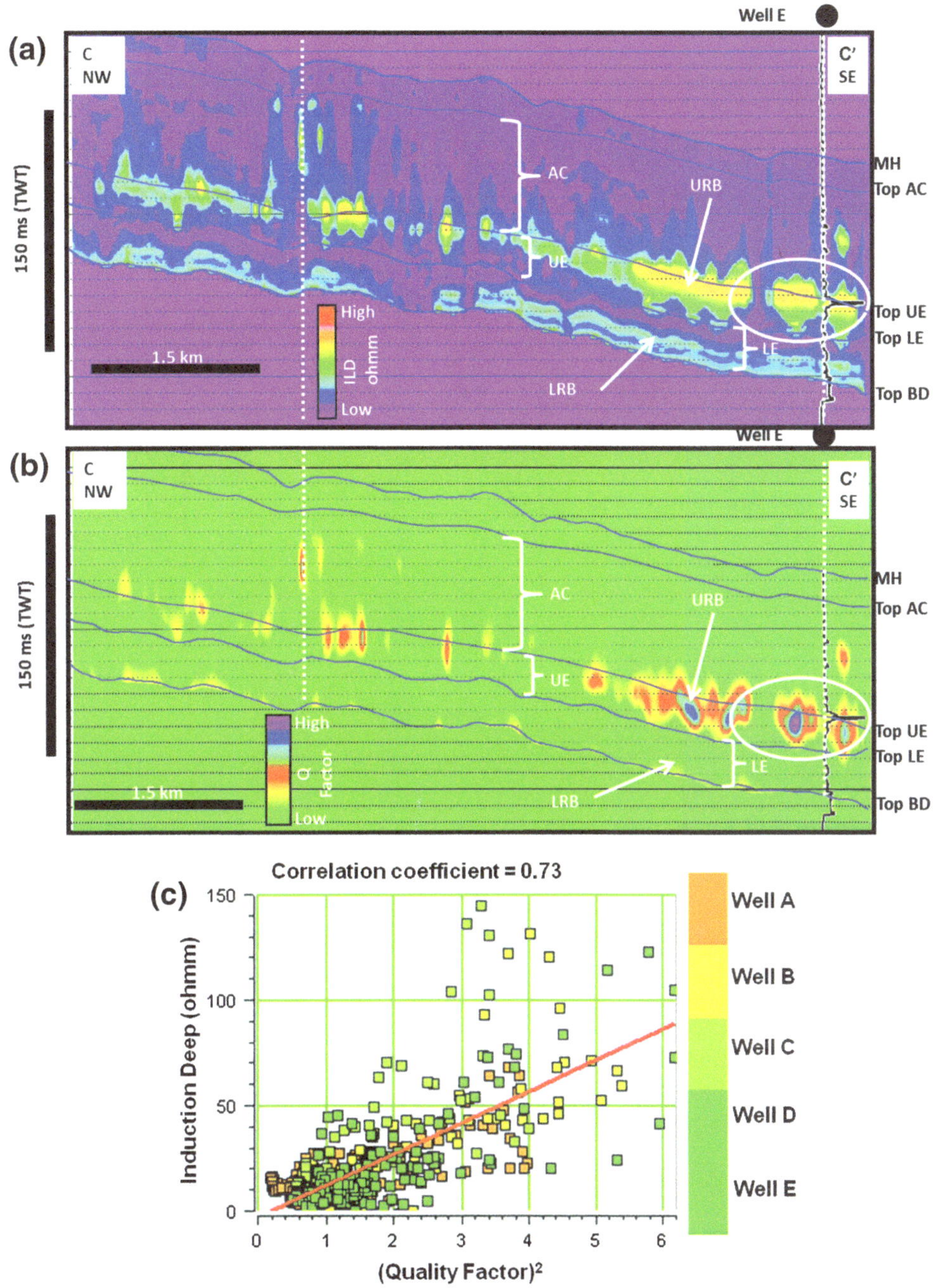

Fig. 12 Comparison of resistivity (ILD) with quality factor (Q) along transect C–C': **a** ILD transect and **b** Q transect. Note inserted ILD curve (*black*) for Well E in both figures; highest resistivity values in **a** occur in zone of highest Q in **b** (*white ellipses*). *White dotted vertical lines* connect corresponding ILD and Q. **c** Plot of Q^2 versus ILD showing linear relationship between both properties—increasing Q suggests increasing resistivity. *LRB* lower resistive bed, *URB* upper resistive bed

variable to determine TOC is Δ Log R multiplied by a constant. Because of the dependence of TOC computation on ILD, the computed TOC log motif is similar to that of ILD (Fig. 5), and the most significant attribute for predicting TOC is also Q^2. Because high Q values—which are indicative of less absorptive, or brittle, rocks correlate with high resistivity values, the observed high ILD values might have had some contribution from the resistant bed. That is, the high ILD values are not entirely due to the TOC or hydrocarbon effect but are instead a combination of TOC or hydrocarbon and resistant lithology.

Physically, Q has no relationship with TOC other than the fact that its values are lower in a rock that is TOC-rich than would be the Q values observed in the same rock type that is devoid of TOC. However, studies of some Eagle Ford cored wells (Sondhi 2011) showed that TOC increases as the carbonate content increases, whereas TOC decreases as the clay content increases. Because TOC increases with increasing carbonate content, high Q, which suggests a high percentage of carbonate, would also suggest high TOC. This characteristic means that high resistivity also suggests high TOC, because high Q suggests high resistivity. As can be seen from Figs. 7a, 8a, high resistivity

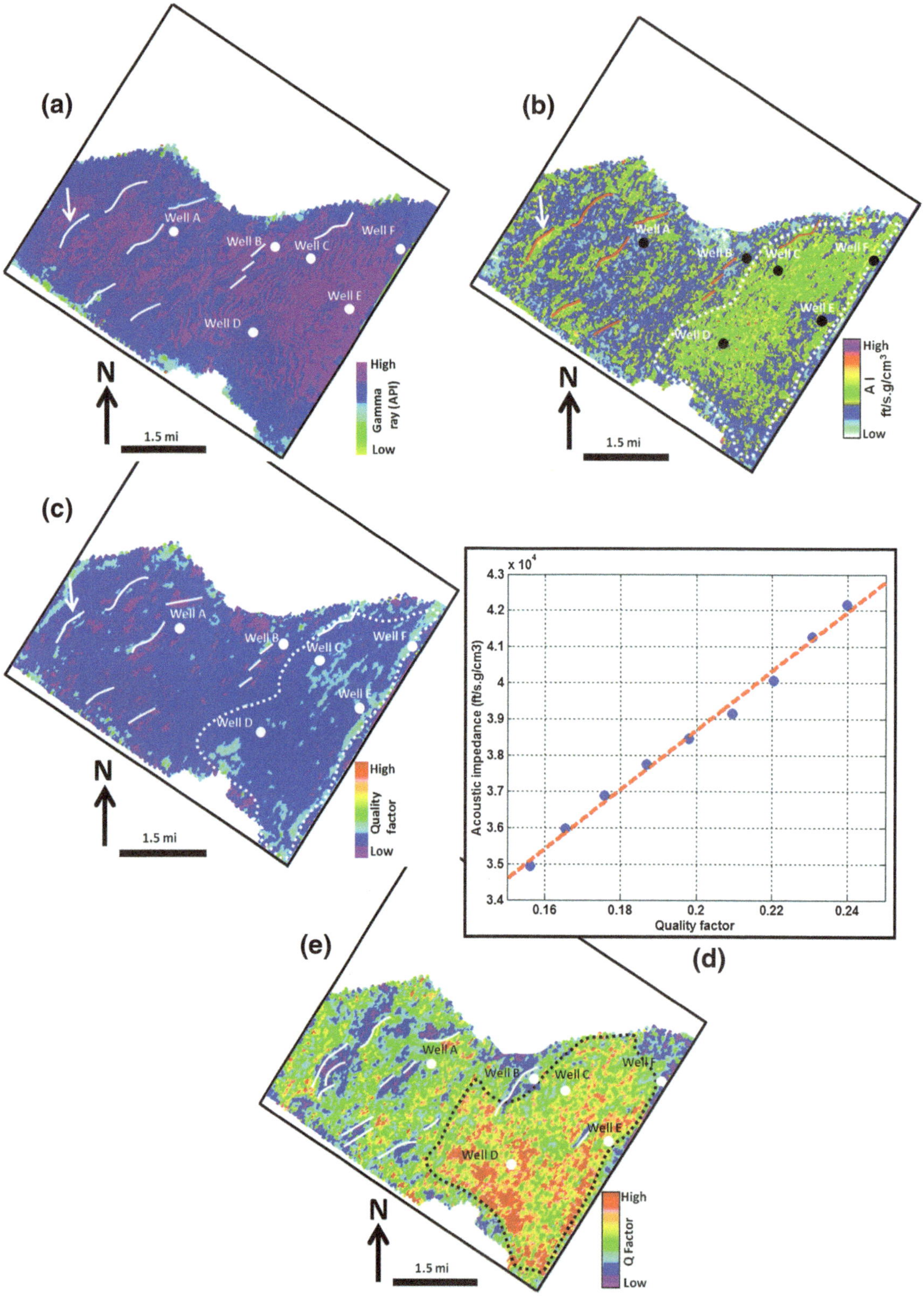

corresponds to high TOC; and from Fig. 12a, high Q corresponds to high resistivity. Hence high Q corresponds to high TOC. In addition, within the lower Eagle Ford Shale Q is low, resistivity is low, and TOC is also low (Figs. 7a, 8a, 12a), suggesting increasing clay content. We would like to emphasize that high Q corresponds with resistant beds (the brittle zones) which are, in this case, carbonate- and TOC-rich beds. Because of the strong correlation between

◀ **Fig. 13** Extracted attribute maps along horizon phantomed 12 m below top lower Eagle Ford showing **a** gamma-ray map, **b** acoustic impedance (AI) map, and **c** quality factor (*Q*) map. Note on gamma-ray map, horizon characterized throughout by high gamma ray (magenta) except in fault zones where low-gamma-ray values are seen (*blue and cyan*). *White lines* fault traces from coherency. Note also almost identical display of AI and *Q* maps, suggesting relationship between AI and *Q*. *White dotted outlines* in both AI and *Q* maps, representing areas of high AI and *Q*, respectively, are almost same. *Red traces* in AI and *white traces* in *Q* maps fault traces from corresponding coherency map. **d** Cross plot of AI versus *Q* in interval containing mapped horizon, showing that relationship between AI and *Q* is linear. **e** *Q* attribute map extracted along horizon phantomed 2 m below upper Eagle Ford, showing area of highest *Q* values (*black dotted outline*). *Dotted outlines* in **b** (*white*), **c** (*white*), and **e** (*black*) are most brittle zones (see text for details)

TOC and carbonate (Sondhi 2011), *Q* was chosen by the multiattribute algorithm as the best attribute to predict TOC. Herein lies the relationship between *Q* and TOC—TOC increases as *Q* increases (Figs. 7a, 8a, 12a). The Eagle Ford Shale is therefore a unique shale—a calcareous marl, in which the TOC content increases with increasing bed resistance. We consider the high-resistant bed a possible barrier to upward hydrocarbon migration in that it could act as a cap rock in some cases, not only trapping hydrocarbons generated within it because it is rich in TOC but also trapping hydrocarbons generated within the TOC-rich lower Eagle Ford, particularly where the bed is more continuous. If the bed were fractured or faulted, trapped hydrocarbons would escape and migrate into the overlying Austin Chalk. Thus, we interpret the sweet spots within the Eagle Ford Shale to lie directly beneath and within the resistant (high-*Q*) beds.

Two such beds/intervals occur within the Eagle Ford Shale in the Maverick Basin: (1) the upper resistant bed (URB) just below the top of the upper Eagle Ford and (2) the lower resistant bed (LRB) within the lower Eagle Ford (Fig. 12a, b). The LRB is thinner, at approximately ~30–95 ft (9–29 m), than the URB, which is about ~68–270 ft (21–82 m) thick. Although the lower Eagle Ford is generally composed of high gamma-ray values (Fig. 13a), the LRB within it exhibits some relatively high AI and *Q* values, suggesting that the LRB could be more calcareous and, therefore, more brittle than the surrounding shale matrix. A comparison of AI and *Q* attribute maps (Fig. 13b, c) shows that both *Q* and AI yield similar results: high-AI and high-*Q* areas are almost identical in both maps. For example, the areas outlined by the white dotted line in the south part of the survey have similar shapes.

Cross plots of AI and *Q* attributes obtained from the lower Eagle Ford at Well D (Fig. 13d) show a good linear relationship between the two attributes—that is, *Q* increases with increasing AI. Because high AI suggests a high elastic modulus, it follows that high *Q* also suggests a high elastic modulus, suggesting increasing brittleness. Therefore, *Q* attributes can be used to identify brittle zones within the Eagle Ford Shale. Because the *Q* attribute is faster and cheaper to compute from 3D poststack seismic

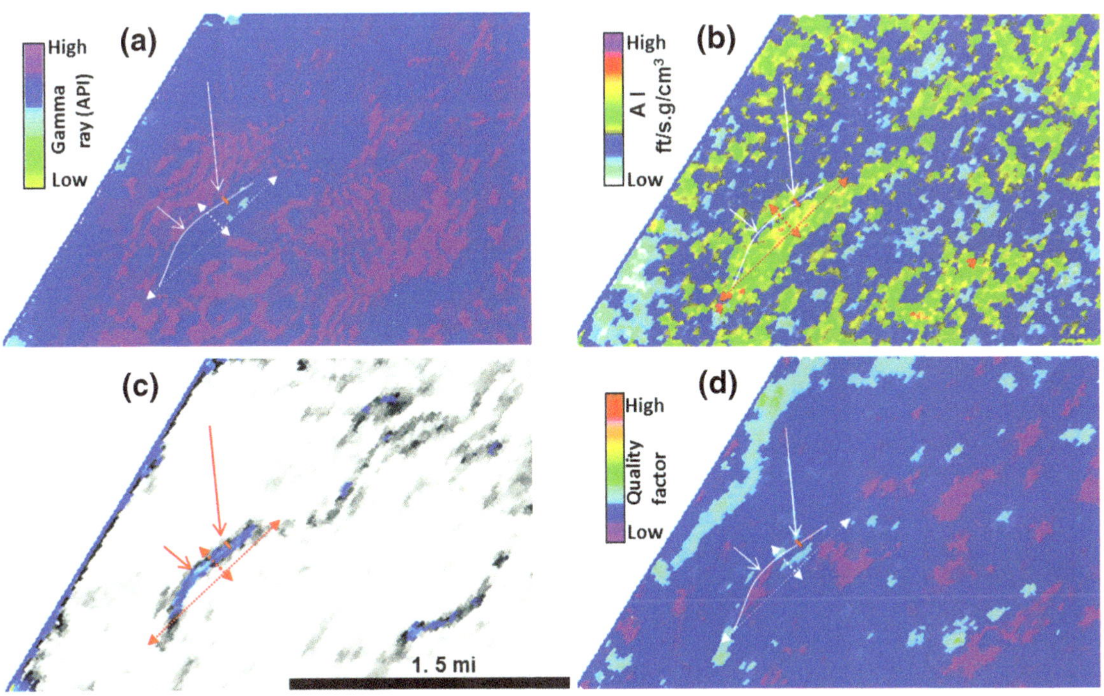

Fig. 14 Enlarged version of parts of Figs. 11a, 13a–c, showing fault zone encased within carbonate-rich (brittle) rock: **a** gamma ray map, **b** AI map, **c** coherency map, and **d** *Q* attribute map. Note *short arrow* points to fault zone and *long arrow* points to width of fault zone indicated by *red short bar* in all four figures

data than other attributes, it could offer the fastest and cheapest method of identifying brittle zones within the Eagle Ford Shale and could assist explorationists in planning horizontal wells. A *Q* attribute map extracted within the URB shows brittle zones (Fig. 13e), and *Q* values in this interval are higher than those within the LRB. As can be seen on the URB map (Fig. 13e), although the most brittle zones (higher *Q*) are in the southeast half of the survey, the entire map area appears to be composed of relatively high *Q*. Areas of low *Q* (magenta) correspond to fault zones. In contrast, within the LRB, the most brittle zones are concentrated mostly in the southeast part of the map (Fig. 13c). Few relatively high *Q* zones, which are linear and trend NE–SW, can be seen in the north half of the survey area. These same linear trends can also be seen in the AI map (Fig. 13b). What Fig. 13b, c suggest is that the rock units in which the faults occur are composed of more brittle (carbonate) materials than the surrounding matrix (predominantly shale). However, areas occupied by the fault traces (i.e., fault zones) have lower *Q* and lower AI compared to the carbonate rock unit that contains the fault zones. For example, consider the linear feature indicated by white arrows in Fig. 13b and c; an enlarged version of it (Fig. 14) shows that it is ~0.96 mi (~1.54 km) long (northeast-trending double-headed arrow) and 1,373 ft (~418 m) wide (southeast-trending double-headed arrow). The fault zone (thin white line indicated by short arrow, Fig. 14a–d) contained by this feature is ~200–300 ft (~61–91 m) wide and is characterized by lower *Q* and lower AI than exhibited by the rest of this rock unit.

Conclusions

In studying the lateral and vertical distribution of TOC and resistivity within the Eagle Ford Shale by analyzing seismic attributes, including AI and well logs, we found that the upper Eagle Ford Shale interval has higher AI values than the lower Eagle Ford Shale because it is more calcareous. Both the upper and the lower Eagle Ford Shale intervals are TOC rich. Horizon slices show that high-TOC and high-resistivity zones are concentrated mostly in the south and southwest parts of the survey and are laterally more continuous there. In the north half of the survey, continuity of these properties is broken by NW-SE-trending faults having throws ranging from 10 to 100 ft (~3–30 m). The high-resistivity and high-TOC zones correlate strongly with high *Q*, an attribute that indicates absorption characteristics and the relative Young's modulus of the beds. Two high-resistivity zones, an upper resistive bed and a lower resistive bed, are identified within the Eagle Ford Shale in the study area. Although both beds exhibit high resistivity, the upper bed is characterized by higher resistivity values. Because high-resistivity zones correlate strongly with the high *Q* attribute, and because the zones are associated with high TOC, the high resistivity values are interpreted to be a combination of resistivity due to TOC and resistant lithology. Additionally, because the *Q* attribute relates to the Young's modulus of the rock and also correlates with AI, it can be used to identify brittle zones. Furthermore, because *Q* is faster and cheaper to compute from the 3D poststack seismic data set than AVO prestack attributes, it could serve as a quicker, alternate method of identifying brittle zones within the Eagle Ford Shale.

Acknowledgments We thank our industry partner, Boilingstone Exploration, for supplying the data. We also thank the Landmark University Program and Hampson and Russell for supplying the software used in this study. We also thank Susie Doenges and Chris Parker for editing the manuscript and John Ames for preparing the figures. Publication authorized by the Director, Bureau of Economic Geology, The University of Texas at Austin.

References

Baldwin OD, Adams JAS (1971) K^{40}/Ar^{40} ages of the alkalic igneous rocks of the Balcones fault trend of Texas. Tex J Sci 22(2 and 3):223–231

Barnes A (1992) The calculation of instantaneous frequency and instantaneous bandwidth (shot note). Geophysics 57:1520–1524

Barnes A (1993) Instantaneous bandwidth and dominant frequency with applications to seismic reflection data. Geophysics 58:419–428

Dawson WC (2000) Shale microfacies: Eagle Ford Group (Cenomanian–Turonian) North-Central Texas outcrops and subsurface equivalents. Gulf Coast Assoc Geol Soc Trans 50:607–622

Donovan AD, Staerker TS (2010) Sequence stratigraphy of the Eagle Ford (Boquillas) Formation in the subsurface of South Texas and outcrops of West Texas. Gulf Coast Assoc Geol Soc Trans 60:861–899

Eastwood RL, Hammes U (2010) Log model development for the Bossier and Haynesville Shales. SPWLA 52nd annual logging symposium, pp 1–17

Ewing TE, Caran SC (1982) Late Cretaceous volcanism in south and central Texas—stratigraphic, structural, and seismic models. Gulf Coast Assoc Geol Soc Trans 32:137–145

Faust ILY (1953) A velocity function including lithologic variations. Geophysics 18:271–287

Goodway B, Chen T, Downton J (1998) AVO and prestack inversion. CSEG annual convention

Grabowski GJ Jr (1981) Source-rock potential of the Austin Chalk, Upper Cretaceous, Southeastern Texas. Gulf Coast Assoc Geol Soc Trans 31:105–113

Grant FS, West GF (1965) Interpretation theory in applied geophysics. McGraw-Hill, New York

Hamilton EL (1972a) A correlation between Qs of shear waves and rigidity. Mar Geol 13:27–30

Hamilton EL (1972b) Compressional-wave attenuation in marine sediments. Geophysics 37:620–646

Hammes U, Hamlin HS, Ewing TE (2011) Geologic analyses of the Upper Jurassic Haynesville Shale in east Texas and west Louisiana. AAPG Bull 95(10):1643–1666

Hampson D, Schuelke J, Quirein J (2001) Use of multiattribute transform to predict log properties from seismic data. Geophysics 66:220–236

Hentz TF, Ruppel SC (2011) Regional lithostratigraphy of the Eagle Ford shale: Maverick Basin to East Texas Basin. Gulf Assoc Geol Soc Trans 60:325–337

Hildred G, Ratcliffe K, Schmidt K (2011) Application of inorganic whole-rock geochemistry to shale resource plays: an example from the Eagle Ford Shale, Texas. http://www.chemostrat.com/wp-content/uploads/2013/08/Application-of-Inorganic-Whole-Rock-Geochemistry-to-Shale-Resource-Plays-an-Example-from-the-Eagle-Ford-Shale-Texas.pdf. Accessed 17 Dec 2013

Hill RJ, Jarvie DM, Zumberge J, Henry M, Pollastro RM (2007) Oil and gas geochemistry and petroleum systems of the Fort Worth basin. AAPG Bull 91:445–473

IHS Energy Information (2012) https://penerdeq.ihsenergy.com/thin2/secure/home/home.jsf. Accessed 16 Aug 2012

Johnston DH, Toksoz MN (1981) Seismic wave attenuation. Geophysics reprint series No. 2. In: Toksoz MN, Johnston DH (eds) Definitions and terminology., pp 1–5

Johnston DH, Toksoz MN, Timur A (1979) Attenuation of seismic waves in dry and saturated rocks: II. Mechanisms. Geophysics 44:691–711

Koesoemadinata A, El-Kaseeh G, Banik N, Dai J, Egan M, Gonzalez A, Tamulonis K (2011) Seismic reservoir characterization in Marcellus shale. Presented at San Antonio SEG annual meeting, 3700–3704

Kuich N (1989) Seismic fracture identification and horizontal drilling: keys to optimizing productivity in fractured reservoir, Giddings field, Texas. Gulf Coast Assoc Geol Soc Trans 39:153–158

Laubach SE, Olson JE, Gross MR (2009) Mechanical and fracture stratigraphy. AAPG Bull 93(11):1413–1426

Lewis JO (1977) Stratigraphy and entrapment of hydrocarbons in the San Miguel sands of Southwest Texas. Gulf Coast Assoc Geol Soc Trans 27:90–98

Luttrell PE (1977) Carbonate diagenesis and facies distribution of the Anacacho Limestone associated with a Late Cretaceous volcano in Elaine Field, Dimmit County, Texas. Master's thesis, The University of Texas at Austin

Ogiesoba OC (2010) Porosity prediction from seismic attributes of the Ordovician Trenton-Black River groups, Rochester field, southern Ontario. AAPG Bull 94(11):1673–1693

Ogiesoba OC, Eastwood R (2013) Seismic multiattribute analysis for shale gas/oil within the Austin Chalk and Eagle Ford Shale in a submarine volcanic terrain, Maverick Basin, South Texas. Interpretation 1(2):61–83

Passey QR, Creaney S, Kulla JB, Maretti FJ, Stroud JD (1990) A practical model for organic richness from porosity and resistivity logs. AAPG Bull 74:1777–1794

Passey QR, Bahacs KM, Esch WL, Kimentidis R, Sinha S (2010) From oil-prone source rock to gas producing shale reservoir—geologic and petrophysical characterization of unconventional shale-gas reservoirs. Soc Pet Eng 13150:29

Rickman R, Mullen M, Petre E, Grieser B, Kundert D (2008) A practical use of shale petrophysics for stimulation design optimization: all shale plays are not clones of the Barnett Shale. SPE Paper No. 115258

Robinson EA, Treitel S (2008) Digital imaging and deconvolution: the ABCs of seismic exploration and processing. Geophysical references series No. 15. Society of Exploration Geophysicists, Hyderabad

Sena A, Castilo G, Chesser K, Voisey S, Estrada J, Carcuz J, Carmona E, Hodgkins P (2011) Seismic reservoir characterization in resource shale plays: stress analysis and sweet spot discrimination. Lead Edge 30:758–764

Slatt RM, Abousleiman Y (2011) Merging sequence stratigraphy and geomechanics for unconventional gas shales. Lead Edge 30(3):274–282

Slatt RM, O'Brien NR, Romero AM, Rodriguez HH (2012) Eagle Ford condense section and its oil and gas storage and flow potential. http://www.searchanddiscovery.com/documents/2012/80245slatt/ndx_slatt.pdf. Accessed 17 Dec 2013

Sondergeld CH, Newsham KE, Comisky JT, Rice MC, Rai CS (2010) Petrophysical considerations in evaluating and producing shale gas resources. SPE Paper No. 131768

Sondhi N (2011) Petophysical characterization of Eagle Ford Shale. Master Thesis, University of Oklahoma, Norman, Oklahoma, USA. http://mpge.ou.edu/research/documents/2011%20thesis/Namrita%20Sondhi_thesis.pdf. Accessed 11 Dec 2013

Spencer AB (1969) Alkalic igneous rocks of the Balcones province, Texas. J Petrol 10(2):272–306

Toksoz MN, Johnston DH, Timur A (1979) Attenuation of seismic waves in dry saturated rocks: I. Laboratory measurements. Geophysics 44:681–690

Treadgold G, Campbell B, McLain B, Sinclair S, Nicklin D (2011) Eagle Ford shale prospecting with 3D seismic data within a tectonic and depositional framework. Lead Edge 30:48–53

7

A sensitivity study of FILTERSIM algorithm when applied to DFN modeling

Rouhollah Ahmadi · Mohsen Masihi · Mohammad Reza Rasaei · Kiomars Eskandaridalvand · Reza Shahalipour

Abstract Realistic description of fractured reservoirs demands primarily for a comprehensive understanding of fracture networks and their geometry including various individual fracture parameters as well as network connectivities. Newly developed multiple-point geostatistical simulation methods like SIMPAT and FILTERSIM are able to model connectivity and complexity of fracture networks more effectively than traditional variogram-based methods. This approach is therefore adopted to be used in this paper. Among the multiple-point statistics algorithms, FILTERSIM has the priority of less computational effort than does SIMPAT by applying filters and modern dimensionality reduction techniques to the patterns extracted from the training image. Clustering is also performed to group identical patterns in separate partitions prior to simulation phase. Various practices including principal component analysis, discrete cosine transform and different data summarizers are used in this paper to investigate a suitable way of reducing pattern dimensions using outcrop maps as training image. Because of non-linear nature of patterns present in fracture networks, linear clustering algorithms fail in determining the borders between the actual partitions; non-linear approaches like Spectral methods, however, can act more efficiently in diagnosing the right clusters. A complete sensitivity analysis is performed on FILTERSIM algorithm regarding search template dimension, type of filtering technique and the number of clusters for each clustering approach described above. Interesting results are obtained for each parameter that is changed during analysis.

Keywords Discrete fracture network · Geometrical modeling · Multiple-point geostatistics · Dimensionality reduction technique · Clustering

R. Ahmadi · M. Masihi
Department of Chemical and Petroleum Engineering, Sharif University of Technology, Tehran, Iran

M. Masihi

M. R. Rasaei
Institute of Petroleum Engineering, Tehran University, Tehran, Iran

K. Eskandaridalvand
DeGolyer and MacNaughton Petroleum Consulting Company, Dallas, TX, USA

R. Shahalipour
Petroiran Development Company, Tehran, Iran

Introduction

Naturally fractured reservoirs are one of the most complex structures in petroleum geology. They represent great heterogeneities in all facets from the geological structure to the flow responses so that characterization and modeling of these reservoirs have been a real challenge for petroleum engineers. Presence of fracture networks plays an important role in flow of hydrocarbon fluids toward the wellbores. They may act as barriers, hindering the production and thus causing different scenarios for economical production. Although the fracture system can increase the porosity and hence the initial oil in place, this causes the drilling fluid lost circulation. Mud invasion through the fracture system gives rise to the cost and time of drilling operation; therefore modeling of fracture system is crucial for assessing the connectivity of the reservoir. In general, familiarity of the fracture system model is important for the development of oil fields, economical issues of geothermal reservoirs and production from groundwater sources. This knowledge can benefit petroleum engineers for better estimates of reserves, production forecasts and optimization of oil recovery factor (Tran et al. 2002; Tran 2004).

Modeling of naturally fractured formations starts from the geological model, ending to the flow simulation; therefore modeling of fracture network requires two steps: fracture geometry characterization and simulation of its dynamic behavior. Geometrical fracture modeling is thus the first step in describing the fractured reservoir.

Characterization of highly fractured regions is one of main purposes of reservoir management, and it is after that one can design the well locations and their trajectories to cut through the fractures. Therefore, construction of the facture intensity distribution models sounds important. In addition, other properties of the individual fractures as well as of the network are necessary to be modeled, since fractures with different properties (like size, dip and connectivity) show different effects on the fluid flow behavior; more fluids flow through the longer and wider fractures. Fracture orientation controls the anisotropy and permeability of the reservoir at different directions. For example subparallel fractures cannot be connected to each other to form a continuous and connected network for fluid flow; in this case, the well trajectory must be deviated to pass through the fractures to increase the production recovery. Using discrete fracture networks facilitates the numerical simulation of fluid flow in the reservoir. For example, where geometric fracture properties (location, orientation, size and width) are available, finite element or boundary element methods can be used to obtain the permeability tensor of the discrete network analytically (Oda et al. 1987; Lee et al. 1999; Teimoori et al. 2004). Cacas et al. (1990) used geometrical fracture statistics instead of permeability tensor to calculate the large-scale hydraulic conductivity of the fractured rock.

How to use DFN models in the flow simulation of fractured reservoirs and the estimation of hydraulic properties are discussed by Bourbiaux et al. (1997), Baker et al. (2001) and Dershowitz et al. (2002).

This paper aims at geometrical modeling of natural fracture networks observed on the outcrops. Since the MPS approaches are capable of reproducing the complex and curvilinear geological features such as natural fractures (Strebelle 2002; Caers and Arpat 2005; Journel and Zhang 2006), this technique is adopted to be used in this work. Utilizing reservoir outcrop maps taken at different locations and structures as training images, fracture networks are modeled under geological and tectonically conditions. We have focused on reproducing the multiple-point statistics incident to an initial training image through the FILTERSIM algorithm. Flow simulation results are finally used to validate the generated realizations by comparing their production recoveries with that of the prior model. FILTERSIM has the priority of fast computations over previously defined algorithms, like SIMPAT, through the classification of extracted patterns in separate clusters (Journel and Zhang 2006; Wu 2007). To reduce dimensionality of pattern space, FILTERSIM has been designed to classify structural patterns using selected filter statistics as specific linear combinations of pattern pixel values that represent different properties. Instead, patterns could be abbreviated in a small set of statistical measures called summaries. Data summaries are also used in this paper to do filtering in FILTERSIM. A full description of clustering algorithms and their applications in FILTERSIM algorithm is provided in the following sections. Furthermore, a new classification of dimensionality reduction practices that could be effectively used in FILTERSIM is provided. Several solutions would also be developed to improve different parts of the available FILTERSIM algorithm in line with the simulation of naturally fracture networks. A complete sensitivity analysis would be finally performed on FILTERSIM parameters including search template size, kind of filtering practice, type of clustering approach and the number of clusters used in partitioning the pattern space. Useful results are induced after this analysis that would help us with improving the FILTERSIM algorithm in further applications.

Discrete fracture network models

DFN models simulate the individual fracture properties, predicting the fluid flow behavior using the fracture geometry and conductivity. These models can be obtained either by deterministic or stochastic approaches. Due to deficiencies inherent to deterministic methods when applying to modeling of complex fractured media, most discrete fracture models make use of probabilistic and stochastic simulation concepts to determine the distribution and orientation of fractures having access to the structural growth stress and tectonic history information which can be obtained with difficulty.

In two-dimensional models (Long et al. 1982; Smith and Schwartz 1984; Mukhopadhyay and Sahimi 1992), fractures are represented as lines distributed randomly in a plane. The center of lines of fractures follows a Poisson distribution while the length and orientation of a fracture can be selected from any given distribution. Hestir and

Long (1990) analyzed various two-dimensional models to determine fracture connectivity and relate the parameters of fracture networks to standard percolation network parameters.

More complicated three-dimensional models were developed (Long and Witherspoon 1985; Charlaix et al. 1987; Pollard et al. 1976) to allow for realistic representation of fracture networks. Fractures are represented as flat planes of finite dimensions and in most of the cases as circular discs.

Fracture data (such as orientation, width and density) are primarily available at well locations. These data are analyzed statistically to estimate the distribution of fracture properties at inter-well locations.

DFN stochastic simulation has many advantages over the geomechanical and mathematical deterministic methods (either continuum or discrete). Dealing with these models, the approximations incident to the continuum models are eliminated by employing the probability distribution functions. The fracture information from different data sources can be used to construct the systems with individual fracture detailed properties in this approach. Stochastic simulation can be conditioned to identified fractures as well as dynamic data by taking the variable uncertainties into consideration. This improves the validity of fracture network models with respect to the reservoir condition.

In general, there are two stochastic approaches for modeling of fracture networks (1) object-based approach in which the fractures are depicted by the closed form objects with clean edges and (2) pixel-based approach, which simulates the nodal values one by one and then by combining these values, final fracture network is constructed. There are other algorithms such as tessellation and Voronoi grid methods which can be attributed to this class. However, due to deficiencies for applying these approaches, they could not be used in practice for modeling natural fracture networks.

In object-based models, fractures are defined as entities with specific center, shape, size and orientation, randomly positioned in space. Random disk models represent the fractures as two-dimensional convex circular disks. The radii of disks can be obtained by a log-normal distribution whose parameters are inferred from trace length distribution observed on the outcrops. Distribution and location of the disks, radii and their orientations are assumed uncorrelated from one fracture to another. Modeling of fracture clusters is difficult with the assumption of random fracture locations. Clustering is performed by parent and daughter model in which a spatial density function is used for cluster modeling. Seed locations of the fractures are determined by this density function and the fracture clusters are simulated over a predefined volume near the initial seed locations (Long et al. 1987).

Fractures are not circular disks; they have elliptical shapes in reality. Based on Long et al.'s (1987) and Hestir et al.'s (1987) findings, if there is not enough data to determine the real shape of the fractures, it is a logical assumption to imagine the fractures as disks. This assumption has many disadvantages and outcrop data do not confirm that.

Many drawbacks are inherent to these models; independence assumption between the parameters, difficulties in inferring the statistical distribution parameters, problems in soft data conditioning, and convex and planar assumption of individual fractures are limitations of this method.

While object-based techniques construct the features with clean edges, however, pixel-based algorithms simulate the nodal values on a discrete network. Useful algorithms such as SISIM, SGSIM or PFSIM are developed for modeling continuous (petrophysical data) and discrete (facies) parameters traditionally. One-point or two-point statistics are not able to reproduce the objects like fractures.

Traditional pixel-based algorithms use the primary data values at their locations as well as univariate statistical distributions and bivariate moments as the required information in the constructed models. This causes the curvilinear and continuous structure of geological entities to be reproduced in a poor manner through the variogram-based geostatistical algorithms.

Two-point statistical moments like variogram and covariance between simulation variables take only the two-point spatial relationships into consideration at one time; the reproduction of such complex geological images requires a relationship between more than two points at the same time and therefore the development of multiple-point simulation approaches seems to be necessary.

MPS techniques are not only capable of handing the prior maximum entropy models and performing multivariate Gaussian algorithms, but they can act very effectively in using the prior structural models of low entropy having more compatibility with hard conditioning data or prior geological models.

Guardiano and Srivastava (1993) developed the first MPS algorithm based on the Journel's extended normal equations. In the following section, a short review of available MPS algorithms with an emphasis on FILTERSIM is introduced.

A review of multiple-point geostatistical simulation algorithms

Various algorithms have been developed for multiple-point simulation of complex geological phenomena. To overcome the problems inherent to the two-point statistics, the issue of multiple-point geostatistics was introduced by Deutsch (1992) and Guardiano and Srivastava (1993). Later, development of a practical MPS simulation algorithm, SNESIM (Strebelle 2002), has led to the wide adoption of MPS simulation for reservoir facies or element modeling. SNESIM draws the patterns of a training image according to the well-defined probabilities and conditioned to the available hard and soft data. This algorithm utilizes a sequential simulation approach in its modeling scheme as in traditional pixel-based simulations: (1) the unknown nodal values of the simulation grid are visited along a predefined random path separately in time; (2) in each unknown cell, the conditional probability is estimated regarding the neighborhood data and the previously simulated values, and (3) a value is randomly selected from the constructed probability distribution and assigned to the current simulation cell. In SNESIM, the conditional probability is generated from the training image, considering the frequency of replicas of the present data event (Strebelle 2000, 2002).

In SNESIM, the training image is scanned only once; for each particular search template, all the conditional proportions present in the training image are stored in a smart search tree data structure to be retrieved as fast as possible in the next steps. SNESIM is composed of two main parts: (1) construction of search tree to save all probability proportions, and (2) simulation section where these proportions are read from the tree and used for simulation of unknown values.

SNESIM can handle only categorical variables such as facies distributions or discrete fracture networks, and is not able to use continuous data. This algorithm takes a large amount of memory especially in the case of large training images containing many various classes of different patterns; thus SNESIM can make only use of a limited number of classes (<5). This algorithm appears suitable for modeling discrete fracture networks. However, using the pattern-based algorithms instead can result in more accurate final realizations. In addition, using quite large scanning templates to capture the large-scale geological features increases the levels of constructed search tree and hence more memory will be required.

Another limitation of SNESIM code is the cost of retrieving the conditional probability when the conditional data event is not fully informed (having missing data locations). At each such missing data location, the SNESIM algorithm will traverse and trace all its children branches in the search tree for all possible data values. This cumbersome tree search is extremely memory and CPU consuming. Indeed, any SNESIM simulation is very slow at the beginning, and gets faster when more and more previously simulated nodes become conditioning data (Hu and Chugunova 2008; Straubhaar 2008; Mariethoz 2010). Non-stationarity problems are the other major drawbacks of the SNESIM algorithm (Chugunova and Hu 2008; DeVries 2009).

To solve the shortcomings of SNESIM algorithm, Caers and Arpat (2005) presented a new approach based on the statistical pattern recognition concepts, named as SIMPAT, in which the simulation is performed using the extracted patterns from the training image (Caers and Arpat 2005, 2007).

Firstly, the training image *TI* is scanned through the search template T to capture all the available patterns present. A filter may be applied to avoid keeping undesired or very similar extracted patterns. The filtered patterns are then stored in a database to be retrieved in the next simulation steps.

During the sequential simulation employed by SIMPAT, each unknown node u is randomly visited and the local conditioning data event $\mathrm{dev}_T(u)$ is recorded simultaneously. Afterward, $\mathrm{dev}_T(u)$ is compared to all available patterns in the database according to a predefined similarity criterion; the most common of which are distance functions. The purpose is to find out the most similar training pattern pat^*_T to the local data event $\mathrm{dev}_T(u)$. In other words, the algorithm evaluates the distance d for all existing patterns and picks up the pattern with the smallest distance as pat^*_T. Having selected the most similar pattern, $\mathrm{dev}_T(u)$ is replaced by pat^*_T such that part of the values inside pat^*_T is pasted onto the simulation grid centered at the current location u (Caers and Arpat 2005).

Due to global search performed on the training patterns database to find the most similar pattern to the data event, SIMPAT suffers from a high computational load (Honarkhah and Caers 2010). This is because of the large number of stored patterns and their high-dimensional feature space in the case of using large scanning templates. A new algorithm known as FILTERSIM was presented by Zhang in 2006 to alleviate the standard SIMPAT's computational intense (Journel and Zhang 2006). The default FILTERSIM starts to decompose each training pattern to a small number of linear filter scores (6 in 2D and 9 in 3D) each of which is a weighted average of the pattern features. These transformed training patterns will then be classified in the new filter space; the aim of clustering is to group the identical patterns into separate classes so that the similarity between them is defined as their distances to each other. The clustering is performed only once for each TI prior to the simulation stage. The sequential simulation approach is adopted later on to see the unknown nodes in a random manner. For each unknown node, the local conditioning data event is compared to the training cluster prototypes in order to select the closest one based on the calculated distance measure. A training pattern is picked up from the selected class in a random manner or conditioned to the available soft data; an inner part of the pattern (called "patch") is then fixed at the simulation grid centered at the current node; their location addresses are then removed from the predetermined random path and the remaining nodes will be replaced by the future evaluated values. Like all the geostatistical algorithms, hard data conditioning is achieved through freezing the corresponding nodal values at their locations not to be changed ever during the simulation step; these primary values are preferentially used in estimating the dissimilarity distances and therefore ensure us that no pattern involving the structural discontinuities will be selected (Wu 2007).

Contrary to the first multiple-point simulation algorithm, SNESIM, in which the *single nodal values* are modeled using the pattern data at each simulation step, SIMPAT and FILTERSIM evaluate *patterns* from the available pattern data. Compared to SNESIM, FILTERSIM is not limited to categorical variables and can be run with continuous values as well. FILTERSIM looks like the SIMPAT algorithm in many aspects, except that the latter does not perform any clustering on the extracted patterns, but searches for the pattern with the most similarity with the multiple-point data event.

Clustering in FILTERSIM

Clustering methods haves gained a great application in many areas such as data mining, document retrieval, image segmentation and pattern classifications. The purpose is to group the patterns according to a similarity (or dissimilarity) criterion so that each cluster represents a set of similar patterns. Each pattern is denoted by a set of system characteristics. There are several books published on classical data clustering (Anderberg 1973; Jain and Dubes 1988; Duda et al. 2001). Clustering algorithms have also been extensively studied in data mining and machine learning books (Han and Kamber 2000; Tan et al. 2005; Bishop 2006). A good survey of different clustering algorithms could be found in Xu and Wunsch (2005), Jain (2010).

After implementing the preprocessing step of FILTERSIM algorithm, it is required to utilize the appropriate similarity criteria between the patterns; the most common similarity measure of standard representation is "distance" (Perner 2007). The distance function used in this paper is of Euclidean type. Other functions could be also applied in this manner as fully described in Ahmadi (2011).

Clustering algorithms try to reorganize the set of training patterns in the database in the form of several separate classes based on a predefined similarity measure. Dimensionality of patterns may have been reduced prior to the clustering. Having accomplished the clustering stage, the patterns in a particular partition are much similar to each other than the patterns in other groups. K-means with a rich and diverse history as it was independently discovered in different scientific fields (Steinhaus 1956; Lioyd, proposed in 1957 but published in 1982; Ball and Hall 1965; MacQueen 1967) is one of the most commonly used clustering algorithms that was applied by Journel and Zhang (2006) in the multiple-point simulation approaches to partition the obtained normal filter scores. Standard partitional clustering methods (such as K-means, fuzzy C-means, Self Organized Maps (SOM) and neural gas) consider the linear borders between the formed clusters and are not able to separate the complex and non-linear data. Other methodologies have been presented by the ability of constructing the separable hypersurfaces. These algorithms are classified into two main categories: kernel-based and Spectral clustering methods.

In the case of kernel-based algorithms, several clustering techniques such as K-means, C-means and SOM have been widely used with the aid of kernel functions. Using kernels, data will be transformed into a new feature space of higher dimension where data could be linearly separated. Separability is performed easier in the higher-dimensional space. This change of coordinate system maintains the structural principles inside data; but it alters the way in which they are represented, so that the principles will be discriminated more straightforwardly. In other words, the non-linear relationship between the information inside data is able to be captured via non-linear mapping of data to a new higher-dimensional space. Correspondingly, the classical clustering algorithms such as K-means acts better in the new space and a more accurate and powerful clustering is achieved henceforth (Shawe and Cristianini 2004; Filippone et al. 2008).

Using the kernel-based clustering methods, however, requires the prior determination of the kernel function and its defining parameters. As an instance, in the case of choosing the Gaussian kernel as the base function, we should try in finding the optimal value of the width; since this parameter which is equivalent to the Gaussian standard deviation indicates the openness of the Gaussian function and the neighborhood area affected by the defined kernel that will influence the final clustering results significantly. However, due to unsupervised nature of clustering algorithms, optimality of kernel parameters does not appear so easily; with the manifold assumption of data, the spectral algorithms could therefore be applied in order for the non-linear pattern clustering.

Spectral clustering algorithms take advantage of graph theory concepts. The main idea is to construct a weighted graph from the initial data set; each node represents a pattern while the weighted edges are simply the similarity degrees between the two patterns on the heads. In this framework, the clustering could be considered like a graph cut problem which may be solved through the spectral graph theory. The fundamental idea is the eigenvalue decomposition of the Laplacian matrix of the weighted graph obtained from data; in reality, there is a close relationship between the second smallest eigenvalue of the Laplacian and the graph cut problem (Chung 1997; Cristianini et al. 2001).

Dimensionality reduction in FILTERSIM

Scanning a training image using a large template size brings about a high dimensionality of the extracted patterns; a preprocessing step would therefore be required to reduce data dimension as much as possible and select part of the features optimally; running clustering algorithms on such a high-dimensional dataset demands for a large computational burden and time.

Dimensionality reduction may be achieved in FILTERSIM through the following classifications:

1. Using image statistics and application of predefined filters to the training patterns,
2. Extraction of unique features from the training patterns or using feature transformations such as features generated from discrete Fourier power spectrum of each pattern,
3. Using common dimensionality reduction techniques such as principal component analysis (PCA), linear discriminant analysis (LDA) and discrete cosine transform (DCT).

The main contribution of this paper is to make use of (1) K-Nearest Neighbor Index (a simple but very common filter, used in the pattern recognition and spatial data mining tasks (Cover and Hart 1967; Toussaint 2005); introduced from the first classification above), (2) PCA (a very common technique in data analysis practices (Jackson 1991; Jolliffe 2002); proposed from the third classification above), and finally (3) DCT (a common transform used in digital image applications (Jain 1989; Pennebaker and Mitchell 1993); also brought from the third classification above). Other methods that are not tried in this paper may be the subject of future works.

Filters

A filter is composed of a set of weights relevant to the search template nodes T_J. The filter value or the weight of node u_j of the template is denoted by f_j; therefore, a J-node filter is defined as $\{f(h_j); j = 1, 2, 3, \ldots, J\}$, where h_j represents the coordinates of the search template T_J with respect to the central node u. The filter is applied to each local training pattern, *pat*, centered at node u in the training image. Application of F filters simultaneously would result in constructing a vector of F filter scores that summarizes any of available patterns in F new dimensions:

$$S_T(u) = \sum_{j=1}^{n} f_i(h_j).\text{pat}(u + h_j); i = 1, 2, \ldots.., F \quad (1)$$

where, $S_T(u)$ denotes the computed filter score and h_j is the offset of each template node from the central node u.

Representation of a pattern with its relevant filter score vector causes a severe dimensionality reduction from the initial pattern size $(n_x \times n_y)$ to F dimensions. Figure 1 shows the process of filtering a 2D training pattern and construction of the corresponding filter score value (Wu 2007).

Instead of averaging the values inside the training patterns through a specific filter, we may summarize each pattern by performing mathematical and statistical operations on their contents. This would be done through the application of *data summaries*.

In general, each pattern extracted from a fracture network training image could be interpreted in three ways: (1) a digital image, (2) a spatial phenomenon, or (3) a fracture phenomenon; regarding each perspective, the appropriate data summary might be designed for the FILTERSIM algorithm:

1. As a digital image, patterns could be defined by their statistical moments of different orders; mean as the first moment and variance with respect to the mean as the second one are common statistics of a random phenomenon. Summarizing each pattern in a small set of its statistical measures might have the potential of reducing dimensionality of the existing patterns in an appropriate and unique manner so that patterns of different occurrences of available categories at their different nodal locations are easily distinguished from each other in the new coordinate system of the selected summaries. No overlapping is expected for the originally separated patterns in the new dimensionality space. Where required, using a more number of efficient measures may benefit to the accuracy of summarizing practice.
2. From the spatial data mining views, the spatial correlation indices such as nearest neighbors index (NNI) (Cover and Hart 1967; Toussaint 2005) and Moran-I parameter (Moran 1950) are suggested to summarize the patterns as the spatial phenomena. To achieve a better and more accurate summarization of training patterns in this mode, it may be practical to

Fig. 1 Filtering process of a training pattern (Wu 2007)

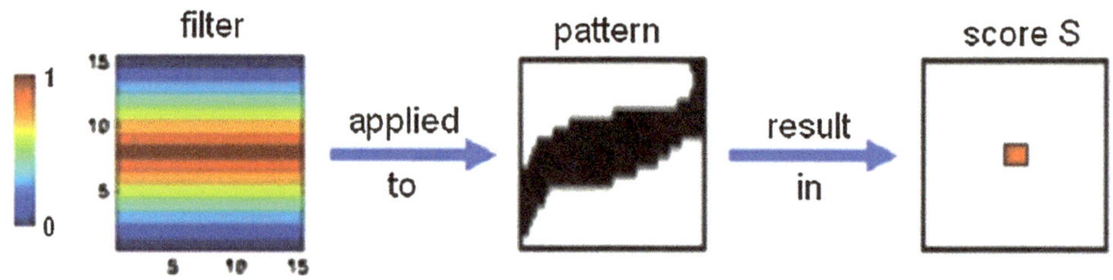

Fig. 2 Four user-defined 2D filters of size 11 × 11 (Wu 2007)

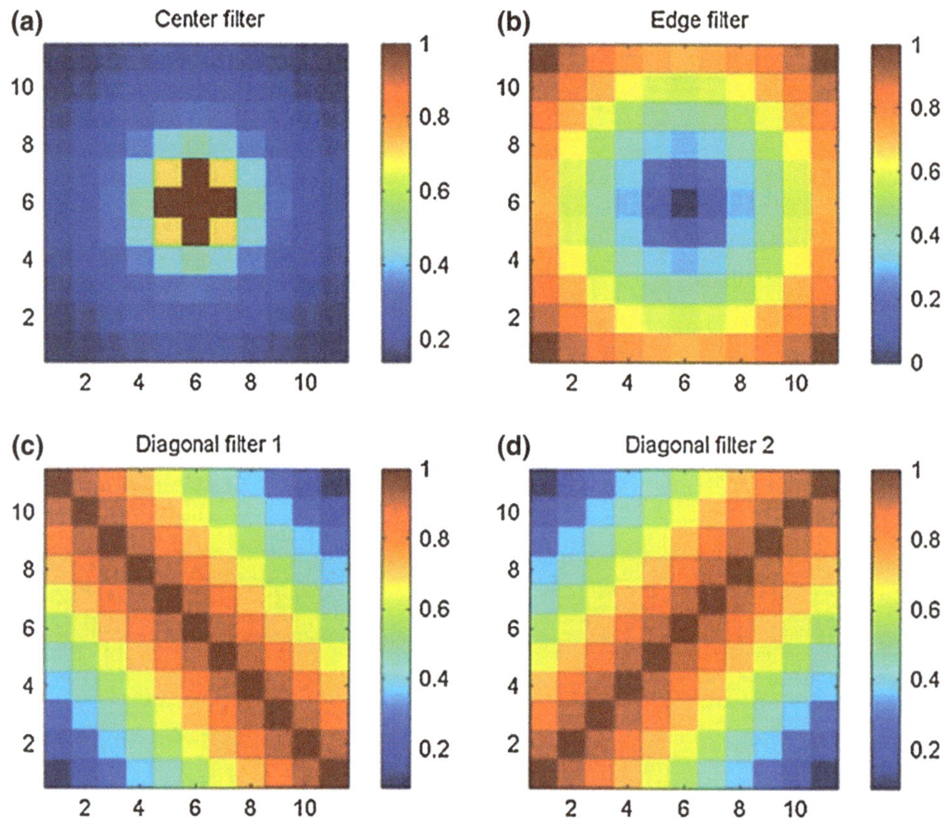

use farther points in the neighborhood of fracture cells inside each pattern.

3. As a fracture network system, it is reasonable to define some new summaries based on the fractures connectivity and the orientation of the fractures along the straight lines.

Default FILTERSIM algorithm makes use of three filters (average, gradient and curvature) in each coordinate direction (Journel and Zhang 2006). The user could instead design its own filters depending on the features of interest in the training image that should be reproduced in the simulation. For instance, the isotropic or oriented filters can have great applications in the case of particular geological structures involving different orientations, such as fracture networks. Examples are central, edge, diagonal I and diagonal II filters shown in Fig. 2 (Wu 2007).

In this paper, we have designed two specific data summaries exclusive to fracture networks including connectivity and direction of individual fractures to obtain a more accurate geological system involving more realistic and connected fractures.

New data summaries for fracture networks modeling

One of the most important and challenging issues in modeling fracture networks and other geological structures like channel reservoirs is the connectivity of final realization so that disconnectivity of simulated nodes belonging to a particular structure is minimal.

Two schematic fracture networks of different connectivities are shown in Fig. 3. Based on a new definition, connectivity of any specific fracture node in the template is defined as the number of fracture nodes to which it is connected immediately. Degree of connectivity of the whole fracture network would then be equal to the sum of connectivities of all fracture nodes existing in the network.

Fig. 3 Two 5 × 5 fracture pattern networks with different connectivities; the blue nodes are in fracture and the *green color* denotes the matrix

The application of connectivity criterion is ranking and selection of the best patterns. Amongst the extracted patterns, the one with the larger degree of connectivity will be won to be selected. How to calculate the connectivity of each pattern network and use the estimated scores consequently as a selection criterion is demonstrated below for the two typical patterns illustrated in Fig. 3. The left-hand-side calculations denote the connectivity of each fracture node in the left pattern, while the right-hand-side computations evaluate the connectivity of fracture nodes for the right pattern:

$$N\{TI(1,1)\} = 1 \quad N\{TI(1,1)\} = 0$$

$$N\{TI(2,2)\} = 2 \quad N\{TI(2,3)\} = 1$$

$$N\{TI(3,3)\} = 2 \quad N\{TI(3,3)\} = 2$$

$$N\{TI(4,4)\} = 2 \quad N\{TI(4,4)\} = 2$$

$$N\{TI(5,5)\} = 1 \quad N\{TI(5,5)\} = 1$$

$$N_{G1} = \sum_{i,j \in F} N\{TI(i,j)\} = 8 \, N_{G2} = \sum_{i,j \in F} N\{TI(i,j)\} = 6$$

where F denotes the set of nodes belonging to the fracture class; TI is the value of current node in the training image TI, and N equals the number of neighborhood nodes in the fracture. For example, the fracture node with the coordination (2,2) in the left pattern in Fig. 3 connects immediately to fracture nodes of coordinations (1,1) and (3,3) in the same pattern. So connectivity of the mentioned fracture node is evaluated as 2. In Fig. 3, since connectivity measure (N_G) of the fracture pattern in the left is greater than that of the right network, the former is selected as the preferable pattern; as it is also clear from Fig. 3, connectivity of the left pattern is visually more than the right one.

Now consider the third fracture pattern in Fig. 4; although it seems visually to be less proper than the pattern in the left part of Fig. 3, however, they have the same degree of connectivity measures mathematically.

$$N_{G3} = N_{G1} = 8$$

Fig. 4 **a** 5 × 5 pattern of different disconnectivity than the pattern in the left side of Fig. 3; the *blue nodes* are in fracture and the *green color* denotes the matrix

To solve this problem, a new measure could be designed by estimating deviation of the fracture nodes from a straight line. This really investigates orientation of fracture pixels along a specific extension (including all diagonals of the pattern matrix). Regarding the disconnectivity quantity, the left pattern in Fig. 3 will be selected rather than the network in Fig. 4, because the fracture nodes in the first pattern are located only along one diagonal (the main diagonal), while in the second pattern fractures are placed alongside the two diagonals and hence have a more disconnectivity than the first one does. To do a right comparison in the case of all possible fracture patterns, it is better to consider the total number of network diagonals with at least a fracture node on those.

Simulation and results

General FILTERSIM algorithm could be broken into four steps including (1) training image scanning, (2) extracted patterns filtering, (3) pattern space clustering, and (4) simulation stage. Simulation is performed through the sequential simulation algorithm as described previously. A sensitivity analysis is, however, performed on the first three steps of FILTERSIM algorithm as explained below:

1. The training image is scanned four times using four square templates of different sizes including 17, 21, 25 and 29 grids on each edge.
2. Clustering algorithms require a relatively large computational time when applied to the pattern database of a very high dimension (depending on the size of search template). It is therefore strongly suggested to perform a preprocessing step on the pattern database and reduce the patterns dimensionality as much as possible. Dimensionality reduction of the extracted patterns, prior to the clustering stage, has been achieved through the application of three different techniques including PCA, KNN and DCT. DCT is applied to each training pattern, while the other two techniques are executed once to the whole set of database. It should be noted that in the case of using 2D DCT as a dimensionality reduction technique, those coefficients will be selected as the transformed features that contain 99 % of the total energy of the whole pattern image. To see the effect of filtering on both running times and accuracy of the simulation algorithm, the models obtained through the above filtering techniques will be compared with the cases where no filtering has been done on the patterns (labeled as FILTERLESS case).
3. It is a common practice in FILTERSIM to group the identical patterns in several clusters prior to the distance minimization step. Using two common clustering methods, K-means and Spectral techniques, partitioning the hard training patterns database and then the modeling tasks are elaborated. Three different numbers of clusters including 1/200, 1/400 and 1/800 of the total number of training patterns extracted from the training image have been used in the clustering algorithm.

Therefore, a complete sensitivity analysis will be performed on the FILTERSIM algorithm, regarding size of the scanning template (17, 21, 25 and 29), type of clustering technique (K-means and Spectral), the number of clusters (1/200, 1/400 and 1/800 of the total number of training patterns) and type of filtering method (PCA, KNN, DCT and Without-Filter). As a result, 48 cases of FILTERSIM runs will be implemented for each clustering method.

Figure 5 shows an outcrop map of a naturally fractured reservoir in south west region of Iran. This site is located in the flank of an anticline structure, next to its nose. The major feature of this site is a large transverse fault observed at the right side of the map. In addition, there are several normal faults as well as many conjugate fractures more or less parallel to the local fold axis. The site is affected by two main joint sets perpendicular to the bedding, oriented at NNE and NS directions. The sensitivity analysis on FILTERSIM algorithm is performed using this training image.

Fig. 5 An outcrop map of a naturally fractured reservoir in south west region of Iran

As an important notation, K-means clustering algorithm may be stopped at the middle iterations due to convergence issues posed by improper initial random assignment of data points to the clusters. Two methods have been adapted to choose the initial cluster centroid positions, sometimes known as seeds. The first approach selects k observations from the dataset X at random. Selecting k points uniformly at random from the range of X is the second method used in the present K-means clustering. Another approach is to perform a preliminary clustering phase on a random 10 % subsample of X. This phase is itself initialized using the first technique. This method is, however, not used in the present study.

In addition, due to dimensionality reduction incident to the Spectral clustering algorithms themselves (where a limited number of eigenvectors of the affinity matrix is to be selected and partitioned), they are expected to have a higher computational speed with respect to the K-means algorithm and thus appear to be more suitable in modeling large fracture networks in this view. This will be tracked during the sensitivity analysis too.

Results comparison

Simulation results could be evaluated through a comparison between the spatial patterns based on several quantitative ways such as MPS, flow simulations or even semivariograms. There are many different types of MPS measures such as connectivity function, lacunarity, the multiple-point density function (MPDF), transition probabilities and distribution of runs (Boisvert et al. 2008; Pyrcz et al. 2008). Flow simulation curves of oil recovery factor have been adapted in this work to be used in verification of results.

A five-spot pattern water flooding with 4 injectors at the corners and 1 producer at the center of model is simulated to obtain the oil recovery factor curve as model performance. Such a flow simulation is designed for each realization as well as the original training image. The simulation is done using a commercial ECLIPSE100 simulator, while it is linked to the MATLAB programming environment to read the images, do the required preprocessing, pass them to the simulator and analyze and post-process the output results. An E-type is generated using 10 FILTERSIM realizations for each case of simulations. Each continuous E-type is converted to a binary image by a threshold of 0.5. The final image to be passed to the flow simulator is therefore a two-class fracture network. The nodes in the fractures are devoted 1,000-millidarcy permeability and full porosity of 1, while the background facies (i.e., non-fracture locations) are assumed to have 10-millidarcy permeability and porosity of 0.1. The training image has a size of 69 × 104 cells in XY coordinates (Fig. 5). The four injection wells are drilled in (1, 1), (69, 1), (1, 104) and (69, 104) coordinates, while the production well is located at (32, 68) coordinate. The wells are allowed to inject or produce for 7,500 days from the beginning. The mean squared error (MSE) values are evaluated between the recovery factor of the original training image and that of each FILTERSIM case.

Description of figures and tables

The final results of sensitivity analysis include the computational times as well as the MSE values (as the measures of accuracy) for the implemented simulations. These data are provided in Tables 1, 2, 3 for the cases that have used K-means clustering method and in Tables 4, 5, 6 for the simulations performed through the Spectral clustering techniques. Simulation times of different parts of FILTERSIM algorithm (including scanning, filtering, clustering and simulation procedures) and MSE of recovery data are shown in these Tables. The errors are displayed in the last column of Table 1 for the realizations obtained through application of K-means clustering technique and in the last column of Table 4 for those realizations modeled using Spectral clustering method. It should be noted that the simulation parameters as in the second columns of the mentioned Tables indicate the order of "filtering method, search template size, ratio of cluster numbers", respectively; for example, the term "PCA, 25, 400" means a FILTERSIM case where filtering is done by PCA, search template size is 25 in each direction on a squared grid and the number of clusters used is equal to 1/400 of the total number of training patterns.

Ranking of different simulation cases based on running times of different sections (including scanning, filtering, clustering and simulation procedures) for realizations obtained through application of K-means clustering technique is displayed in Table 2. The same results, but in the case of applying Spectral technique, are provided in Table 5. Each column of Tables 2 and 5 indicates the index of a specified running time in an ascending order. The numbers indicate the row index of each case as specified in Tables 1 and 4. The last column displays label of simulation case corresponding to ranking results of total time of simulation.

In addition, Tables 3 and 6 show ranking results of different FILTERSIM cases based on MSE values of production recoveries in the case of K-means and Spectral clustering, respectively.

Figure 6 illustrates the most and the least accurate results of sensitivity analysis for the cases generated by applying K-means and Spectral clustering techniques. Figure 6a shows the binary training image that has been used in the FILTERSIM simulations for the analysis. It is actually the digital form of the analog outcrop map displayed in Fig. 5. Red nodes indicate fracture locations whereas blue nodes are in background facies (i.e., non-fracture locations). It has a size of 69 × 104 grids on a two-dimensional coordinate. The modeling area has been built through selection of 300 data points from the training image at random. Figure 6b displays the least accurate E_type obtained through application of K-means clustering algorithm during sensitivity analysis. As it is clear in Table 3, the least accurate result of K-means cases is obtained using KNN as filtering method, a search template of size 17 × 17, and a number of clusters equal to 1/800 of the total number of training patterns (i.e., a total number of 7,176 scanned patterns). Figure 6c illustrates the most accurate E_type when using K-means clustering. This is achieved without any filtering, using a search template of size 21 × 21, and a number of clusters equal to 1/200 of the total number of training patterns. The least accurate E_type of FILTERSIM using Spectral clustering method is demonstrated in Fig. 6d where it comes from KNN as filtering method, a search template of size 29 × 29, and a number of clusters equal to 1/200 of the total number of training patterns. Finally, the most accurate E_type of Spectral cases is shown in Fig. 6e; using PCA as filtering method, a search template of size 17 × 17, and a number of clusters equal to 1/200 of the total number of

Table 1 Simulation times of different sections (including scanning, filtering, clustering and simulation procedures) and mean squared errors of recovery data for realizations obtained through application of K-means clustering technique

Row	Simulation parameters (label)	Scanning time (s)	Filtering time (s)	Clustering time (s)	Simulation time, (s)	Total time (s)	Recovery MSE
1	pca 17,200	114.6	1.4	148.7	119.3	384.1	5.00E−04
2	pca 17,400	111.5	1.1	104.3	93.6	310.5	5.00E−03
3	pca 17,800	110.7	1.7	98.8	71.8	282.9	9.00E−04
4	pca 21,200	134.7	3.2	185.5	140.4	463.9	4.70E−03
5	pca 21,400	128.5	3.1	164.9	103.1	399.6	2.00E−04
6	pca 21,800	126.8	3.0	96.5	93.0	319.3	4.00E−04
7	pca 25200	168.5	6.8	238.5	185.8	599.6	1.00E−04
8	pca 25,400	168.5	6.8	160.1	180.9	516.2	1.00E−04
9	pca 25,800	170.9	6.9	112.0	115.5	405.4	3.00E−04
10	pca 29,200	219.0	14.5	331.1	302.6	867.1	2.90E−03
11	pca 29,400	217.9	14.4	235.6	212.7	680.7	5.90E−03
12	pca 29,800	218.7	15.0	263.8	222.9	720.3	3.00E−04
13	knn 17,200	104.3	13.5	60.8	61.1	239.7	1.40E−03
14	knn 17,400	113.0	13.4	69.2	65.7	261.4	4.90E−03
15	knn 17,800	111.7	12.9	71.3	68.0	264.0	6.40E−03
16	knn 21,200	130.4	21.8	65.4	64.9	282.5	8.00E−04
17	knn 21,400	134.4	22.2	69.1	63.6	289.3	5.10E−03
18	knn 21,800	130.2	22.5	67.6	62.3	282.6	1.00E−03
19	knn 25,200	172.7	29.2	84.4	76.9	363.1	1.00E−04
20	knn 25,400	169.3	29.1	82.6	75.3	356.2	5.00E−04
21	knn 25,800	168.6	29.0	83.5	76.3	357.5	2.00E−04
22	knn 29,200	224.4	40.1	107.5	97.7	469.7	5.90E−03
23	knn 29,400	226.4	40.0	108.0	98.1	472.5	2.80E−03
24	knn 29,800	222.9	40.0	107.0	97.3	467.2	7.00E−04
25	dct 17,200	112.2	9.5	69.1	64.0	254.7	3.50E−03
26	dct 17,400	106.8	9.5	67.0	63.0	246.3	7.00E−04
27	dct 17,800	112.1	9.4	67.7	64.4	253.6	5.00E−04
28	dct 21,200	132.9	15.7	67.5	62.3	278.4	5.00E−04
29	dct 21,400	138.3	15.9	67.8	62.9	284.9	4.00E−04
30	dct 21,800	138.1	15.8	68.0	62.5	284.4	5.00E−04
31	dct 25,200	172.9	23.3	83.7	75.9	355.8	6.00E−04
32	dct 25,400	178.5	23.2	85.8	78.4	365.9	2.00E−04
33	dct 25,800	172.2	25.2	85.8	78.2	361.4	3.00E−04
34	dct 29,200	259.3	34.0	120.5	108.2	522.0	1.00E−04
35	dct 29,400	226.8	33.6	106.1	96.3	462.7	5.10E−03
36	dct 29,800	249.6	34.9	117.7	105.4	507.6	2.00E−04
37	filterless 17,200	105.8	0.0	180.4	214.9	501.1	4.00E−04
38	filterless 17,400	112.0	0.0	185.3	146.5	443.7	9.00E−04
39	filterless 17,800	127.6	0.0	111.1	115.1	353.8	9.00E−04
40	filterless 21,200	143.7	0.0	304.5	255.2	703.5	0.00E+00
41	filterless 21,400	128.4	0.0	176.0	172.7	477.2	2.00E−04
42	filterless 21,800	126.4	0.0	117.2	98.3	341.9	4.50E−03
43	filterless 25,200	173.1	0.0	359.5	361.4	894.0	1.00E−04
44	filterless 25,400	172.6	0.0	291.6	314.3	778.6	2.90E−03
45	filterless 25,800	171.8	0.0	172.3	187.7	531.7	8.00E−04
46	filterless 29,200	220.9	0.0	542.4	484.9	1,248.2	2.90E−03
47	filterless 29,400	217.4	0.0	470.6	235.8	923.9	2.50E−03
48	filterless 29,800	215.5	0.0	351.5	177.7	744.7	2.90E−03

Table 2 Ranking of different simulation cases based on running times of different sections (including scanning, filtering, clustering and simulation procedures) for realizations obtained through application of K-means clustering technique

Scanning time	Filtering time	Clustering time	Simulation time	Total time	Simulation method
40	26	22	22	1	pca 17,200
39	1	21	20	2	pca 17,400
14	46	17	16	3	pca 17,800
9	10	36	34	4	pca 21,200
7	15	23	23	5	pca 21,400
41	3	31	31	6	pca 21,800
11	33	11	45	7	pca 25,200
32	18	37	48	8	pca 25,400
8	14	39	42	9	pca 25,800
23	36	48	10	10	pca 29,200
12	21	7	11	11	pca 29,400
46	19	44	12	12	pca 29,800
13	37	13	13	13	knn 17,200
2	48	18	29	14	knn 17,400
15	41	27	26	15	knn 17,800
27	40	30	25	16	knn 21,200
6	2	20	3	17	knn 21,400
25	43	25	27	18	knn 21,800
30	27	35	35	19	knn 25,200
28	7	6	32	20	knn 25,400
17	8	3	6	21	knn 25,800
33	28	1	9	22	knn 29,200
44	30	8	1	23	knn 29,400
45	12	34	39	24	knn 29,800
3	47	28	30	25	dct 17,200
37	38	16	28	26	dct 17,400
26	39	26	18	27	dct 17,800
38	42	29	17	28	dct 21,200
42	45	15	15	29	dct 21,400
1	44	14	14	30	dct 21,800
16	4	33	33	31	dct 25,200
29	25	24	24	32	dct 25,400
4	9	2	2	33	dct 25,800
48	32	38	8	34	dct 29,200
20	11	42	36	35	dct 29,400
43	17	41	41	36	dct 29,800
31	16	45	38	37	filterless 17,200
21	13	9	5	38	filterless 17,400
18	5	32	19	39	filterless 17,800
10	20	12	37	40	filterless 21,200
19	29	5	4	41	filterless 21,400
5	6	19	21	42	filterless 21,800
35	24	43	44	43	filterless 25,200
22	34	10	40	44	filterless 25,400
47	31	4	7	45	filterless 25,800
34	22	46	46	46	filterless 29,200
36	23	47	43	47	filterless 29,400
24	35	40	47	48	filterless 29,800

Table 3 Ranking of different simulation cases based on flow simulation recoveries for realizations obtained through application of K-means clustering technique

Simulation method	MSE
filterless 21,200	0.00E+00
dct 29,200	1.00E−04
pca 25,400	1.00E−04
filterless 25,200	1.00E−04
pca 25,200	1.00E−04
knn 25,200	1.00E−04
pca 21,400	2.00E−04
dct 25,400	2.00E−04
knn 25,800	2.00E−04
filterless 21,400	2.00E−04
dct 29,800	2.00E−04
pca 29,800	3.00E−04
pca 25,800	3.00E−04
dct 25,800	3.00E−04
filterless 17,200	4.00E−04
pca 21,800	4.00E−04
dct 21,400	4.00E−04
knn 25,400	5.00E−04
dct 21,200	5.00E−04
pca 17,200	5.00E−04
dct 21,800	5.00E−04
dct 17,800	5.00E−04
dct 25,200	6.00E−04
knn 29,800	7.00E−04
dct 17,400	7.00E−04
filterless 25,800	8.00E−04
knn 21,200	8.00E−04
filterless 17,400	9.00E−04
pca 17,800	9.00E−04
filterless 17,800	9.00E−04
knn 21,800	1.00E−03
knn 17,200	1.40E−03
filterless 29,400	2.50E−03
knn 29,400	2.80E−03
filterless 29,800	2.90E−03
filterless 29,200	2.90E−03
pca 29,200	2.90E−03
filterless 25,400	2.90E−03
dct 17,200	3.50E−03
filterless 21,800	4.50E−03
pca 21,200	4.70E−03
knn 17,400	4.90E−03
pca 17,400	5.00E−03
knn 21,400	5.10E−03
dct 29,400	5.10E−03
knn 29,200	5.90E−03
pca 29,400	5.90E−03
knn 17,800	6.40E−03

training patterns has led to the most accurate result obtained through application of Spectral clustering. Figure 6 provides only a visual inspection into sensitivity analysis performed. Other figures, however, exhibit the complete results including running times and MSE values evaluated.

Running times of different simulation stages and mean squared errors of production recoveries of different realizations are plotted in Figs. 7 and 8, respectively, in the case of K-means clustering technique and in Figs. 9 and 10 for models generated by Spectral method. The vertical axes in Figs. 7 and 9 show time of simulation in seconds, for each quarterly stages; while the horizontal axes denote numeric index of each FILTERSIM case as labeled in Tables 1 and 4.

The vertical axes on Figs. 8 and 10, on the other hand, show MSE values for each simulation stage; numeric index of each FILTERSIM case as labeled in Tables 1 and 4 is displayed on the horizontal axes, however. These two graphs are simultaneously plotted on Fig.11 with the same axes description as before for a better comparison between K-means and Spectral clustering methods.

Results and discussions

As observed from tables of data (Tables 1, 4) and Figs. 7 and 9, scanning process takes the most running time in a simulation procedure. Clustering and simulation sections stand on the second and third stages, respectively, in this view point. Filtering process, however, takes a small time compared to other sections and does not have a substantial impact on total running time. Yet, the filter type utilized may have a considerable effect on clustering and simulation running times as well as on the accuracy of final realizations. This is a valid statement in all cases including use of different filtering and clustering methods through application of different-sized scanning templates.

All running times of scanning, filtering, clustering and simulation procedures increase as search template size is growing. This could be easily followed up by looking into the Figs. 7 and 9.

The number of clusters used in Spectral clustering scheme does not have a considerable effect on any parts of running time, even on clustering section; on the contrary, an increase in the number of clusters decreases computational load of K-means clustering either when the training patterns stored in the database have been filtered through PCA method or they have not been filtered anyway. Accordingly, total time of simulation decreases as a lower number of partitions is used to group the training patterns. Running times of simulations utilizing DCT or KNN techniques in filtering section are

Table 4 Simulation running times of different sections (including scanning, filtering, clustering and simulation) and mean squared errors of recovery data for realization obtained through application of Spectral clustering technique

Row	Simulation parameters	Scanning time (s)	Filtering time (s)	Clustering time (s)	Simulation time (s)	Total time (s)	Recovery MSE
1	pca 17,200	88.9	1.0	89.0	80.7	259.6	4.60E−05
2	pca 17,400	91.4	1.0	90.8	82.1	265.4	1.12E−04
3	pca 17,800	91.1	1.1	90.0	81.4	263.7	3.42E−03
4	pca 21,200	122.8	2.4	112.0	100.7	337.9	2.11E−04
5	pca 21,400	126.2	2.6	114.3	103.5	346.6	1.18E−04
6	pca 21,800	126.1	2.5	113.9	102.8	345.3	1.68E−04
7	pca 25,200	168.4	6.2	145.2	130.5	450.3	3.18E−04
8	pca 25,400	166.2	6.2	143.4	129.3	445.1	2.24E−04
9	pca 25,800	168.2	6.1	143.6	129.1	447.1	1.09E−04
10	pca 29,200	212.0	13.1	177.6	159.5	562.2	4.76E−05
11	pca 29,400	215.9	13.1	180.8	161.4	571.2	1.53E−03
12	pca 29,800	223.4	14.8	184.3	165.2	587.7	1.37E−03
13	knn 17,200	92.3	13.8	71.0	65.0	242.1	2.31E−04
14	knn 17,400	90.5	12.9	71.3	65.6	240.3	5.77E−05
15	knn 17,800	91.3	13.2	71.6	65.6	241.7	5.68E−05
16	knn 21,200	127.3	18.9	85.8	78.1	310.0	1.28E−04
17	knn 21,400	128.6	18.1	86.0	78.5	311.1	2.28E−03
18	knn 21,800	131.2	18.6	86.6	79.0	315.3	2.26E−04
19	knn 25,200	171.1	24.2	103.2	93.4	391.9	2.29E−04
20	knn 25,400	167.1	24.0	101.1	91.3	383.4	3.57E−04
21	knn 25,800	167.3	24.2	101.1	91.5	384.0	4.74E−04
22	knn 29,200	214.7	33.4	121.4	110.1	479.7	3.75E−03
23	knn 29,400	212.9	33.4	121.7	109.6	477.6	2.55E−03
24	knn 29,800	213.4	33.4	121.6	109.9	478.3	2.41E−04
25	dct 17,200	91.2	8.9	72.5	66.0	238.5	1.32E−04
26	dct 17,400	90.4	8.9	72.8	66.6	238.8	8.69E−05
27	dct 17,800	90.1	9.0	71.6	65.5	236.1	2.34E−03
28	dct 21,200	124.9	13.8	87.3	79.1	305.1	7.36E−05
29	dct 21,400	124.6	13.7	87.0	79.1	304.4	1.49E−04
30	dct 21,800	124.3	13.8	86.6	78.6	303.3	1.38E−04
31	dct 25,200	166.2	20.5	105.4	95.0	387.1	1.19E−04
32	dct 25,400	165.7	20.5	105.5	95.5	387.2	4.40E−04
33	dct 25,800	166.6	20.5	105.1	95.4	387.6	8.14E−05
34	dct 29,200	215.0	29.6	126.8	114.3	485.7	2.17E−03
35	dct 29,400	214.9	29.4	127.2	114.9	486.4	5.96E−05
36	dct 29,800	214.2	29.4	126.6	114.3	484.4	5.77E−05
37	filterless 17,200	89.7	0.0	101.0	91.5	282.2	1.34E−04
38	filterless 17,400	90.1	0.0	100.9	91.5	282.5	2.33E−03
39	filterless 17,800	90.6	0.0	100.8	91.5	282.9	2.50E−03
40	filterless 21,200	124.6	0.0	131.8	118.6	375.0	4.65E−05
41	filterless 21,400	124.1	0.0	131.8	118.4	374.3	1.69E−03
42	filterless 21,800	124.4	0.0	131.3	118.5	374.2	6.66E−05
43	filterless 25,200	167.2	0.0	170.0	152.7	489.8	8.33E−05
44	filterless 25,400	166.3	0.0	170.3	152.8	489.4	1.04E−03
45	filterless 25,800	169.6	0.0	174.4	156.1	500.1	1.16E−03
46	filterless 29,200	224.0	0.0	222.2	197.1	643.3	2.41E−04
47	filterless 29,400	192.8	0.0	207.1	182.9	582.8	3.59E−04
48	filterless 29,800	218.7	0.0	213.5	190.7	622.9	1.13E−04

Table 5 Ranking of different simulation cases based on running times of different sections (including scanning, filtering, clustering and simulation procedures) for realizations obtained through application of Spectral clustering technique

Scanning time	Filtering time	Clustering time	Simulation time	Total time	Simulation method
39	47	16	16	1	pca 17,200
25	46	18	30	2	pca 17,400
3	41	17	17	3	pca 17,800
28	9	21	38	4	pca 21,200
5	7	19	19	5	pca 21,400
6	8	20	37	6	pca 21,800
9	12	35	35	7	pca 25,200
43	28	36	36	8	pca 25,400
21	13	34	34	9	pca 25,800
35	35	10	10	10	pca 29,200
34	36	11	11	11	pca 29,400
48	23	47	47	12	pca 29,800
14	40	26	26	13	knn 17,200
38	42	15	15	14	knn 17,400
26	45	25	25	15	knn 17,800
42	4	39	20	16	knn 21,200
40	6	38	21	17	knn 21,400
29	5	37	39	18	knn 21,800
20	30	23	22	19	knn 25,200
32	14	4	4	20	knn 25,400
31	10	6	6	21	knn 25,800
19	16	40	40	22	knn 29,200
7	17	42	41	23	knn 29,400
45	18	41	42	24	knn 29,800
37	39	14	27	25	dct 17,200
27	48	27	14	26	dct 17,400
1	38	13	13	27	dct 17,800
30	3	2	2	28	dct 21,200
41	2	3	3	29	dct 21,400
4	1	1	1	30	dct 21,800
8	11	5	5	31	dct 25,200
44	15	22	23	32	dct 25,400
33	29	24	24	33	dct 25,800
10	33	9	8	34	dct 29,200
23	31	7	7	35	dct 29,400
47	32	8	9	36	dct 29,800
15	43	30	18	37	filterless 17,200
2	44	29	29	38	filterless 17,400
13	37	28	28	39	filterless 17,800
18	27	32	32	40	filterless 21,200
17	26	31	33	41	filterless 21,400
16	25	33	31	42	filterless 21,800
36	21	44	44	43	filterless 25,200
24	20	43	43	44	filterless 25,400
22	19	45	45	45	filterless 25,800
46	22	46	46	46	filterless 29,200
11	34	12	12	47	filterless 29,400
12	24	48	48	48	filterless 29,800

Table 6 Ranking of different simulation cases based on flow simulation recoveries for realizations obtained through application of Spectral clustering technique

Simulation method	MSE
pca 17,200	4.60E−05
filterless 21,200	4.65E−05
pca 29,200	4.76E−05
knn 17,800	5.68E−05
dct 29,800	5.77E−05
knn 17,400	5.77E−05
dct 29,400	5.96E−05
filterless 21,800	6.66E−05
dct 21,200	7.36E−05
dct 25,800	8.14E−05
filterless 25,200	8.33E−05
dct 17,400	8.69E−05
pca 25,800	1.09E−04
pca 17,400	1.12E−04
filterless 29,800	1.13E−04
pca 21,400	1.18E−04
dct 25,200	1.19E−04
knn 21,200	1.28E−04
dct 17,200	1.32E−04
filterless 17,200	1.34E−04
dct 21,800	1.38E−04
dct 21,400	1.49E−04
pca 21,800	1.68E−04
pca 21,200	2.11E−04
pca 25,400	2.24E−04
knn 21,800	2.26E−04
knn 25,200	2.29E−04
knn 17,200	2.31E−04
filterless 29,200	2.41E−04
knn 29,800	2.41E−04
pca 25,200	3.18E−04
knn 25,400	3.57E−04
filterless 29,400	3.59E−04
dct 25,400	4.40E−04
knn 25,800	4.74E−04
filterless 25,400	1.04E−03
filterless 25,800	1.16E−03
pca 29,800	1.37E−03
pca 29,400	1.53E−03
filterless 21,400	1.69E−03
dct 29,200	2.17E−03
knn 21,400	2.28E−03
filterless 17,400	2.33E−03
dct 17,800	2.34E−03
filterless 17,800	2.50E−03
knn 29,400	2.55E−03
pca 17,800	3.42E−03
knn 29,200	3.75E−03

not sensibly affected by cluster numbers while using K-means clustering method.

Amongst the filtering approaches used in this study, PCA takes the smallest time to do a reduction in problem dimensionality than KNN and DCT techniques. KNN has slightly brought more running time compared to DCT practice. Considering total simulation time, however, Figs. 7 and 9 verify that DCT, KNN, PCA and FILTERLESS practices do the filtering in the lowest time, respectively. This trend is more regular in the case of using Spectral technique rather than K-means method.

Obviously, time of clustering depends, to some extent, on the outputs of filtering process. This could be easily incepted in Tables 1 and 4. Usage of KNN and DCT techniques in filtering process leads to smaller (and nearly the same) running times of clustering. Utilizing PCA method or skipping filtering (FILTERLESS case) increases running time subsequently. The same conclusion could be induced about running time of simulation process itself. Total time of FILTERSIM, from scanning process through simulation stage, obeys the same trend with the filter type.

The graphs shown in Figs. 8 and 10 are displayed on a single graph in Fig. 11 for a better comparison of recovery MSE values for both K-means and the Spectral clustering techniques. As it is clear from these figures, the realizations obtained through application of the Spectral clustering method have shown a better accuracy than those acquired by applying K-means clustering. This is related to the non-linear behavior of the pattern database and the fact that Spectral methods are able to partition the pattern space through the non-linear boundary, whereas K-means is only capable of regarding linear borders between the clusters. The two graphs exhibit nearly the same trend, although some deviations exist from what we expected.

Another important conclusion is that skipping the filtering section during the simulation procedure traces a better accuracy within the simulated realizations obtained through application of either Spectral or K-means clustering method in comparison with the cases where dimensionality reduction has been implemented prior to clustering section. Leaving extracted patterns without passing them to filtering section, however, brings about more running times in clustering stage and hence in total simulation procedure, as observed in Tables 1 and 4. Moreover, when patterns are filtered through PCA or DCT method, generated realizations have quite a reliable agreement with original training image from the reservoir characteristic point of view, with application of both clustering methods. The simple KNN method, however, does fail in reproducing the connected patterns carefully to capture the flow behavior of the system in a reasonable way when patterns are clustered through K-means method. However, when Spectral method is used in clustering scheme, KNN approach does the filtering more satisfactorily.

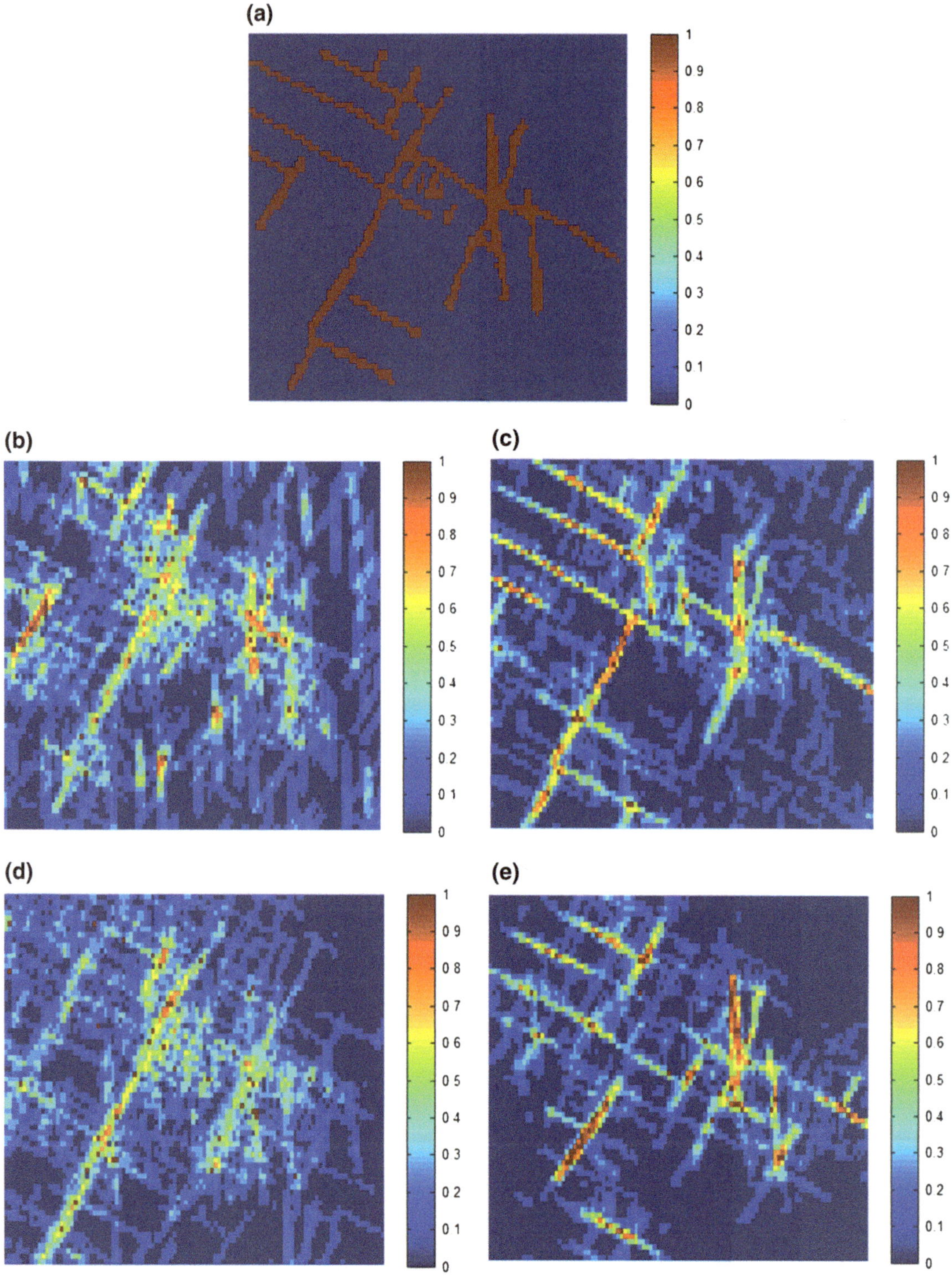

Fig. 6 Some results of FILTERSIM sensitivity analysis. **a** Training image of size 69 × 104 grids used in sensitivity analysis; **b** The least accurate E_type of FILTERSIM cases obtained through application of K-means clustering algorithm; **c** The most accurate E_type of FILTERSIM cases obtained through application of K-means clustering algorithm; **d** The least accurate E_type of FILTERSIM cases obtained through application of Spectral clustering algorithm; and **e** The most accurate E_type of FILTERSIM cases obtained through application of Spectral clustering algorithm. *NB* The colorbar next to each figure represents degree of fracture evidence in it. The more-*red* the nodes, the more the probability that the specified node is located in fracture; on the other hand, more-*blue nodes* denote less-fractured areas of the system

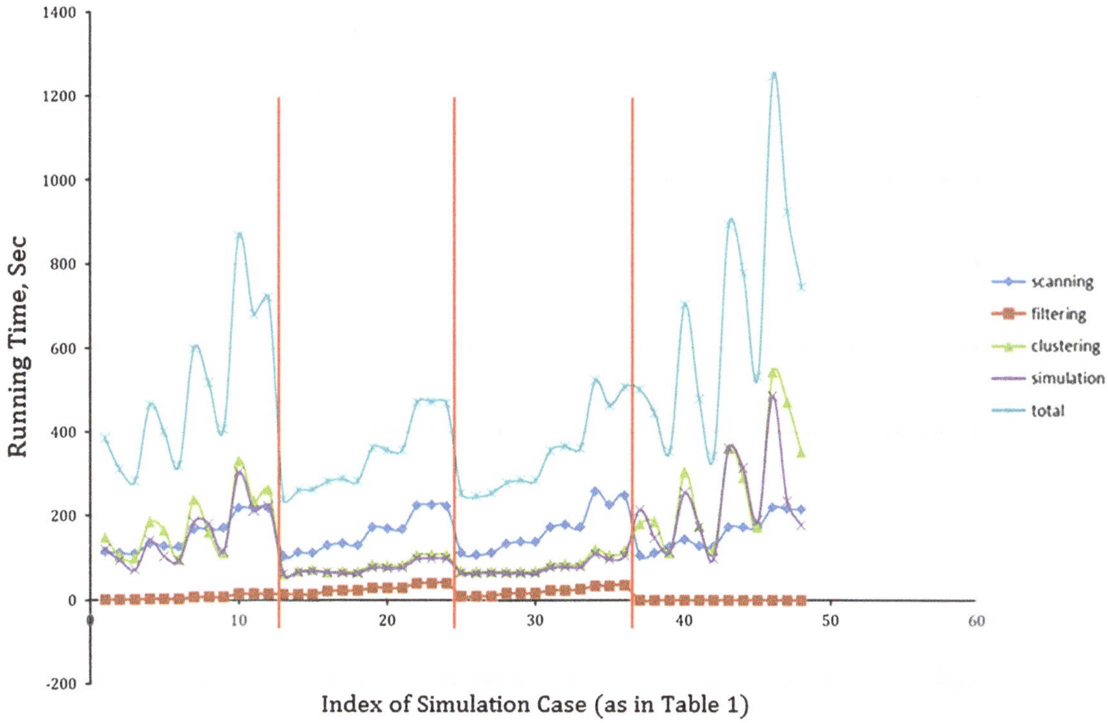

Fig. 7 Running times of different simulation stages (including scanning, filtering, clustering and simulation procedures) for realizations obtained through application of K-means clustering technique. The *vertical axis* is in seconds, showing running times of each simulation stage; while the *horizontal axis* denotes number of each FILTERSIM case as labeled in Table 1. *Vertical solid lines* in *orange* indicate the boundary of each filtering method; from *left* to *right*, the regions are exclusive to PCA, KNN, DCT and FILTERLESS techniques

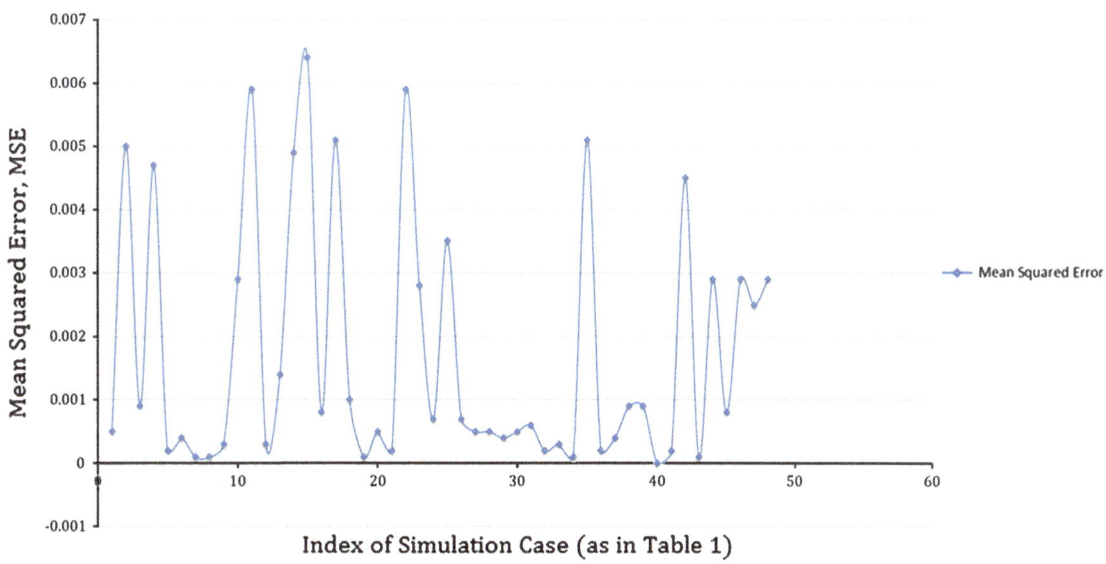

Fig. 8 Mean squared errors of production recoveries of different realizations obtained through application of K-means clustering technique. The *vertical axis* shows MSE value for each simulation stage; while the *horizontal axis* denotes number of each FILTERSIM case as labeled in Table 1

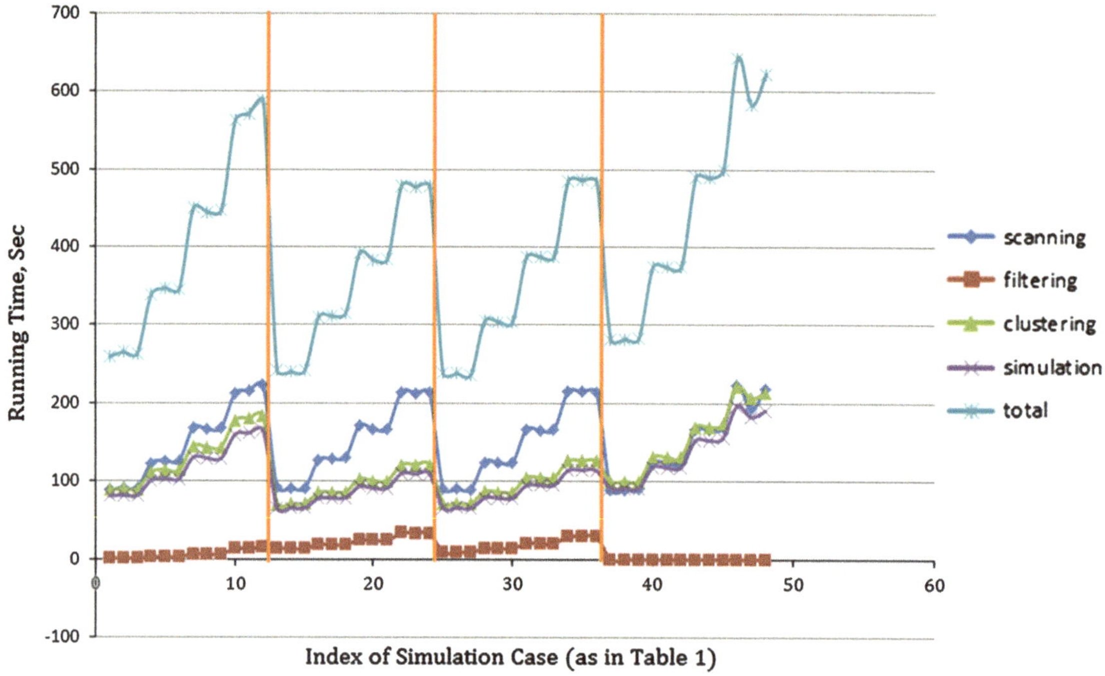

Fig. 9 Running times of different simulation stages (including scanning, filtering, clustering and simulation procedures) for realizations obtained through application of Spectral clustering technique. The vertical axis is in seconds, showing running times of each simulation stage; while the horizontal axis denotes number of each FILTERSIM case as labeled in Table 4. *Vertical solid lines* in *orange indicate* the boundary of each filtering method; from *left* to *right*, the regions are exclusive to PCA, KNN, DCT and FILTERLESS techniques

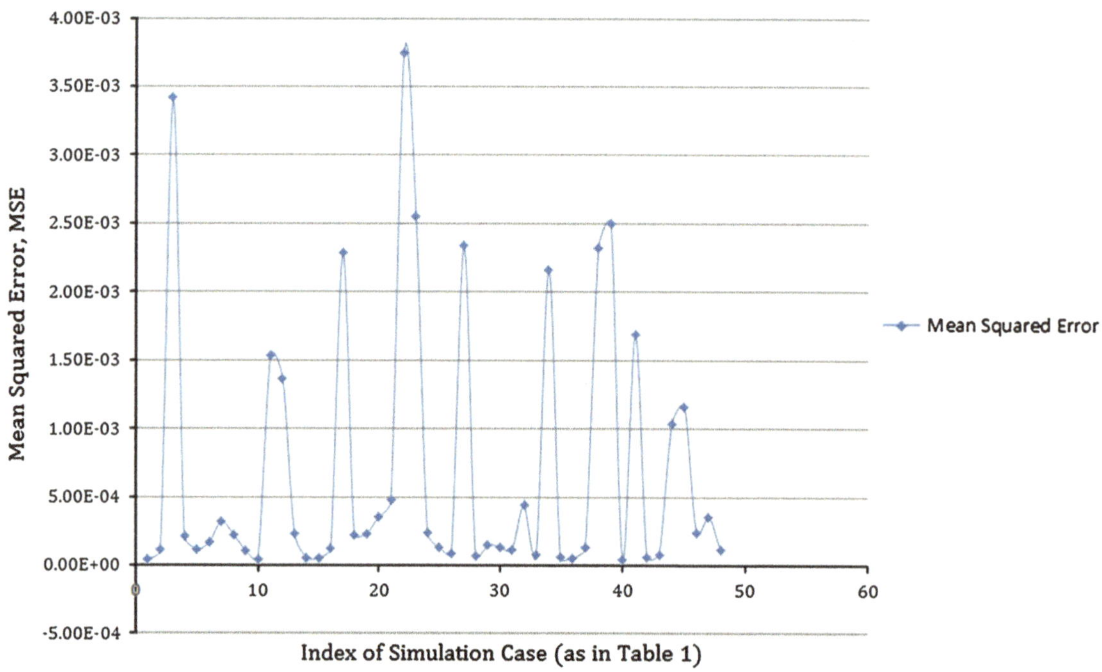

Fig. 10 Mean squared errors of production recoveries of different realizations obtained through application of Spectral clustering technique. The *vertical axis* shows MSE value for each simulation stage; while the *horizontal axis* denotes number of each FILTERSIM case as labeled in Table 4

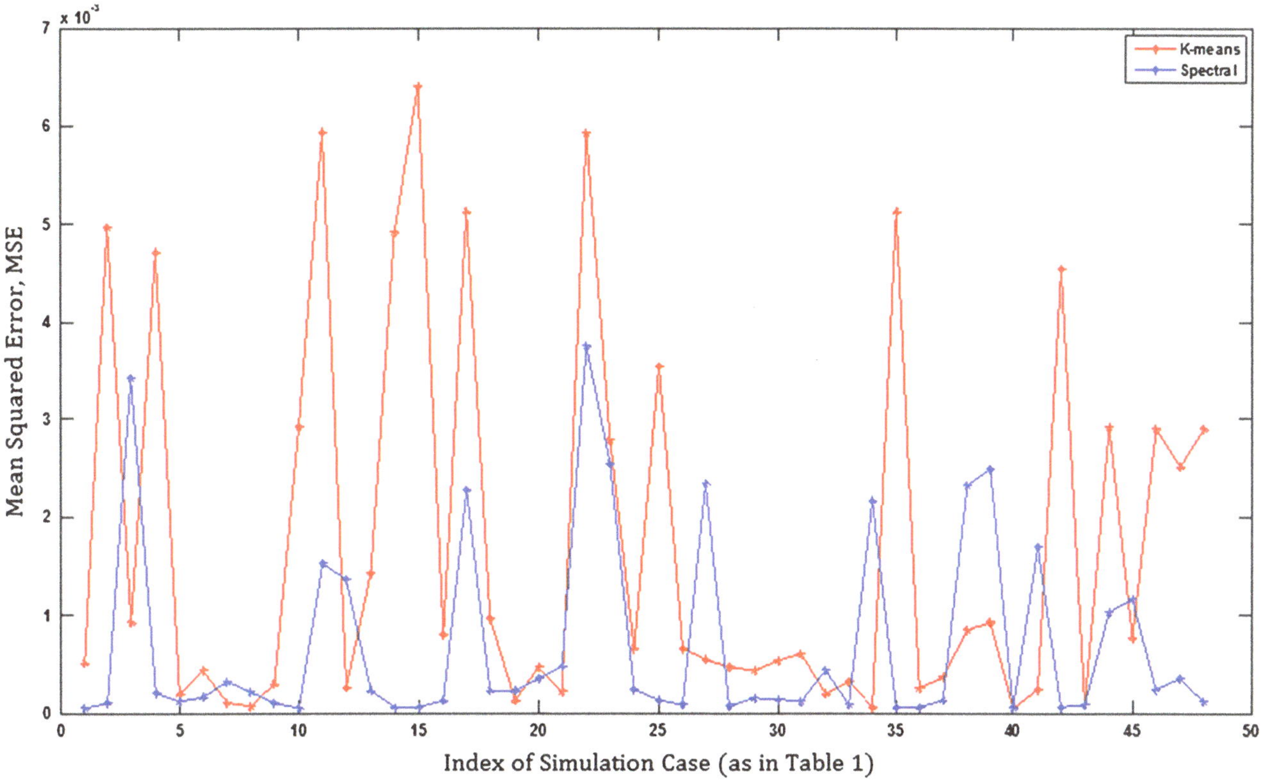

Fig. 11 Mean squared errors of production recoveries of different realizations obtained through application of **a** K-means clustering, and **b** Spectral clustering technique. As shown on legends, *red lines* denote K-means results and *blue lines* indicate Spectral results. The *vertical axis* shows MSE value for each simulation stage; while the *horizontal axis* denotes number of each FILTERSIM case as labeled in Table 1

Conclusions

1. MPS simulation algorithms are effectively able to model complex, heterogeneous and curvilinear geological structures of discrete fracture networks. This is achieved through capturing the multiple-point patterns of training images and considering the spatial relationship between more than two points in the model simultaneously.
2. Amongst the four steps of FILTERSIM algorithm, scanning process, clustering section and simulation procedure take the most part of total simulation time, respectively. Filtering process, however, takes a small time compared to other sections and does not have a significant effect on total running time.
3. Growing search template dimensions will lead to a corresponding increase in all running times during scanning, filtering, clustering and simulation procedures.
4. Changing the number of partitions used in Spectral clustering algorithm does not affect any parts of total running time considerably, even including the clustering section. Using a less number of clusters, however, raises the computational speed of K-means clustering technique either when the training patterns stored in the database have been filtered using PCA method or they have not been filtered anyway (FILTERLESS case).
5. PCA takes the smallest time to do a reduction in problem dimensionality than KNN and DCT techniques which bring nearly the same filtering times. Regarding FILTERSIM computational time, however, DCT, KNN, PCA and FILTERLESS practices do the filtering in the lowest time, respectively.
6. Application of KNN and DCT as filtering techniques leads to a smaller clustering time. Utilizing PCA method or skipping filtering (FILTERLESS case) increases running time subsequently. The same conclusion could be induced about running time of simulation process itself as well as total time of FILTERSIM.
7. Realizations obtained through application of Spectral clustering method show a better accuracy than those generated by K-means clustering.
8. Filtering patterns prior to clustering them may decrease the accuracy of final realizations, though they a have a great effect on the reduction of simulation times. Nevertheless, PCA and DCT

methods may lead to quite acceptable results with both clustering approaches, while at the same time total simulation time decreases substantially.

References

Ahmadi R (2011) Geometrical fracture modeling within multiple-point statistics framework. MS Dissertation, Sharif University of Technology, Department of Chemical and Petroleum Engineering, Iran

Anderberg MR (1973) Cluster analysis for applications. Academic Press, New York

Baker RO, Schechter DS, McDonald P, Knight WH, Leonard P, Rounding C (2001) Development of a fracture model for spraberry field, Texas USA. SPE 71635. In: Annual technical conference and exhibition, Louisiana

Ball G, Hall D (1965) ISODATA, a novel method of data analysis and pattern classification. Technical report NTIS AD 699616. Stanford Research Institute, Stanford, CA

Bishop CM (2006) Pattern recognition and machine learning. Springer, Berlin

Boisvert JB, Pyrcz MJ, Deutsch CV (2008) Multiple-point statistics for training image selection. Nat Resour Res 16(4):313–321

Bourbiaux BJ, Cacas MC, Sarda S, Sabathier JC (1997) A fast and efficient methodology to convert fractured reservoir images into a dual-porosity model. SPE 38907. In: Annual technical conference and exhibition, Texas, USA

Cacas MC, Ledoux E, de Marsily G, Tillie B (1990) Modeling fracture flow with a stochastic discrete fracture network: calibration and validation. Water Resour Res 26:479–489

Caers J, Arpat GB (2005) A multiple-scale, pattern-based approach to sequential simulation. PhD dissertation, Stanford University, Department of Petroleum Engineering, Stanford, USA

Caers J, Arpat GB (2007) Conditional simulation with patterns. Math Geol 39(2):177–203

Charlaix E, Guyon E, Roux S (1987) Permeability of an array of fractures with randomly varying apertures. Transp Porous Med. 2:31–43

Chugunova T, Hu L (2008) Multiple-point simulations constrained by continuous auxiliary data. Math Geosci 40(2):133–146

Chung FRK (1997) Spectral graph theory. In: CBMS regional conference series in mathematics 92, American Mathematical Society, Providence, RI

Cover TM, Hart PE (1967) Nearest neighbor pattern classification. IEEE Trans Inf Theory 13(1):21–27

Cristianini N, Taylor JS, Kandola JS (2001) Spectral kernel methods for clustering. In: NIPS, pp 649–655

Dershowitz W, Shuttle D, Parney R (2002) Improved oil sweep through discrete fracture network modelling of gel injections in the South Oregon Basin Field, Wyoming. In: SPE 75162, energy thirteenth improved oil recovery symposium, Oklahoma, USA

Deutsch CV (1992) Annealing techniques applied to reservoir modeling and the integration of geological and engineering (well test) data. PhD Dissertation, Stanford University, Stanford

DeVries L (2009) Application of multiple point geostatistics to non-stationary images. Math Geosci 41(1):29–42

Duda R, Hart P, Stork D (2001) Pattern Classification, 2nd edn. Wiley, New York

Filippone M, Camastra F, Masulli F, Rovetta S (2008) A survey of kernel and spectral methods for clustering. Pattern Recognit. 41:176–190

Guardiano F, Srivastava R (1993) Multivariate geostatistics: beyond bivariate moments. Geostatistics-troia. Kluwer Academic Publications, Dordrecht, pp 133–144

Han J, Kamber M (2000) Data mining: concepts and techniques. Morgan Kaufmann, San Francisco

Hestir K, Long JCS (1990) Analytical expressions for the permeability of random two-dimensional Poisson fracture networks based on regular lattice percolation and equivalent media theories. J Geophys Res 95(B13):565–581

Hestir K, Long JC, Chilks JP, Billaux (1987) Some techniques for stochastic modeling of three-dimensional fracture networks. In: Proceedings of the conference on geostatistical, sensitivity, and uncertainty methods for ground -water flow and radionuclide transport modeling, San Fransisco, CA, pp 495–519

Honarkhah M, Caers J (2010) Stochastic simulation of patterns using distance-based pattern modeling. Math Geosci 42(5):487–517

Hu L, Chugunova T (2008) Multiple-point geostatistics for modeling subsurface heterogeneity: a comprehensive review. Water Resour Res 44(W11413).

Jackson JE (1991) A user's guide to principal components. Wiley, NewYork, p 592

Jain AK (1989) Fundamentals of digital image processing. Prentice-Hall, Englewood Cliffs

Jain AK (2010) Data clustering: 50 years beyond K-means. Pattern Recognit Lett 31:651–666

Jain AK, Dubes RC (1988) Algorithms for clustering data. Prentice Hall, Upper Saddle River

Jolliffe IT (2002) Principal component analysis, 2nd edn. Springer, Berlin

Journel A, Zhang T (Dec.2006) The necessity of a multiple-point prior model. Math Geol 38(5)

Lee SH, Jensen CL, Lough MF (1999) An efficient finite difference model for flow in a reservoir with multiple length-scale fractures. In: SPE 56752, annual technical conference and exhibition, Texas

Lioyd S (1982) Least squares quantization in PCM. IEEE Trans Inf Theory 28:129–137 (Originally as an unpublished Bell laboratories Technical Note (1957))

Long JCS, Witherspoon PA (1985) The relationship of the degree of interconnections to permeability in fracture networks. J Geophys Res 90(B4):3087–3098

Long JCS, Remer JS, Wilson CR, Witherspoon PA (1982) Porous media equivalents for networks of discontinuous fractures. Water Resour Res 18(3):645–658

Long JC, Billaux D, Hestir K, Chilks JP (1987) Some geostatistical tools for incorporating spatial structure in fracture network modeling. Proc Congr Int Soc Rock Mech 6(1):171–176

MacQueen JB (1967) Some methods for classification and analysis of multivariate observations, vol 1. In: Proceedings of 5th Berkeley symposium on mathematical statistics and probability, Berkeley, University of California Press, pp 281–297

Mariethoz G (2010) A general parallelization strategy for random path based geostatistical simulation methods. Comput Geosci 37(7):953–958

Moran PAP (1950) Notes on continuous stochastic phenomena. Biometrika 37(1/2):17–23. (Biometrika Trust)

Mukhopadhyay S, Sahimi M (1996) Scaling properties of a percolation model with long-range correlations. Phys Rev E 54(4):3870–3880

Oda M, Hatsuyama Y, Ohnishi Y (1987) Numerical experiments on permeability tensor and its application to jointed granite at stripa mine, Sweden. J. Geophys Res 92:8037–8048

Pennebaker WB, Mitchell JL (1993) JPEG Still image data compression standard, Chap 4. Van Nostrand Reinhold, New York, NY

Perner P (2007) Machine learning and data mining in pattern recognition. In: Proceedings of the 5th international conference, MLDM 2007, Leipzig, Germany

Pollard DD (1976) On the form and stability of open hydraulic fractures in the Earth's crust. Geophys Res Lett 3(9):513–516

Pyrcz MJ, Boisvert JB, Deutsch CV (2008) A library of training images for fluvial and deepwater reservoirs and associated code. Comput Geosci 34(5):542–560

Shawe TJ, Cristianini N (2004) Kernel methods for pattern analysis. Cambridge University Press, Cambridge

Smith L, Schwartz FW (1984) An analysis of the influence of fracture geometry on mass transport in fractured media. Water Resour Res 20(9):1241–1252

Steinhaus H (1956) Sur la division des corp materiels en parties. Bull Acad Polon Sci IV (C1.III):801–804

Straubhaar J (2008) Optimization issues in 3D multipoint statistics simulation. In: Geostats, Santiago, Chile

Strebelle S (2000) Sequential simulation drawing structures from training images. PhD Dissertation, Stanford University, Stanford, CA

Strebelle S (2002) Conditional simulation of complex geological structures using multiple-point statistics. Math Geol 34(1):1–21

Tan PN, Steinbach M, Kumar V (2005) Introduction to data mining. Addison-Wesley Longman Publishing Co., Inc., Boston

Teimoori A, Tran NH, Chen Z, Rahman SS (2004) Simulation of fluid flow in naturally fractured reservoirs with the use of effective permeability tensor. In: Annual proceeding of geothermal resources council (GRC), Australia

Toussaint GT (2005) Geometric proximity graphs for improving nearest neighbor methods in instance-based learning and data mining. Int J Comput Geom Appl 15(2):101–150

Tran NH (July 2004) Characterization and modeling of naturally fractured reservoirs. PhD Dissertation, The University of New South Wales, School of Petroleum Engineering, Sydney, Australia

Tran NH, Rahman MK, Rahman SS (2002) A nested neuro-fractal-stochastic technique for modeling naturally fractured reservoirs. In: SPE 77877, Asia Pacific oil and gas conference and exhibition, Melbourne, Australia

Wu J (2007) 4D seismic and multiple-point pattern data integration using geostatistics. PhD Dissertation, Stanford University, Department of Energy Resources Engineering

Xu R, Wunsch D (2005) Survey of clustering algorithms. IEEE Trans Neural Netw. 16(3):645–678

8

Numerical simulation of two-phase fluid flow

José M. Carcione · Stefano Picotti ·
Juan E. Santos · Ayman Qadrouh ·
Hashim S. Almalki

Abstract We simulate two-phase fluid flow using a stress–strain relation based on Biot's theory of poroelasticity for partial saturation combined with the mass conservation equations. To uncouple flow and elastic strain, we use a correction to the stiffness of the medium under conditions of uniaxial strain. The pressure and saturation differential equations are then solved with an explicit time stepping scheme and the Fourier pseudospectral method to compute the spatial derivatives. We assume an initial pressure state and at each time step compute the wetting- and non wetting-fluid pressures at a given saturation. Then, we solve Richards's equation for the non wetting-fluid saturation and proceed to the next time step with the updated saturations values. The pressure and saturation equations are first solved separately and the results compared to known analytical solutions showing the accuracy of the algorithm. Then, the coupled system is solved. In all the cases, the non-wetting fluid is injected at a given point in space as a boundary condition and capillarity effects are taken into account. The examples consider oil injection in a water-saturated porous medium.

J. M. Carcione · S. Picotti
Instituto Nazionale di Oceanografia e di Geofisica Sperimentale (OGS), Borgo Grotta Gigante 42c, 34010 Sgonico, Trieste, Italy

J. E. Santos
CONICET, Instituto del Gas y del Petróleo, Facultad de Ingeniería, Universidad de Buenos Aires, Buenos Aires, Argentina

J. E. Santos
Universidad Nacional de La Plata, La Plata, Argentina

J. E. Santos
Department of Mathematics, Purdue University, West Lafayette, IN 47907-2067, USA

A. Qadrouh · H. S. Almalki
SAC, KACST, PO Box 6086, Riyadh 11442, Saudi Arabia

Keywords Two-phase flow · Pressure · Saturation · Diffusion · Richards equation · Fourier method

Introduction

Diffusion equations can be obtained in poroelasticity at low frequencies, when the inertial terms are neglected (e.g. Carcione 2007). Chandler and Johnson (1981) have shown the equivalence between quasi-static fluid flow and Biot's diffusive wave (see also Carcione 2007). Hence, fluid flow and pressure diffusion are phenomena described by the same differential equation. In hydrology and hydrocarbon exploration, diffusion equations are mainly used to model fluid flow in reservoir rocks (Peaceman 1977). Shapiro et al. (2002) describe the phenomenon of micro-seismicity caused by fluid injection in boreholes, while Müller (2006) provides a detailed physical analysis of the pore pressure induced by a fluid mass point source. His results support the hypothesis that the diffusive slow P-wave is mainly responsible for the triggering of microearthquakes.

Carcione and Gei (2009) simulated fluid-pressure diffusion in inhomogeneous media by considering a single fluid. The purpose of the present work is to generalise that approach to more than one fluid and use a similar method to solve the equations governing the flow. Possible applications involve simulation of fluid depletion and injection of hydrocarbons, geomechanical analysis of reservoirs (Settari and Mourits 1994) and micro-seismicity, as mentioned above. A convenient equation to define the stress–strain relation involving two fluids can be obtained from Biot's theory of poroelasticity (Santos et al. 1990a, b; Ravazzoli et al. 2003). The diffusion can be fully uncoupled from the elastic deformations by neglecting the strain, but a less stringent condition can be obtained under uniaxial strain conditions, where the effect of the elastic deformations requires the modification of the rock stiffness involved in the diffusion equation. Uniaxial-strain conditions hold for the case of a fluid-injection source in a borehole, where the porous medium deforms uniaxially only in the vertical direction and the horizontal strains are zero. This approximation is valid for geological formations whose lateral dimensions are large compared to their thickness. In this case, the lateral deformations are small compared to the horizontal dimensions, and hence the lateral strain is approximately zero (e.g., Gutierrez and Lewis 2002).

In this work, we combine Biot's poroelastic equations and the fluid mass conservation equations averaged to obtain a system of differential equations for the pore pressures and saturations of the wetting and non-wetting phases. The theory used here to define the stress–strain relations is a generalization of Biot's theory and has been developed by Santos et al. (1990a, b) (see also Carcione et al. 2004). It is important to note here that the constitutive equations introduced by Santos et al. (1990a, b) have never been used to simulate fluid-flow phenomena, but only to describe wave propagation. Biot's wave propagation theory, neglecting the acceleration terms, is equivalent to the consolidation theory if the infinitesimal stresses and pressures are assumed to be their absolute counterparts, since the stress–strain relations can be interpreted as a relation between incremental fields, where stress and strain are increments with respect to a reference stress and strain—the case of wave propagation—or, as relations between the absolute fields (consolidation theory; Biot 1941). Other models accounting for the coupling between the mechanical behaviour of the matrix and the fluid dynamics include the Hassanizadeh model (Hassanizadeh et al. 2002) and the Barenblatt–Biot model (Barenblatt and Gilman 1987).

The poroelastic stress–strain relation for the two fluids can be recasted as a first-order differential equation in time and allows us to obtain the fluid pressures and total flow velocity. On the other hand, the mass conservation equation can be recasted to obtain the advection–diffusion Richards equation which is coupled to the fluid pressures through the flow velocity. The diffusion term is determined by the capillarity effects. We first solve the pressure and saturation equation separately and test the results with known analytical solutions for the advection and diffusion equation in 1D and 2D space. Then, we investigate the effects of partial saturation and capillary forces on the pressure and saturation profiles and finally solve the coupled system, which consists in computing the flow velocity and then Richards's equation for the fluid saturations. The algorithm is fully explicit and based on the Euler–Picard time stepping and on the Fourier pseudospectral method to compute the spatial derivatives (e.g., Carcione 2007). It has been shown that the Picard iteration method demonstrated perfect mass balance in fluid-flow equations when used with direct-grid methods such as the finite difference and finite element approximations in space (Zarba 1988). It is expected that the same performance can be achieved with the Fourier pseudospectral method. In this case, the model is discretized on a mesh the spatial derivatives are calculated with the fast Fourier transform. The spatial derivatives computed with the Fourier method have spectral accuracy and require coarser grids compared to finite-difference methods. The use of this spectral method overcomes two drawbacks: low accuracy and stringent stability conditions, since the error in time decays exponentially. The method has proved to be efficient when solving diffusion equations. (Carcione 2006, 2010; Carcione and Gei 2009).

Darcy's equations

We consider a porous medium (a rock for instance) saturated with two immiscible fluids and denote with the subscripts (and superscripts) w and n quantities related to the wetting and non-wetting phases, respectively. Let v^s, $\bar{v}^n$, and $\bar{v}^w$ denote the interstitial particle-velocity vectors of the solid grains, non-wetting fluid and wetting fluid, respectively. First of all, we assume that the rock deforms slowly compared to multiphase flow. In this case, we have the relative particle-velocity components $v_i^m = \phi(\bar{v}_i^m - v_i^s) \approx \phi\bar{v}_i^m, i = 1, 2, 3$, $m = n, w$, where ϕ is the porosity. Then, Darcy's equations relating Biot's relative velocities and pressures p_m are

$$S_m v_i^m = -\lambda_m (p_{m,i} - \rho_m g \delta_{iz}), \quad \lambda_m = \frac{\kappa_m}{\eta_m}, \quad m = n, w \tag{1}$$

(Gai 2004, see Eqs. (2.12, 2.16) ; Santos et al. 1990a), where S, ρ, η and κ denote saturation, density, viscosity and permeability, respectively, g is the gravity constant, i indicates the spatial variables $x(1)$, $y(2)$ and $z(3)$, δ_{ij} is Kronecker's delta and the subindex "i" denotes spatial differentiation. Darcy's velocities, as given in Gai (2004, Eq. 2.16), are $\phi N_\alpha \bar{v}_i^m = N_\alpha v_i^m$, where N_α are the so-called (dimensionless) concentrations.

Stress–strain relations

Following Santos et al. (1990a, b), we define

$$\dot{\xi}_m = -v_{i,i}^m, \tag{2}$$

where ξ is the variation of fluid content and a dot above a variable denotes time differentiation. Let ϵ_{ij} denote the strain components of the solid grains let $\vartheta = \epsilon_{11} + \epsilon_{22} + \epsilon_{33}$ denote the dilatation field. Then, the total stress components and fluid stresses are

$$\begin{aligned}\tau_{ij} &= 2N\epsilon_{ij} + \delta_{ij}(\lambda_c\vartheta - B_1\xi_n - B_2\xi_w),\\ \tau_n &= -(S_n + \beta + \zeta)p_n + (\beta + \zeta)p_w = B_1\vartheta - M_1\xi_n - M_3\xi_w,\\ \tau_w &= -(S_w + \zeta)p_w + \zeta p_n = B_2\vartheta - M_3\xi_n - M_2\xi_w,\end{aligned} \tag{3}$$

where

$$\lambda_c = K_c - \frac{2}{3}N, \tag{4}$$

N is the dry-rock shear modulus and K_c is the undrained (closed) bulk modulus. It is

$$K_c = \frac{K_s(K_m + G)}{K_s + G}, \tag{5}$$

where K_s is the bulk modulus of the solid grains, K_m is the dry-rock bulk modulus,

$$G = \frac{K_f(K_m - K_s)}{\phi(K_f - K_s)}, \tag{6}$$

$$K_f = \alpha\left(\frac{\gamma S_n}{K_n} + \frac{S_w}{K_w}\right)^{-1}, \tag{7}$$

$$\alpha = 1 + (S_n + \beta)(\gamma - 1) \tag{8}$$

$$\begin{aligned}B_1 &= \theta K_c[(S_n + \beta)\gamma - \beta + (\gamma - 1)\zeta],\\ B_2 &= \theta K_c[(S_w + (1 - \gamma)\zeta],\end{aligned} \tag{9}$$

$$\theta = \left[\varphi + \phi\left(\frac{1}{K_m} - \frac{1}{K_c}\right)\right]\left\{\alpha\left[\varphi + \phi\left(\frac{1}{K_m} - \frac{1}{K_f}\right)\right]\right\}^{-1}, \tag{10}$$

$$\begin{aligned}M_1 &= -M_3 - B_1/(\varphi K_m),\\ M_2 &= rB_2/q + \zeta/q,\\ M_3 &= -B_2[r/q + 1/(K_m\varphi)] - \zeta/q,\end{aligned} \tag{11}$$

$$\beta = \frac{p_c}{p'_c}, \quad \zeta = \frac{\bar{p}_w}{p'_c}, \tag{12}$$

where K_s is the bulk modulus of the grains, K_n and K_w are the bulk moduli of the non-wetting and wetting fluids, respectively, p_c is the capillary pressure (i.e., the difference between the non-wet and wetting-fluid absolute pressures), $\bar{p}_w$ is an average hydrostatic pressure (assumed to be zero in this work) and p'_c the derivative of p_c with respect to S_n; moreover,

$$\varphi = \frac{1}{K_s} - \frac{1}{K_m}, \tag{13}$$

$$\gamma = \left(1 + \frac{p'_c S_n S_w}{K_w}\right)\left(1 + \frac{p'_c S_n S_w}{K_n}\right)^{-1}, \tag{14}$$

$$r = \frac{S_n + \beta}{K_s} + \frac{1}{K_c - K_m}\left[qB_2 + (S_n + \beta)\left(1 - \frac{K_c}{K_s}\right)\right], \tag{15}$$

$$q = \phi\left(\frac{1}{K_n} + \frac{1}{p'_c S_n S_w}\right). \tag{16}$$

Uncoupling fluid flow and deformation

The fluid pressure is coupled with the strain of the matrix in Eq. (3). This fact makes the problem much more difficult to solve, but there are situations where these field variables can be uncoupled. They occur when the displacement field is irrotational or when the fluid is very compressible (e.g., Detournay and Cheng 1993). We may avoid such approximation by using a less stringent one. Let us assume the case of a fluid-injection source in a borehole, uniaxial strain conditions and vertical deformations only. In this case, the only non-zero differential strain is $d\vartheta = d\epsilon_{33}$. Assuming no changes in the vertical stress, we obtain from Eq. (3):

$$d\tau_{33} = 0 \simeq (\lambda_c + 2N)d\vartheta - B_1 d\xi_n - B_2 d\xi_w. \tag{17}$$

Using this result, the second and third Eq. (3) become

$$\begin{aligned}\tau_n &= -(S_n + \beta + \zeta)p_n + (\beta + \zeta)p_w = a_1\xi_n + a_3\xi_w,\\ \tau_w &= -(S_w + \zeta)p_w + \zeta p_n = a_3\xi_n + a_2\xi_w,\\ a_1 &= B_1^2/(\lambda_c + 2N) - M_1,\\ a_2 &= B_2^2/(\lambda_c + 2N) - M_2,\\ a_3 &= B_1 B_2/(\lambda_c + 2N) - M_3.\end{aligned} \tag{18}$$

More general approaches involving the coupling of fluid flow and deformation based on Biot's equations are given in Gutierrez and Lewis (2002).

Mass conservation equations

The system of equations are complemented with the balance of mass

$$\phi\dot{S}_m + \nabla\cdot(S_m\mathbf{v}^m) = s_m, \quad m = n, w \tag{19}$$

(Gai 2004, Eqs. 2.17–2.19; Peaceman 1977; Santos et al. 1990a), where $\mathbf{v}^m = (v_1^m, v_2^m, v_3^m)^T$, s_m is the source or sink term and we have assumed constant volume factors.

The pressure and Richards equations

Replacing the flow velocities from Eq. (1) into the time derivative of Eq. (18) and into Eq. (19), we obtain

$$\begin{aligned}\dot{\tau}_n &= a_1\nabla\cdot b_n(\nabla p_n - \rho_n\mathbf{g}) + a_3\nabla\cdot b_w(\nabla p_w - \rho_w\mathbf{g}),\\ \dot{\tau}_w &= a_3\nabla\cdot b_n(\nabla p_n - \rho_n\mathbf{g}) + a_2\nabla\cdot b_w(\nabla p_w - \rho_w\mathbf{g}),\\ \dot{S}_n &= \tfrac{1}{\phi}[s_n + \nabla\cdot(\lambda_n(\nabla p_n - \rho_n\mathbf{g}))],\\ \dot{S}_w &= \tfrac{1}{\phi}[s_w + \nabla\cdot(\lambda_w(\nabla p_w - \rho_w\mathbf{g}))],\\ b_m &= \tfrac{\lambda_m}{S_m}, \quad m = n, w,\end{aligned} \tag{20}$$

where $\mathbf{g}$ is the gravity vector. These equations are subject to the constraints

$$p_c = p_n - p_w, \quad S_n + S_w = 1. \tag{21}$$

From Eq. (18) we have

$$p_n = [\zeta\tau_n - (\beta + \zeta)\tau_w]/Q, \quad p_w = [(S_w + \zeta)\tau_n - (S_n + \beta + \zeta)\tau_w]/Q, \quad Q = \beta S_w + \zeta(S_w - S_n) \tag{22}$$

and the first constraint Eq. (21) becomes

$$p_c = -(S_w\tau_n + S_n\tau_w)/Q. \tag{23}$$

In particular, we set

$$\tau_n = \frac{1}{S_w}[(1 - S_w)\tau_w - Qp_c], \quad S_n = 1 - S_w. \tag{24}$$

From Eq. (20), we obtain the system to be solved:

$$\begin{aligned} \dot{\tau}_w &= a_3\nabla \cdot b_n(\nabla p_n - \rho_n\mathbf{g}) + a_2\nabla \cdot b_w(\nabla p_w - \rho_w\mathbf{g}), \\ \tau_n &= \frac{1}{S_w}[(1 - S_w)\tau_w - Qp_c], \\ \phi\dot{S}_n &= \nabla \cdot (h_n\nabla S_n) - \mathbf{c}^t \cdot f_n'\nabla S_n - \nabla \cdot (\lambda_w f_n(\rho_n - \rho_w)\mathbf{g}) \\ &\quad -(f_n - 1)s_n - f_n s_w, \\ \mathbf{c}^t &= -(\lambda_n\nabla p_n + \lambda_w\nabla p_w) + (\lambda_n\rho_n + \lambda_w\rho_w)\mathbf{g}, \end{aligned} \tag{25}$$

where the third (Richards's) equation is obtained in "Appendix A" $\mathbf{c}^t$ is the total Darcy velocity. This system is complemented with Eqs. (18) and (22), subject to the initial condition (p_{w0}, S_{w0}). In the limit $S_w \rightarrow 1$ (assuming the residual saturations $S_{rw} = S_{rn} = 0$), $a_2 \rightarrow -M$, where M is given in Eq. (10) of Carcione and Gei (2009) and the first Eq. (25) becomes Eq. (9) of that paper.

The algorithm to solve these equations is explicit in the time variable, $t = j\ \mathrm{d}t$, where $\mathrm{d}t$ is the time step and $j = 0, 1, \ldots$. It uses the forward Euler method to approximate the time derivatives and the Fourier pseudospectral method to compute the spatial derivatives (e.g., Carcione 2006, 2007; Carcione and Gei 2009). We assume an initial state (p_{w0}, S_{w0}), compute τ_w at $j + 1$ (first equation) assuming the saturations obtained at j on the r.h.s. of the equations. Then, we compute τ_n, the pressures with Eq. (22) and solve the last two equation for S_n at $j + 1$ proceed to the next time step. To improve the Euler method we solve the saturation Eq. (25) by using the Picard method (Pang 2006) (see "Appendix B").

Fourier analysis

The characteristics of the pressure diffusion can be analysed with a Fourier analysis. Let us assume a kernel of the form $p_w = \exp[i(\omega t - \mathbf{k} \cdot \mathbf{x})]$, where ω is the angular frequency, $\mathbf{k}$ is the complex wavenumber vector and $\mathbf{x}$ is the position vector. Assuming homogeneous properties, we have $\mathbf{k} = k\ (l_1, l_2, l_3)^T$, where $k = \mathrm{Re}(k) - i\bar{\alpha}, \bar{\alpha}$ is the attenuation factor l_i are the direction cosines defining the propagation direction, where Re takes the real part.

From Eqs. (18) and (22) we have $\tau_w = -S_w p_w + \zeta p_c$ and $\dot{\tau}_w = -S_w\dot{p}_w$ at constant saturation. Substituting the above kernel into the first Eq. (25), in the absence of source, gives the dispersion equation

$$i\omega = (a_3\lambda_n + a_2\lambda_w)k^2, \tag{26}$$

We define the complex velocity as

$$v_c = \frac{\omega}{k} = (1 + i)\sqrt{-\frac{\omega(a_3b_n + a_2b_w)}{2}} \tag{27}$$

(the quantity inside the square root is positive). The same kinematic concepts used in wave propagation (acoustics and electromagnetism) are useful in this analysis (see Carcione 2007, Chap. 8). For $S_w \rightarrow 1$, we obtain the complex velocity of Carcione and Gei (2009).

The phase velocity of the pressure front and respective attenuation factor can be obtained from the complex velocity as

$$v_p = \left[\mathrm{Re}\left(v_c^{-1}\right)\right]^{-1} \quad \text{and} \quad \bar{\alpha} = -\omega \mathrm{Im}(v_c^{-1}), \tag{28}$$

respectively, where Im takes imaginary part. The skin depth is the distance d for which $\exp(-\bar{\alpha}d) = 1/e$, where e is Euler's number. It is usually taken as the effective distance of penetration of the signal. Then, $d = 1/\bar{\alpha}$.

Numerical simulations

Tests are performed in the cases where analytical solutions exist. First, we consider the first Eq. (25) in the homogeneous case, without gravity, $S_w = 1$ and $p_{w0} = 0$. In this case, $\tau_w = -p_w$, $a_3 = 0$ and $a_2\ b_w = -M\kappa/\eta_w$, where M is given in Eq. (10) of Carcione and Gei (2009). We obtain the equation

Table 1 Material properties

Grain bulk modulus, K_s	40 GPa
Shear modulus, μ_s	38 GPa
Density, ρ_s	2,600 kg/m^3
Frame bulk modulus, K_m	12 GPa
Shear modulus, N	8 GPa
Porosity, ϕ	0.25
Permeability, κ	0.8 D
Brine density, ρ_w	1,040 kg/m^3
Viscosity, η_w	0.001 Pa s
Bulk modulus, K_w	2.25 GPa
Oil density, ρ_n	800 kg/m^3
Viscosity, η_n	0.1 Pa s
Bulk modulus, K_n	1.3 GPa

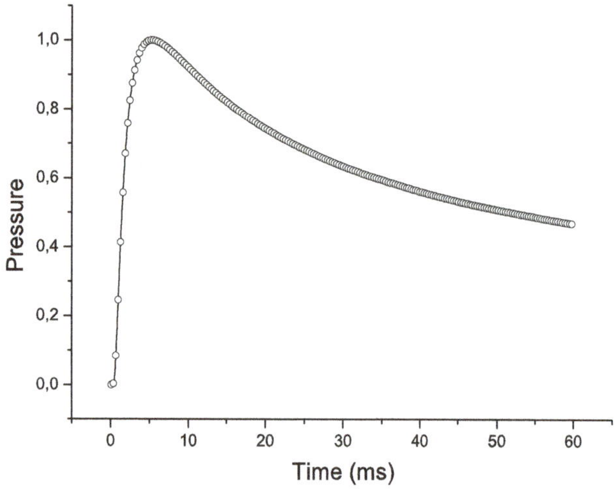

Fig. 1 Comparison between the numerical (*symbols*) and analytical (*solid line*) solutions of the pressure field (normalized) at 100 % wetting-fluid saturation. The receiver is located at 24 cm from the source

$$\dot{p}_w = \frac{\kappa M}{\eta_w}\Delta p_w + \delta(\mathbf{x})\delta(t), \quad (29)$$

where Δ is the Laplacian, κ is the absolute permeability and here δ is the Dirac delta function, representing the source. The analytical solution is the Green's function given by Eq. (35) in Carcione and Gei (2009). We consider the 1D case, the material properties shown in Table 1, a number of grid points $n_x = 165$ and a grid spacing $dx = 4$ m/n_x. Figure 1 compares numerical and analytical time histories (normalized) at 24 cm from the source computed with a time step $dt = 10$ μs. The agreement is excellent.

In the absence of sources and gravity effects and $\mathcal{D} = h_n/\phi$ (constant), an analytical solution is known for the 1D version of the third Eq. (25), i.e.,

$$\dot{S}_n = \mathcal{D}\partial_{xx}S_n - cf_n'\partial_x S_n, \quad c = \frac{c^t}{\phi}, \quad (30)$$

where $\mathcal{D}$ is a diffusivity, for the initial condition $S_n(x,0) = 0$ and boundary conditions $S_n(x_0, t) = 1$ and $S_n(\infty, t) = 0$. It is given by

$$S_n = \frac{1}{2}\mathrm{erfc}\left(\frac{x - x_0 - cf_n't}{2\sqrt{\mathcal{D}t}}\right) + \frac{1}{2}\exp(cf_n'(x - x_0)/\mathcal{D})\mathrm{erfc}\left(\frac{x - x_0 + cf_n't}{2\sqrt{\mathcal{D}t}}\right), \quad (31)$$

where erfc is the complementary error function (Peaceman 1977, Eq. 4.35). In 1D space c^t is a constant since $\partial c^t/\partial x = 0$. The velocity at which the saturation front travels is

$$v = cf_n' = \frac{c^t f_n'}{\phi}. \quad (32)$$

Equation (30) with $p_c = 0$ ($\mathcal{D} = 0$) is the Buckley–Leverett (1942) equation. To solve Eq. (30), we consider $\mathcal{D} = 0.01, f_n' = 0.9$ and the velocity field shown in Fig. 2a, where $c_0 = 1.2$ m/s, i.e., $c = -c_0$ for $x \le x_0$ and $c = +c_0$ for $x \ge x_0$, where $S_n(x_0 = 0, t) = 1$. An important parameter is the Courant number C, giving the distance traveled due to advection in a time step dt with respect to the grid spacing, i.e., $C = cf_n' dt/dx$. Being explicit, the algorithm is stable for $C < 1$, which means that one cannot advect the saturation more than one grid cell in a single time step. Accuracy requires a time step smaller than that imposed by

Fig. 2 Velocity field (**a**) and comparison between the numerical (*symbols*) and analytical (*solid line*) solutions corresponding to Richards's Eq. (30) with boundary conditions (**b**)

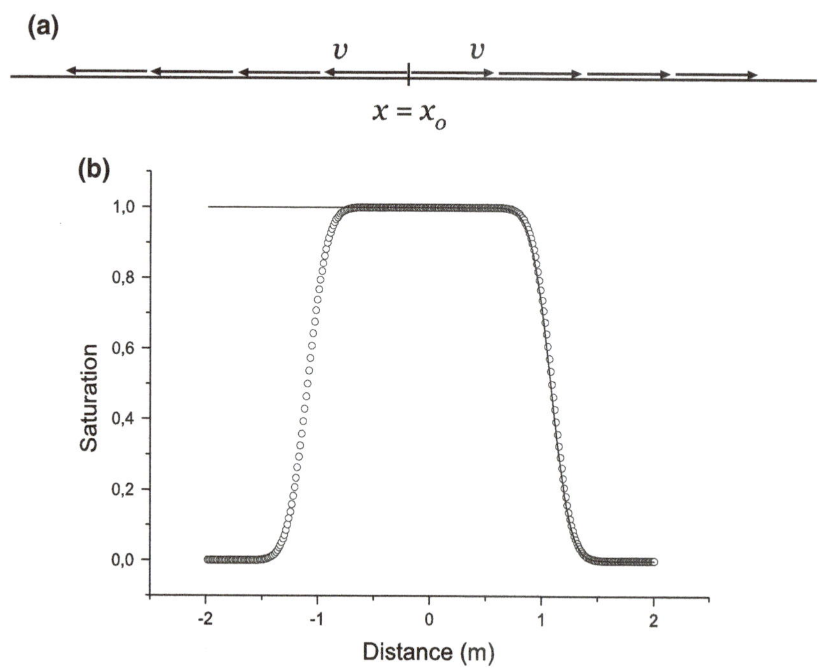

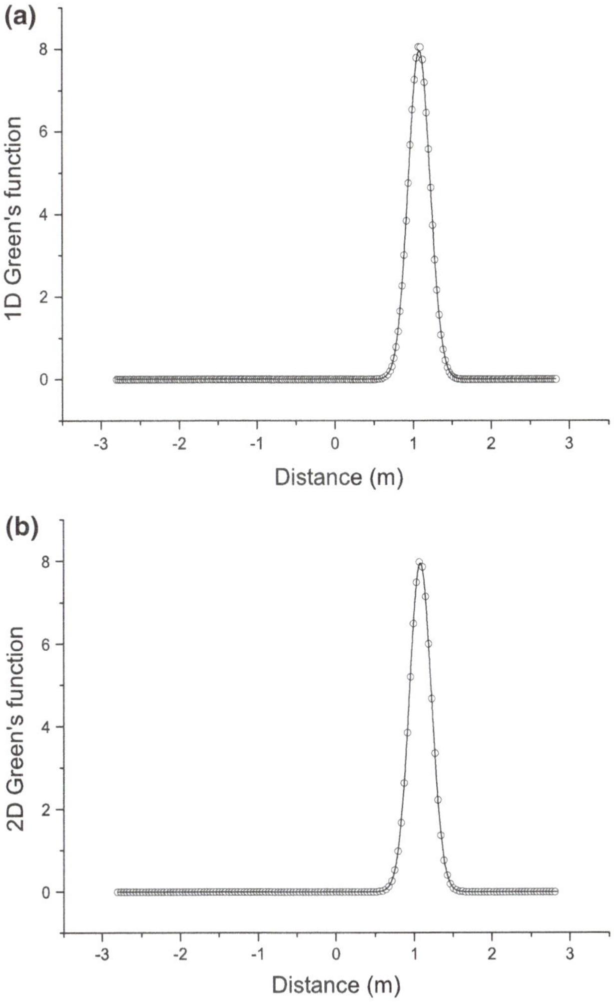

Fig. 3 Comparison between 1D (a) and 2D (b) numerical (*symbols*) and analytical (*solid line*) solutions corresponding to Richards's Eq. (30) with a point source

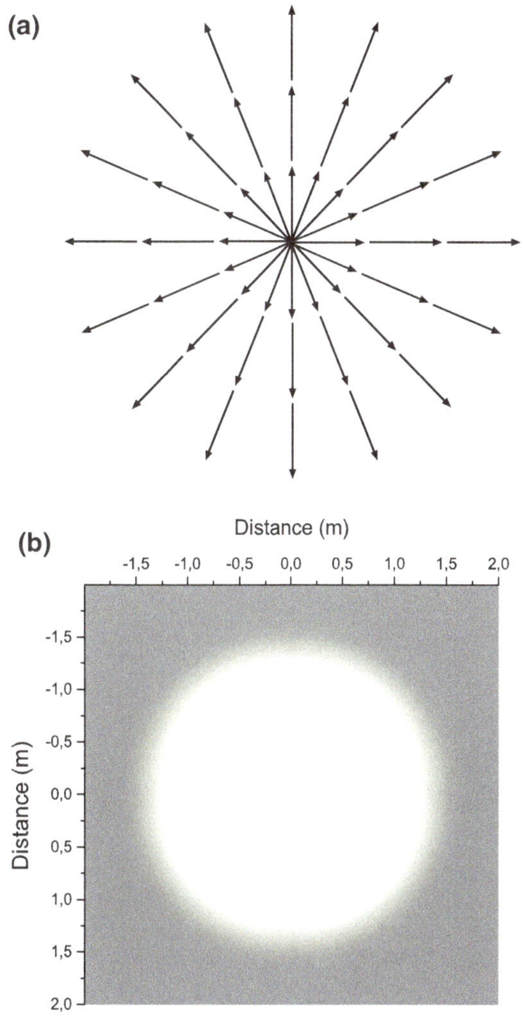

Fig. 4 Velocity field (a) and saturation at t = 1 s corresponding to Richards's Eq. (30) with boundary conditions (b)

stability. Figure 2b shows the comparison between the numerical and analytical solutions at 1 s, where $n_x = 429$, $dx = 4\ m/n_x$ and $dt = 0.2$ ms. The analytical solution is shown for $x > x_0$.

In 2D space, assuming diffusivity anisotropy ($\mathcal{D}_x$, $\mathcal{D}_z$) a flow velocity (v_x , v_z) and a point-source injection, Eq. (30) can be generalized as

$$\dot{S}_n = (\mathcal{D}_x \partial_{xx} + \mathcal{D}_z \partial_{zz}) S_n - f_n'(v_x \partial_x + v_z \partial_z) S_n + s_0 \delta(t)\delta(x)\delta(z), \tag{33}$$

where s_0 is the rate of injection per unit area and the solution to this equation yields the Green function. It is well known that this equation can be recast as a pure diffusion equation. Let us use an analogy and consider the x-direction. If we are traveling in a train moving at velocity $f_n' v_x$, we do not see the advection. If the train starts to move at $t = 0$ and $x = 0$, its location is $x = f_n' v_x t$ and we may define a moving coordinate $x' = x - f_n' v_x t$, so that $x' = 0$ always. With respect to those moving coordinates, we have the following diffusion equation:

$$\dot{S}_n = (\mathcal{D}_{x'} \partial_{x'x'} + \mathcal{D}_{z'} \partial_{z'z'}) S_n + s_0 \delta(t)\delta(x')\delta(z'), \tag{34}$$

whose solution, at the original coordinates, is

$$S_n(x,z) = \frac{s_0}{4\pi t\sqrt{\mathcal{D}_x \mathcal{D}_z}} \exp\left[-\frac{(x - f_n' v_x t)^2}{4\mathcal{D}_x t} - \frac{(z - f_n' v_z t)^2}{4\mathcal{D}_z t}\right] \tag{35}$$

(Carslaw and Jaeger 1984; Carcione 2007). If the injection occurs at (x_0, z_0), the solution is obtained by simply replacing t by $t - t_0$, x by $x - x_0$ and z by $z - z_0$. In 1D space the solution is $S_n(x) = (s_0/\sqrt{4\pi t \mathcal{D}_x}) \exp\left[-(x - f_n' v_x t)^2/(4\mathcal{D}_x t)\right]$. We consider $n_x = n_z = 429$, $dx = dz = 4\ m/n_x$ and $dt = 0.2$ ms. Recall that the discrete spatial delta is represented by $1/dx$ in 1D space and $1/(dx\ dz)$ in 2D space. Figure 3 compares 1D (a) and 2D

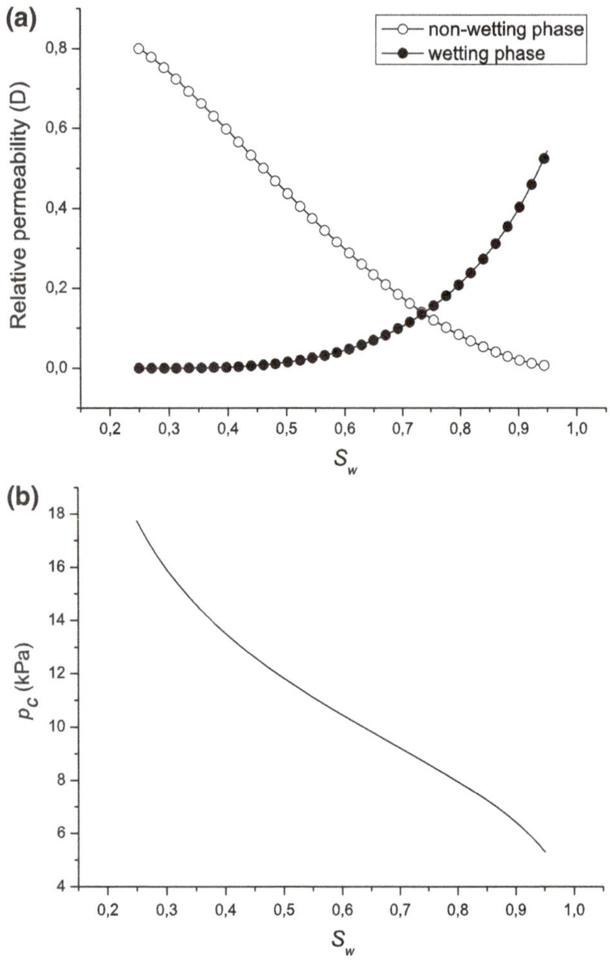

Fig. 5 Permeabilities (**a**) and capillary pressure (**b**) as a function of the wetting-fluid saturation

(b) numerical and analytical solutions at t = 1 s for $D_x = D_z = 0.01$ D, $v_x = v_z = v_0/\sqrt{2}$ with $v_0 = 1.2$ m/s at all the grid points.

Next, we consider the boundary condition S_n (x_0 , t) = 1, a radial velocity flow as shown in Fig. 4a and ($\mathbf{v} = v_0\hat{\mathbf{r}}$), with v_0 =1.2, $f_n' = 1.2$ and the same diffusivity of the previous examples. Figure 4b shows the snapshot of the saturation, where we have taken $n_x = n_z = 1{,}287$, dx = dz = 4 m/n_x and dt = 0.2 ms.

Now we consider the mobilities and the capillary pressure effects. The permeabilities are given by

$$\begin{aligned} \kappa_n &= \kappa\sqrt{1 - S_{ew}}(1 - S_{ew}^{1/p})^{2p}, \\ \kappa_w &= \kappa\sqrt{S_{ew}}[1 - (1 - S_{ew}^{1/p})^p]^2, \\ S_{ew} &= \frac{S_w - S_{rw}}{1 - S_{rw}} \end{aligned} \tag{36}$$

(van Genuchten 1980; Lo et al. 2007), where κ is the absolute permeability, $0 < p < 1$ S_{rn} and S_{rw} are residual saturations. These relations are based on laboratory experiments performed on various porous rocks during imbibition and drainage processes (neglecting hysteresis effects).

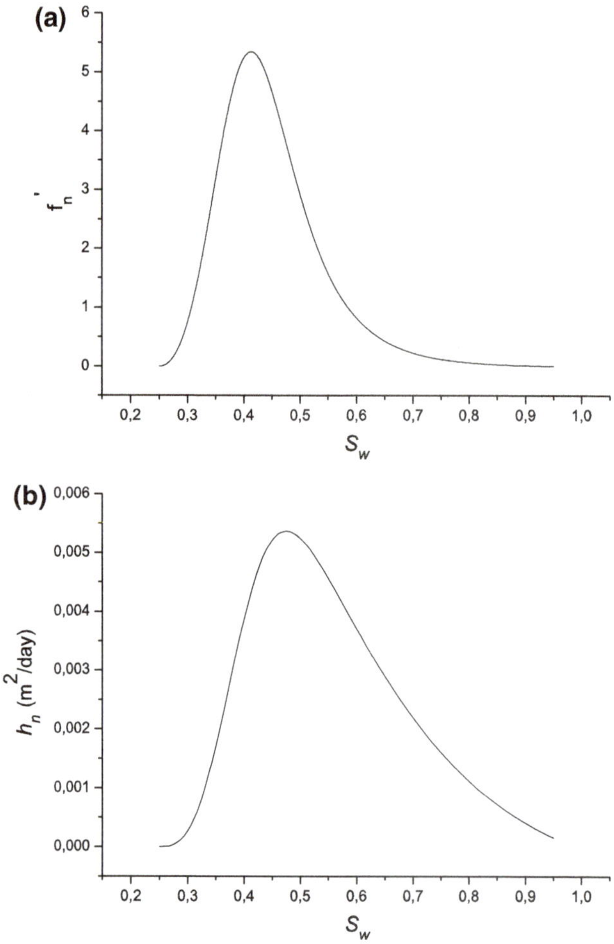

Fig. 6 Advective (**a**) and diffusive (**b**) parameters as a function of S_w

The capillary pressure as a function of saturation is the van Genuchten function

$$\begin{aligned} p_c &= P_0(S_*^{-1/p} - 1)^{1-p}, \\ p_c' &= \frac{1-p}{p}\frac{P_0}{1-S_{rw}}(S_*^{-1/p} - 1)^{-p}S_*^{-1-1/p}, \quad S_* = \frac{S_w - S_{rwv}}{1 - S_{rwv}}, \end{aligned} \tag{37}$$

where P_0 is a reference pressure (e.g., Goumiri et al. 2011) and $S_{rwv} = 0.1\ S_{rw}$. Here, $p_c' = \mathrm{d}p_c/\mathrm{d}S_n$. Relative permeabilities and capillary pressure as a function of the wetting-fluid saturation are shown in Fig. 5, where $\kappa = 0.8$ D, $S_{rw} = 0.25$, $S_{rn} = 0.05$, $P_0 = 10$ kPa and $p = 0.75$. Figure 6 display the advective and diffusive terms f_n' and h_n appearing in the Richards Eq. (25).

We now solve the first Eq. (25) for partial saturation with an initial condition in the wetting-fluid pressure field, without gravity effects and considering the capillary pressure. The equations are

$$\begin{aligned} \dot{\tau}_w &= a_3\nabla\cdot b_n\nabla p_n + a_2\nabla\cdot b_w\nabla p_w, \\ \tau_n &= \frac{1}{S_w}[(1 - S_w)\tau_w - Qp_c], \\ p_w &= [(S_w + \zeta)\tau_n - (S_n + \beta + \zeta)\tau_w]/Q \\ p_n &= [\zeta\tau_n - (\beta + \zeta)\tau_w]/Q. \end{aligned} \tag{38}$$

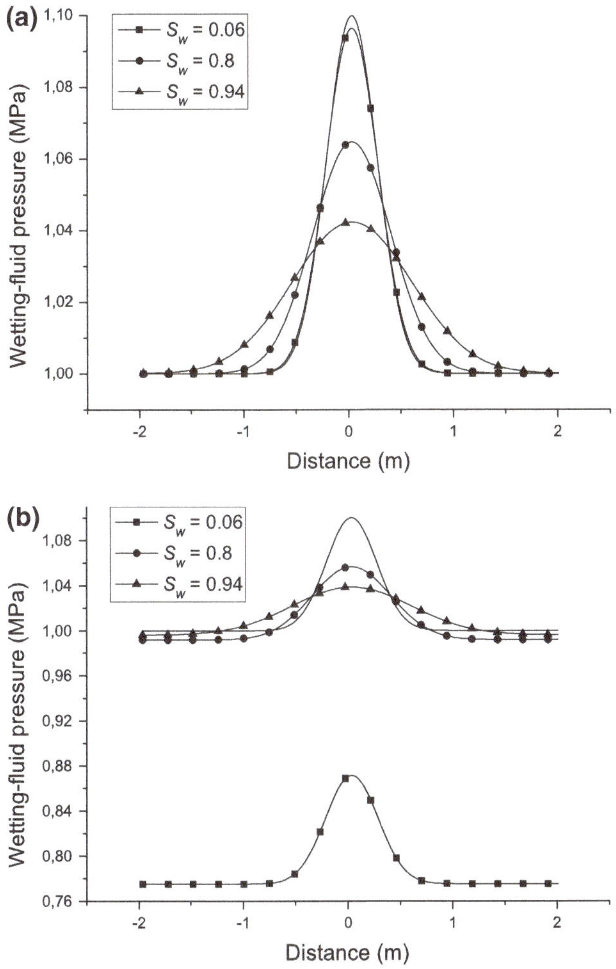

Fig. 7 Snapshots of the wetting-fluid pressure at 60 ms without (**a**) and with (**b**) capillary pressure. The *line without symbols* corresponds to the initial condition

The initial condition is

$$p_w(x,0) = f(x) = p_0 + p_1 \exp[-k_0^2(x-x_0)^2], \tag{39}$$

where $p_0 = 1$ MPa, $p_1 = 0.1$ MPa $k_0 = 3$/m. We consider the properties of Table 1, $S_{rw} = 0.05$, $S_{rn} = 0.05$ and $p = 0.5$. We first test the case where there is an analytical solution, i.e. only the presence of the wetting phase ($S_w = 1$) and an initial condition $f(x) = p_1 \exp[-k_0^2(x-x_0)^2]$. The 1D Green's function is well known (Carslaw and Jaeger 1984; Carcione and Gei 2009) and is given by $G(x) = (4\pi d t_0)^{-1/2} \exp[-(x-x_0)^2/(4dt_0)]$, where $d = \kappa M/\eta_w$ and t_0 is a given time. The numerical solution fits the analytical solution (not shown here), which is given by $p_w(x) = G * f$, where "$*$" denotes spatial convolution. The next simulation considers three saturations with and without the presence of capillary pressure. We take $n_x = n_z = 165$, $dx = dz = 4$ m/n_x and $dt = 50$ µs. The results at 60 ms are shown in Fig. 7, where the reference initial condition is also shown. Increasing wetting-fluid saturation implies

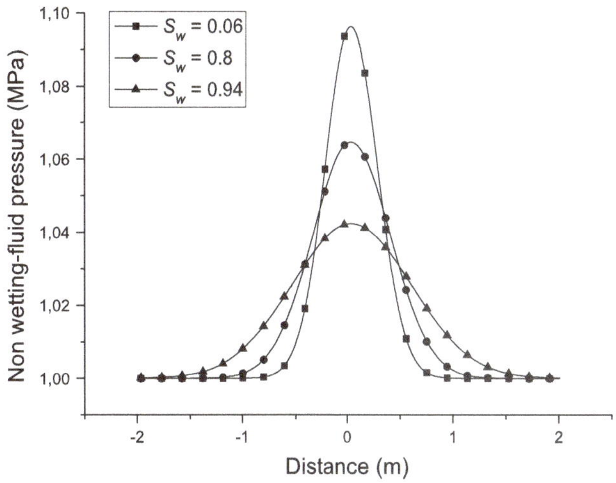

Fig. 8 Snapshots of the non-wetting fluid pressure at 60 ms with capillary pressure

attenuation and dispersion with higher levels at higher wetting-fluid saturations. The presence of capillary pressure produces a shift of the curves downwards. Figure 8 shows the non-wetting fluid pressure, which is not affected by capillary effects.

Now, we compute the numerical solution of Richards's Eq. (25) at different saturations, with and without capillary effects. We ignore sources and gravity. We have

$$\begin{aligned} \phi \dot{S}_n &= \nabla \cdot (h_n \nabla S_n) - \mathbf{c}^t \cdot f_n' \nabla S_n, \\ \mathbf{c}^t &= -(\lambda_n \nabla p_n + \lambda_w \nabla p_w), \end{aligned} \tag{40}$$

where

$$\begin{aligned} f_n' &= \frac{f_n}{(1-S_{rw})\lambda_n}\left[\lambda_n'(1-f_n) - f_n \lambda_w'\right], \\ \lambda_n' &= \frac{\kappa}{2\eta_n}\sqrt{\frac{1-S_{rw}}{1-S_w}}(1-S_w^{1/p})^{2p}\left[1+4\left(\frac{1-S_w}{1-S_{rw}}\right)(S_w^{-1/p}-1)^{-1}S_w^{-1}\right], \\ \lambda_w' &= -\frac{\kappa[1-(1-S_w^{1/p})^p]}{2\eta_w\sqrt{S_w}}\left[1-(1-S_w^{1/p})^p + 4(1-S_w^{1/p})^{p-1}S_w^{1/p}\right]. \end{aligned} \tag{41}$$

We then apply the boundary condition $S_n(x_0, t) = 1 - S_{rw}$. We consider a 1D medium with $c = -c_0$ for $x \le x_0$ and $c = +c_0$ for $x \ge x_0$, where $c_0 = 0.4$ m/s. The grid size is $n_x = 315$, with $dx = 4$ m/n_x. Figure 9 shows a snapshot of the saturation at 15 s as a function of distance, where we have used $P_0 = 10$ kPa, $p = 0.75$, a time step $dt = 0.05$ ms and $K = 10$ iterations of the Picard algorithm (see "Appendix B"). The capillarity pressure has little effect on the shape of the saturation front.

In order to appreciate the capillarity effects, we consider $P_0 = 1$ GPa, $p = 0.5$, $c_0 = 0.4$ m/s, $S_n(x_0, t) = 1 - S_{rw} - 0.01$, the previous mesh and a time step $dt = 0.1$ ms. Figure 10a shows snapshots of the saturation at different propagation times, where the front propagates at a velocity of approximately 3.25 m/s. The velocity actually depends on the fluid saturations. The front velocity Eq. (32) as a function of the wetting fluid saturation is shown in Fig. 10b.

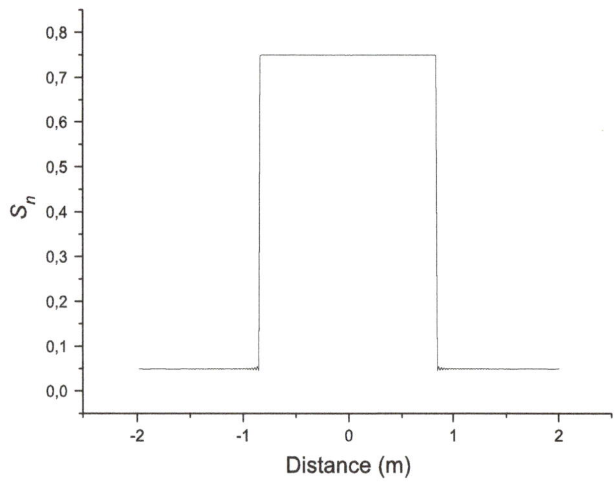

Fig. 9 1D snapshot of the non-wetting saturation at 15 s

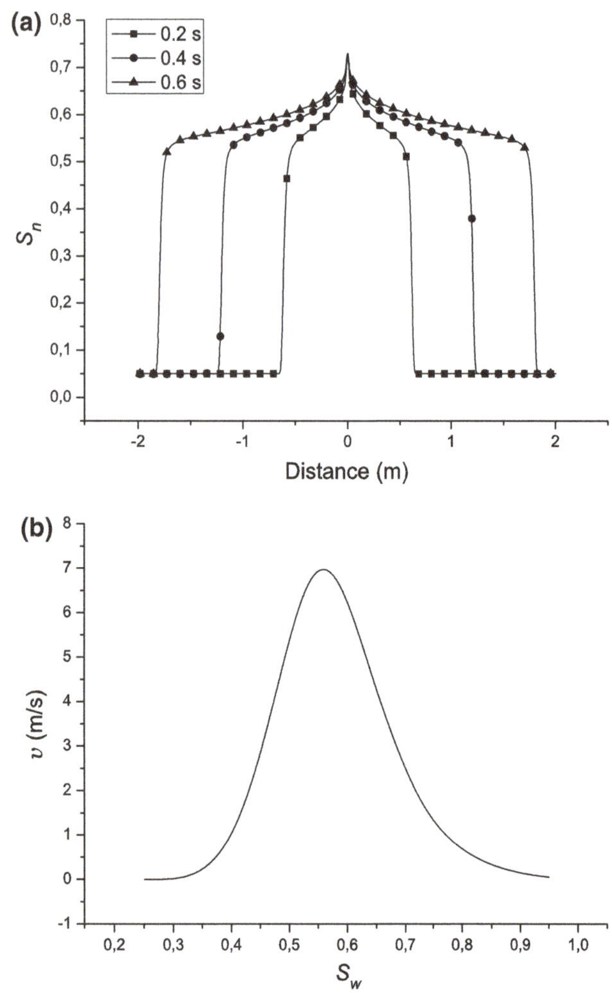

Fig. 10 1D snapshot of the non-wetting saturation at successive propagation times (**a**) and front velocity (**b**). Capillarity effects dominate

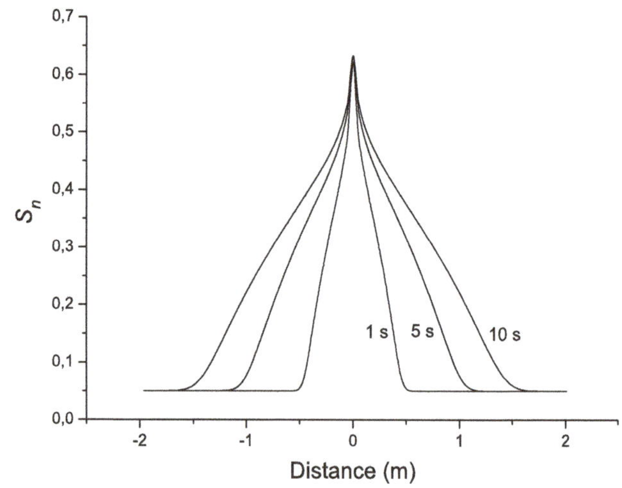

Fig. 11 Snapshots of the non-wetting saturation from the solution of the coupled system of Eq. (25)

Finally, we solve the coupled system Eq. (25) in 1D space, where the flow velocity (last equation) is obtained by solving for the pressures (first and second equations). We set the initial condition Eq. (39), with $p_0 = 1$ MPa, $p_1 = 0.1$ MPa, and $k_0 = 3$/m, $S_{rw} = 0.25$, $S_{rn} = 0.05$ and 95 % water saturation at $t = 0$. Then, we apply the boundary condition $S_n\ (x_0\ , t) = 0.95 - S_{rw}$, with $P_0 = 1$ GPa and $p = 0.5$. We take $n_x = n_z = 165$, $dx = dz = 4\ m/n_x$, $K = 10$ and $dt = 10$ μs. Figure 11 shows the saturations at various propagation times. Diffusion effects due to the capillarity forces dominate in this case.

Conclusions

We have developed a fully explicit algorithm to simulate the injection of a non-wetting fluid into a porous medium saturated with a wetting fluid. The fluid pressures are obtained from the stress–strain relation of the theory of poroelasticity, from which the total (Darcy) flow velocity is computed. This velocity determines the advection term of Richards's saturation equation. Then, the coupled pressure–saturation system of equations is solved with the Euler–Picard time stepping algorithm and the Fourier pseudospectral method. The example assumes oil injection in a water-saturated porous medium. Analytical solutions allow us to verify the accuracy of the results in various situations, where the two unknown variables (pressure and saturation) are uncoupled, with and without capillarity effects based on the van Genuchten functions. Strong capillary forces have to be assumed to see the effects as shown in the simulations of the saturation front. Here, we have considered two fluids. The extension of the theory to include gases, which requires to include pressure–volume–temperature and chemical effects, will be developed in a future work.

Acknowledgments This paper has been partially funded by the CO2Monitor project.

Appendix A: Richards's equation

Let us define the Darcy velocity vector as

$$\mathbf{c}^{\mathrm{m}} = S_{\mathrm{m}}\mathbf{v}^{\mathrm{m}}. \tag{42}$$

Then, from Eq. (1),

$$\mathbf{c}^{\mathrm{m}} = -\lambda_{\mathrm{m}}(\nabla p_{\mathrm{m}} - \rho_{\mathrm{m}}\mathbf{g}). \tag{43}$$

From this equation we obtain

$$\mathbf{c}^{\mathrm{n}} = f_{\mathrm{n}}\mathbf{c}^{\mathrm{t}} - h_{\mathrm{n}}\nabla S_{\mathrm{n}} + \lambda_{\mathrm{w}}f_{\mathrm{n}}(\rho_{\mathrm{n}} - \rho_{\mathrm{w}})\mathbf{g}, \tag{44}$$

where

$$\mathbf{c}^{\mathrm{t}} = \mathbf{c}^{\mathrm{n}} + \mathbf{c}^{\mathrm{w}} \tag{45}$$

is the total Darcy velocity,

$$f_{\mathrm{n}} = \frac{\lambda_{\mathrm{n}}}{\lambda_{\mathrm{n}} + \lambda_{\mathrm{w}}} \tag{46}$$

and

$$h_{\mathrm{n}} = \frac{\lambda_{\mathrm{n}}\lambda_{\mathrm{w}}}{\lambda_{\mathrm{n}} + \lambda_{\mathrm{w}}}p'_{\mathrm{c}}. \tag{47}$$

Here, $p'_{\mathrm{c}} = \mathrm{d}p_{\mathrm{c}}/\mathrm{d}S_{\mathrm{n}}$.

Substituting Eq. (44) into Eq. (19) yields

$$\phi\dot{S}_{\mathrm{n}} + \nabla\cdot(f_{\mathrm{n}}\mathbf{c}^{\mathrm{t}}) - \nabla\cdot(h_{\mathrm{n}}\nabla S_{\mathrm{n}}) + \nabla\cdot(\lambda_{\mathrm{w}}f_{\mathrm{n}}(\rho_{\mathrm{n}} - \rho_{\mathrm{w}})\mathbf{g}) = s_{\mathrm{n}}. \tag{48}$$

The second term is

$$\nabla\cdot(f_{\mathrm{n}}\mathbf{c}^{\mathrm{t}}) = \mathbf{c}^{\mathrm{t}}\cdot\nabla f_{\mathrm{n}} + f_{\mathrm{n}}(s_{\mathrm{w}} + s_{\mathrm{n}}), \tag{49}$$

where we have used Eq. (19). Then,

$$\begin{aligned}\nabla\cdot(h_{\mathrm{n}}\nabla S_{\mathrm{n}}) = {}& \phi\dot{S}_{\mathrm{n}} + \mathbf{c}^{\mathrm{t}}\cdot f'_{\mathrm{n}}\nabla S_{\mathrm{n}} + \nabla\cdot(\lambda_{\mathrm{w}}f_{\mathrm{n}}(\rho_{\mathrm{n}} - \rho_{\mathrm{w}})\mathbf{g}) \\ & + (f_{\mathrm{n}} - 1)s_{\mathrm{n}} + f_{\mathrm{n}}s_{\mathrm{w}},\end{aligned} \tag{50}$$

where $f'_{\mathrm{n}} = \mathrm{d}f_{\mathrm{n}}/\mathrm{d}S_{\mathrm{n}}$. This is a general form of the so-called Richards's (1931) equation.

Appendix B: Euler–Picard method

The Picard method is based on a sequence of functions which converge to the desired solution. It is useful for the error analysis of Eulerian methods. Picard's method starts with a guess solution. The idea is that the process of checking each new solution yields a new guess which, even if it is not the correct solution, is a better approximation of the previous one (Pang 2006).

The saturation Eq. (25) can be re-written as

$$\dot{S} = f(S, t), \tag{51}$$

where $S = S_n$. The forward Euler algorithm is

$$S_{j+1} = S_j + \mathrm{d}tf_j, \quad t = j\mathrm{d}t. \tag{52}$$

Since this method has generally low accuracy, we use a numerical quadrature to integrate Eq. (51),

$$S_{j+1} = S_j + \frac{\mathrm{d}t}{2}(f_j + f_{j+1}) \tag{53}$$

(Pang 2006) and then perform Picard iterations to obtain $f_{j+1} = f(S_{j+1})$. We take $S^0_{j+1} = S_j$ and perform

$$S^{k+1}_{j+1} = S^k_j + \frac{\mathrm{d}t}{2}(f^k_j + f^k_{j+1}), \quad k = 1,\ldots K, \tag{54}$$

compute f_{j+1} and solve Eq. (53).

References

Barenblatt GI, Gilman AA (1987) A mathematical model of non-equilibrium counter-current capillary imbibition. J Eng Phys 52:456–461

Biot M (1941) General theory of three-dimensional consolidation. J Appl Phys 12:155–164

Buckley SE, Leverett MC (1942) Mechanism of fluid displacements in sands. Trans AIME 146:107–116

Carcione JM (2006) A spectral numerical method for electromagnetic diffusion. Geophysics 71:I1–I9

Carcione JM (2007) Wave fields in real media. Theory and numerical simulation of wave propagation in anisotropic, anelastic, porous and electromagnetic media, 2nd edn. Elsevier, Amsterdam

Carcione JM (2010) Simulation of electromagnetic diffusion in anisotropic media. Progr Electromagnet Res B (PIER B) 26:425–450

Carcione JM, Cavallini F, Santos JE, Ravazzoli CL, Gauzellino PM (2004) Wave propagation in partially-saturated porous media. Simulation of a second slow wave. Wave Motion 39:227–240

Carcione JM, Gei D (2009) Theory and numerical simulation of fluid-pressure diffusion in anisotropic porous media. Geophysics 74:N31–N39

Carslaw HS, Jaeger JC (1984) Conduction of heat in solids. Clarendon Press, Oxford, UK

Chandler RN, Johnson DL (1981) The equivalence of quasi-static flow in fluid-saturated porous media and Biot's slow wave in the limit of zero frequency J Appl Phys 52:3391–3395

Detournay E, Cheng AH-D (1993) Fundamentals of poroelasticity. In: Fairhurst C (ed) Comprehensive rock engineering: principles, practice and projects, vol. II, Chap. 5. Pergamon, New York, pp 113-171

Gai X (2004) A coupled geomechanics and reservoir flow model on parallel computers. PhD Thesis, Texas University, Austin

Goumiri IR, Prévost JH, Preisig M (2011) The effect of capillary pressure on the saturation equation of two-phase flow in porous media. Int J Numer Anal Method Geomech 32:352–361

Gutierrez MS, Lewis RW (2002) Coupling of fluid flow and deformation in underground formations. J Engrg Mech 128: 779–787

Hassanizadeh SM, Celia MA, Dahle HK (2002) Dynamic effects in the capillary pressure–saturation relationship and their impacts on unsaturated flow. Vadose Zone J 1:38–57

Lo WC, Yeh CL, Tsai CT (2007) Effect of soil texture on the propagation and attenuation of acoustic wave at unsaturated conditions. J Hydrol 338:273–284

Müller TM (2006) Spatio-temporal pore pressure evolution due to fluid-mass point sources in dynamic poroelasticity. Geophys J Int 165:906–912

Pang T (2006) An introduction to computational physics. Cambridge University Press, Cambridge

Peaceman DW (1977) Fundamentals of numerical reservoir simulation. Elsevier, Amsterdam

Ravazzoli CL, Santos JE, Carcione JM (2003) Acoustic and mechanical response of reservoir rocks under variable saturation and effective pressure. J Acoust Soc Am 113:1801–1811

Richards LA (1931) Capillary conduction of liquids through porous mediums. Physics 1:318–333

Santos JE, Douglas Jr J, Corberó JM, Lovera OM (1990) A model for wave propagation in a porous medium saturated by a two-phase fluid. J Acoust Soc Am 87:1439–1448

Santos JE, Douglas Jr J, Corberó JM (1990) Static and dynamic behaviour of a porous solid saturated by a two-phase fluid. J Acoust Soc Am 87:1428–1438

Settari A., Mourits FM (1994) Coupling of geomechanics and reservoir simulation models. In: Siriwardane H, Zaman M (eds) Computer methods and advances in geomechanics. A. A. Balkema, Rotterdam, pp 2151–2158

Shapiro SA, Rothert E, Rath V, Rindschwentner J (2002) Characterization of fluid transport properties of reservoirs using induced microseismicity. Geophysics 67:212–220

Van Genuchten MT (1980) Closed-form equation for predicting the hydraulic conductivity of unsaturated soils. Soil Sci Soc Am J 44:892–898

Zarba RL (1988) A numerical investigation of unsaturated flow. M.S. thesis, Department of Civil Engineering, Massachusetts Institute of Technology, Cambridge

9

Delineating the fluvial channel system in the Upper Triassic formation of the Elward area in the Syrian Euphrates Graben using 3-D seismic attributes

Mohammad Alsouki · Riad Taifour · Omar Al Hamad

Abstract 3-D seismic data were used to evaluate the reservoir in the Mulussa F formation in the Elward area and to discover new drillable locations by delineating the fluvial channel system and the features of its facies. Seismic attributes such as relative acoustic impedance, coherency volume and reflection strength were considered. The attribute maps revealed different types of channels such as braided, meandering and anastomosing channels. They also revealed fluvial features such as point bars, Ox-Bows and crevasse splays, which are considered potential points of hydrocarbon accumulation due to the abundance of sandy bodies. The map of relative acoustic impedance demonstrated that the sandstone beds have high impedance, whereas the shale beds have low impedance. In fact, well data indicated that the shaly beds were surrounded by sandy beds, and their thickness was in the range of 20–25 m within the Mulussa F formation. The coherency attribute showed some of the meandering tributary channels that were not seen on the relative acoustic impedance map. The abundance of fluvial channels in the Euphrates region indicates a large water discharge during the deposition of the Mulussa F formation because of the wet climate prevailing during the Upper Triassic.

M. Alsouki · O. Al Hamad
Syrian Petroleum Company, Damascus, Syria

O. Al Hamad

R. Taifour
Geology Department, Sciences Faculty, University of Damascus, Damascus, Syria

Keywords Fluvial · Channel · Seismic · Attributes · Euphrates

Introduction

The area of study is located in the middle part of the Euphrates Graben area, between the Elward and the Jido Grabens and within the Euphrates Graben, situated to the east of Syria (Fig. 1a). This Graben is located in the northern part of the Arabian Plate, and it contains significant petroleum reserves. Elward Field is considered one of the most important structural traps that produces oil from the Mulussa F formation, which dates from the Upper Triassic.

The area has accumulated a large amount of sediment from the Palaeozoic to the present. The Mulussa F formation consists of clastic sediments with a narrow thickness range of 280–320 m (Fig. 2), as indicated by the well data from the study area. It is considered the most important reservoir sequence along with the Lower Cretaceous sandy Rutba formation. The Mulussa F and Rutba reservoirs retain 80–90 % of the hydrocarbon volume within the Graben area, as indicated in regional studies. The reservoir characteristics of Mulussa F in Elward Field are fairly good. The net/gross ratios are usually approximately 35 %, and the average porosities are in the range of 20–30 %.

The high-quality 3-D seismic used in this study was a powerful tool that made it possible to interpret the sedimentary features and obtain a better understanding of the depositional environment. Different interpretation techniques, such as conventional seismic interpretation, were used to extract various types of attributes, such as the relative acoustic impedance and coherency attributes that were required for detection of the subtle channel features that constitute the main goal of our study. Thus, the aim of

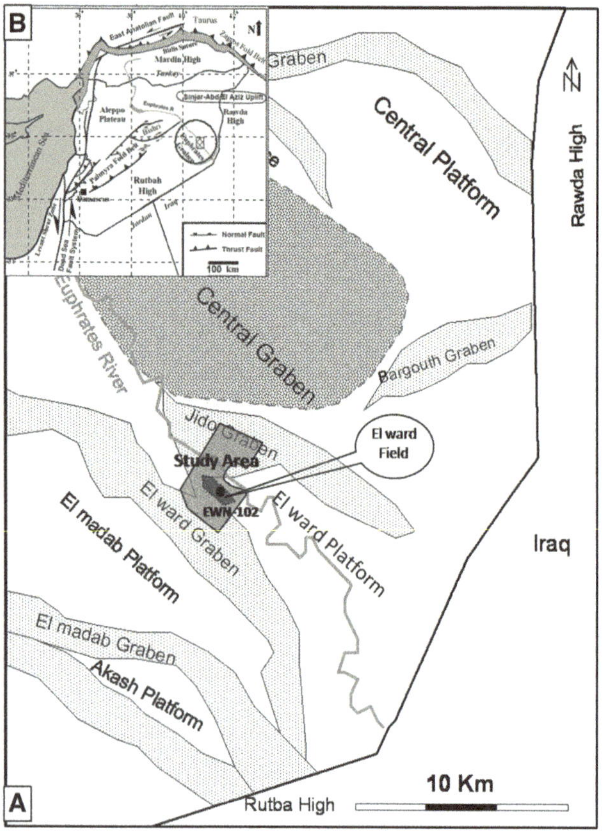

Fig. 1 **a** Location of study area on a tectonic map of the Euphrates Graben. **b** Syrian tectonic map showing the Arabia-Eurasia collision line

this paper is to delineate the fluvial channel system in the Mulussa F formation and to describe the features of its facies, which may form a potential stratigraphic trap.

Geological and tectonic setting of the Euphrates Graben

The Syrian Euphrates Graben is part of the Euphrates fault system, which is located in the northern part of the Arabian plate near the edge of the Eurasia/Arabian plate collision (Brew et al. 2001b). It is characterised by sets of extensional faults that give rise to a series of horst and graben features. The Arabian Plate contacts the Eurasia plate on a roughly north-north westerly direction (McClusky et al. 2000). This collision is obvious because of the active transformation and convergent plate boundaries that are currently proximal to Syria (Fig. 1b). The events that occurred at these boundaries have largely dominated the Paleozoic tectonics of Syria, particularly the Mesozoic–Cenozoic tectonics.

The Euphrates Graben is a fault-bounded failed rift that has been studied extensively by Litak et al. 1997 showed that the Euphrates Fault System is a related, but less-deformed zone of an extension that extends from the Iraqi border in the southeast to the Turkish border in the northwest and includes the Euphrates Graben.

A Turonian–Coniacian age unconformity most likely marked by a pre-rift uplift was extensively developed in the Euphrates Graben, and the underlying limestone was eroded and dolomitised. Coniacian rifting that resulted in approximately 6 km of extension and an indefinite amount of strike-slip movement was accommodated by a distributed system of steep normal faults (Litak et al. 1998). The syn-rift carbonate deposition culminated in the Campanian to early Maastrichtian with the deposition of marly limestones within the graben. Rift-related extension stopped during the Maastrichtian.

The Paleogene was marked by widespread thermal subsidence as the lithosphere re-established thermal equilibrium after rifting (Brew et al. 2001a).

The stratigraphy of the Euphrates Graben generally records a clastic-dominated Palaeozoic section underlying Triassic to Meocene carbonate rocks and evaporites interbedded with lesser amounts of sandstone and shale (Fig. 2). Early Palaeozoic subsidence led to the deposition of Cambro–Ordovician clastic sediments throughout most of the Euphrates Graben along an east-facing continental margin (Best et al. 1993).

Mulussa F was deposited during the tectono-depositional phase of the Palmyrides Basin adjacent to the Euphrates Graben. The tectonic extensional events that took place in the Jurassic period led to complete erosion of Jurassic formations, and partial erosion of the Mulussa F formation in the Euphrates Graben produced an unconformity surface over the Mulussa F Formation called the BKL (Base Cretaceous Lower) (Fig. 2).

Based on well data from the Euphrates Graben, Mulussa F sediments represent a regressive clastic cycle during the Upper Triassic, which consists of marine dolomites and shales in the basal section, passing upward into the coastal plain and fluvial claystones and sandstones. Mulussa F extends laterally from the continental clastic sediment overlying the Doubayat Formations in the southern Euphrates Graben to a 300-m thick layer of regression coastal and fluvial sediments in the study area, and it gradually changes to marine carbonates and clastics in the distal areas near the Palmyride Basin in most of the western Euphrates Graben. The actual thickness of this formation is largely controlled by the rate of erosion at the BKL. This facies distribution suggests that the sediment comes from southern and south western areas in the most interior parts of the Arabian Platform (Brew et al. 2001a).

The time map of the Mulussa F formation top in the study area (Fig. 3) shows three structural units: the Elward Platform and the Jido and El Ward Grabens, which are oriented east–west to northwest–southeast and which are parallel to the central Euphrates Graben. These units form an intracratonic rift system that developed within the northern Arabic

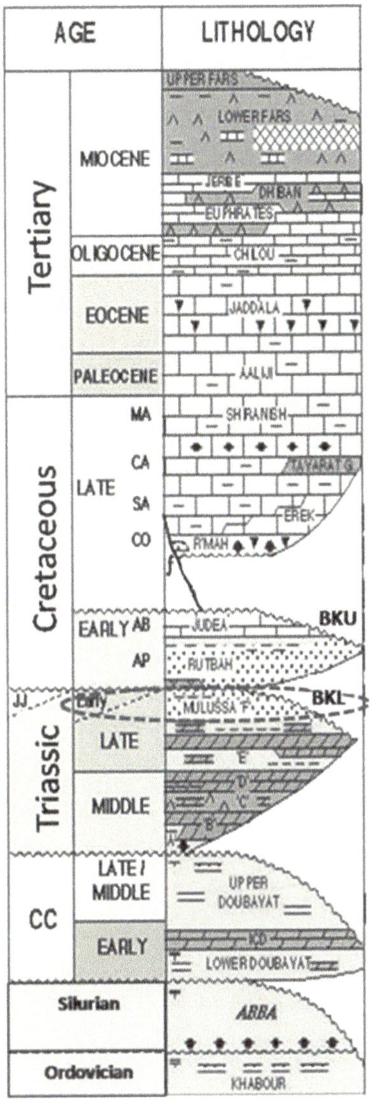

Fig. 2 Mulussa F formation and the unconformity surface (BKL) of the stratigraphic column of the study area

plate in the beginning of late Cretaceous (Coniacian), separating the Rutba High to the south from the Rawda High to the east (Figs. 1, 3). The upper cretaceous formations (Lower and Upper Shiranish) in the aforementioned Grabens are mainly dominated by carbonates with shaly beds. These grabens are considered to be a "kitchen" for the surrounding Jido and Elward Fields, and they form potential stratigraphic traps within the Mulussa F formation.

Methodology

Seismic data set

The seismic data were acquired using a sercel SN368 and a recording system with 24 channel 10 Hz geophones (sensor sm-4). A vibroseis was used as the seismic source. The 3-D seismic PSTM data set covered 215 square kilometres. In general, the quality of the data is good, with a good signal-to-noise ratio that allows the deep horizons to be recognised. The interpretation of the top and bottom of the Mulussa F formation in the majority of the study area is clear. The average thickness of the latter formation is approximately 300 m. The orientation of the inline is NE–SW. The lateral continuity of the internal seismic reflections of the Mulussa F formation is limited due to the depositional fluvial system.

Seismic facies analysis of the Mulussa F sequence

The concepts of seismic stratigraphy and facies analysis (Vail et al.1977) have been applied to the present study to characterise the internal geometry of the Mulussa F formation and interpret the various facies in terms of depositional environment and lithology. The main step is to define the seismic sequence separated by the major surfaces of discontinuity and the correlative continuity surface. These surfaces are interpreted as unconformities. In the stacked seismic sections, the distinct seismic sequence of the Mulussa F formation, and its constituent facies are recognised with the following characteristics:

Mulussa F sequence

The Mulussa F sequence was deposited before the main rifting phase that occurred during the late cretaceous (Coniacian). It is bounded by a truncation surface below, which is the top of Doubayat formation (Fig. 2), and it is bounded on top by a correlative continuity surface called the BKL. Reflections below the surface clearly dip toward the northeast and southwest. From point CDP 500–700, the sequence internal reflections paralleling the upper and lower correlative surfaces dip down steeply towards northeast at an apparent angle of almost 25°, then they rise up gradually to point CDP 700, forming a concave-up geometry (Fig. 4). After that, the sequence reflections dip down gradually until point CDP 840. From point CDP 500–255, the sequence slants increasingly toward the southwest, at an apparent angle of 20° (Fig. 4).

The Mulussa F sequence facies are characterised by a sub-parallel, parallel reflection pattern, poor to moderate continuity and medium amplitude. The amplitude is typically in the range of 70–80 %. This facies pattern is interpreted as a thick sand bed with interbedding shales deposited in a low-medium energy deltaic shelf margin. This pattern is typical of a fluvial environment (Sangree and Widmier 1978). Low-amplitude zones on a seismic section indicate that either the beds are too thin to be resolved by seismic methods or the deposits are uniform.

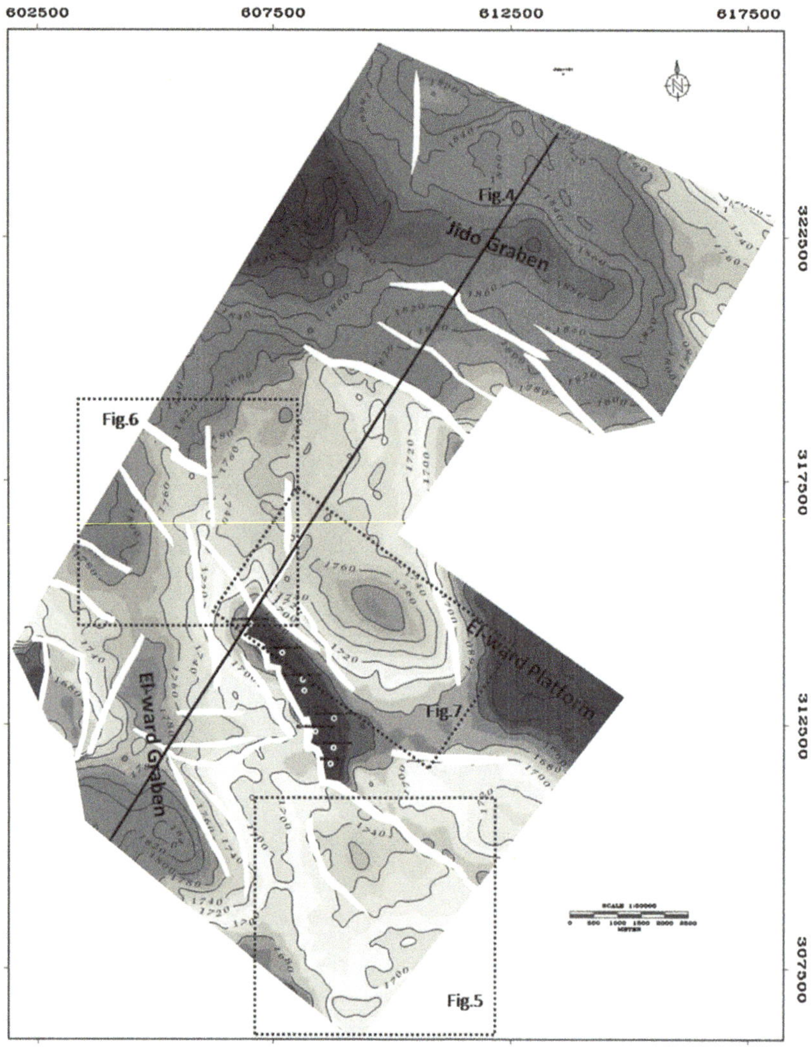

Fig. 3 Time map of the Mulussa F formation including local structural units

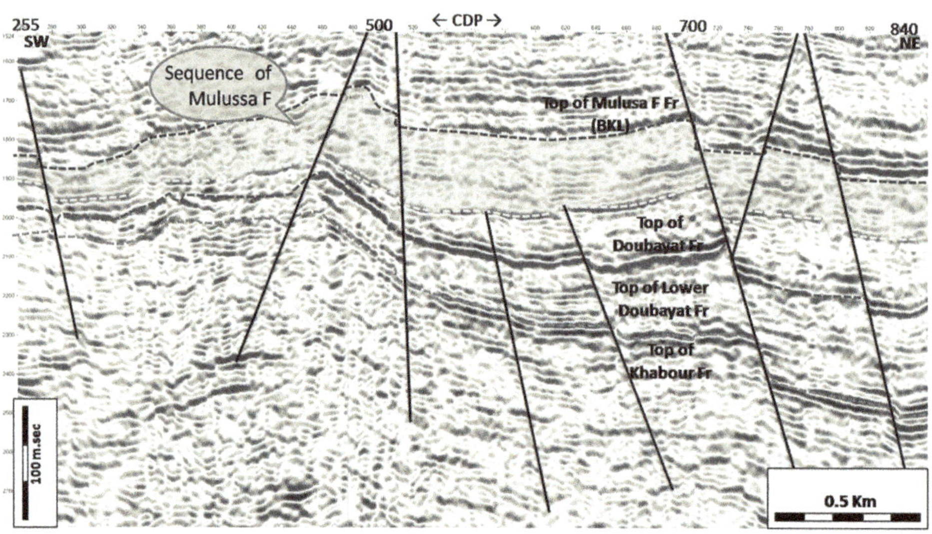

Fig. 4 Seismic section (*inline 2150*) reveals the sequence of Mulussa F and other formations tops

Good reflection amplitudes may represent interbedding of the relatively thick sandstones with the shales and silt-stones. The thickness of the Mulussa F sequence decreases toward the southwest, i.e., in the direction of Rutba high, but it generally increases toward the northeast, i.e., in the direction of the Central Euphrates Graben (Figs. 1, 4).

Delineating the fluvial channel system using 3-D seismic attributes

Seismic attributes extract information from seismic reflection data that can be used for quantitative and qualitative interpretation. Attributes are used by geophysicists and others to delineate depositional features. Some attributes, such as relative acoustic impedance and coherency volume, are directly sensitive to changes in seismic amplitude (Chopra and Marfurt 2005). Seismic attributes are commonly used for two purposes: detecting features and predicting physical properties of the formation of interest (Radovich and Oliveros 1998).

Relative acoustic impedance (RAI)

Delineating the fluvial channel system using the relative acoustic impedance attribute is a technique that relies on intrinsic physical property calculations for interface beds. As the calculations determine the density of the geological formation (ρ) multiplied by the propagation velocity of P-waves (V) in the formation (Eq. 1) (Becquey et al. 1979).

$$AI = \rho v \qquad (1)$$

The relative acoustic impedance has been shown by Taner (2000) to be expressed and computed by integrating the real part of the seismic trace. This method assumes that the input seismic data have been processed to reduce the noise and eliminate multiple contamination. Based on this assumption, the seismic trace represents the band-limited reflective series (Eq. 2) (Schlumberger. 2007a). The integration of the seismic trace provides a band-limited estimate of the natural logarithm of the acoustic impedance (Eq. 3) (Schlumberger 2007a).

$$f(t) = \frac{1\Delta\rho v}{2\rho v} \qquad (2)$$

$$ln\,AI = 2\int_{t=0}^{t=T} f(T)\mathrm{d}t \qquad (3)$$

The relative acoustic impedance map is computed by integrating the attribute across the given time window. For example, the attribute can be integrated between two interpreted horizons. Physically, this is related to the logarithm of the acoustic impedance contrast. Variations in the integrated acoustic impedance attributes map are related by density variation, porosity, lithology and other features. Thus, the relative acoustic impedance map is used for reservoir characterisation (Taner et al. 1979).

Because the acoustic impedance attribute enhances impedance contrast boundaries, it may help to delimit different facies within a fluvial complex. In general, due to the lithological differences between channels and the differences between channel fills and/or flood plains, the amplitudes of seismic reflections are obviously distinct, particularly if the difference in acoustic impedance is large.

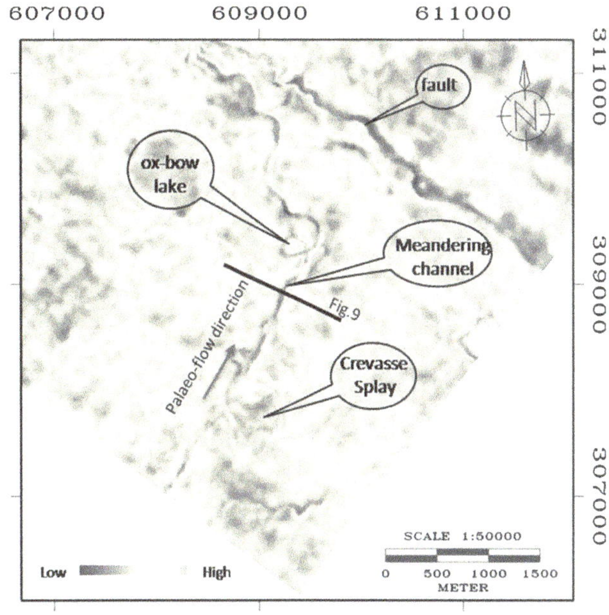

Fig. 5 Relative acoustic impedance attribute map within the Mulussa F sequence. The map shows an Ox-Bow lake, a crevasse splay and a meandering channel. It also highlights the impedance contrast boundaries and the changes in lithology

Hence, we can delineate the fluvial channel system by calculating the acoustic impedance volume on the time interval of the Mulussa F Sequence, i.e., between two horizons: the top of Mulussa F and the top of Doubayat (Figs. 5, 6).

Coherency

The coherency attribute is calculated from 3-D seismic data. It is a quantitative measure of the similarity or dissimilarity of nearby seismic traces based on Eq. 4 (Bahorich and Farmer 1995).

$$\Phi(t,\ d) = \frac{\sum_{k=t+N/2}^{k=t+N/2} G_k H_{k+d}}{\sqrt{\sum_{k=t+N/2}^{k=t+N/2} G_k^2 \sum_{k=t+N/2}^{k=t+N/2} H_{k+d}^2}} \qquad (4)$$

In this equation, Φ is the correlation coefficient at time (*t*) (milliseconds) for a specific geologic dip (*d*). *G* and *H* are the correlated traces, and *N* is the number of samples in the correlation time window.

Because the coherence measurements in three dimensions represent the trace-to-trace similarity, they produce interpretable changes (Chopra et al. 2011). Similar traces are mapped with high coherence coefficients and discontinuities, such as faults and fluvial channels, and subtle stratigraphic features have low coefficients. The regions of seismic traces cut by channels, for example, result in sharp discontinuities in the trace-to-trace coherence, producing delineation of low coherence along the channel profile. The significant details of stratigraphic features such as river channels, levees and submarine canyons are undetectable in seismic data even with close scrutiny. The attribute involves a numeric process, which does not require any interpreted horizons, and so it is free from interpretational bias.

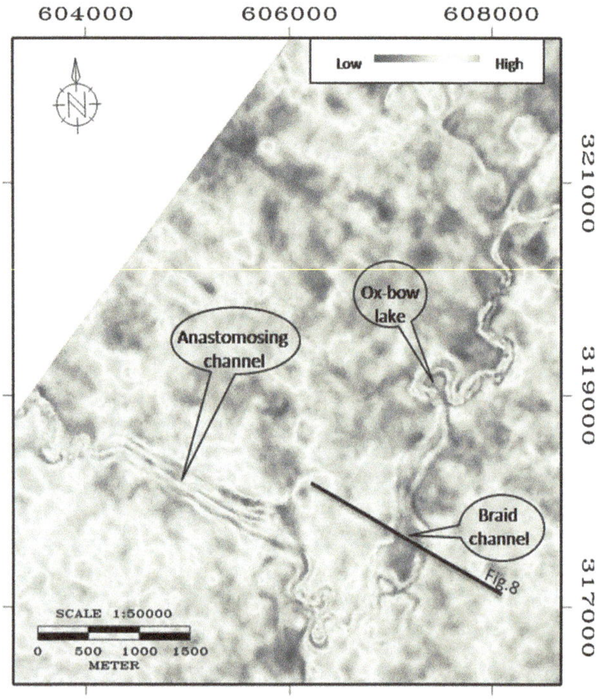

Fig. 6 Relative acoustic impedance attribute map within the Mulussa F sequence. The map shows the anastomosing channel and the braided channel

The coherence attribute map within Mulussa F contains different discontinuities, such as faults and channels. The black areas represent areas of high coherency, and the white areas represent features of low coherency, which likely include stratigraphic information (Fig. 7).

Fluvial channel types within the Mulussa F formation

In this study, the attribute maps illustrate a wide variety of fluvial channel types, dimensions, vertical successions, and lateral relationships. The mapped channels seem to have different sizes, as shown in Figs. 5, 6, 7. The differences are likely due to differences in the substratum erodibility, channel discharge (sediment supply and flow rate), and/or tectonic control. A braided channel is clearly observed in Fig. 6. Such a channel is created when a stream channel is divided into several smaller ones by the accumulation of channel deposits (Fig. 8a). By applying the reflection strength attribute on the previous seismic section in Fig. 8a, sandy beds with high reflection strength (light colour) can clearly be observed enclosed by shales with low reflection strength (grey colour) (Fig. 8b). The braided channel is approximately 2.3 km long in the study area, and the medium single channel is approximately 300–400 m wide.

A meandering fluvial channel is observed in the study area. It forms a sinuous curve, as shown in Figs. 5, 7. It moves in loops, and its velocity is greatest at the outer bank where erosion occurs and lowest at the inner bank where deposition of the point bar sands occurs. The length of the

Fig. 7 Meandering tributary channel mapped through the coherency attribute at time slice 1816 ms within the Mulussa F sequence (the *dotted curve* indicated with *arrows*)

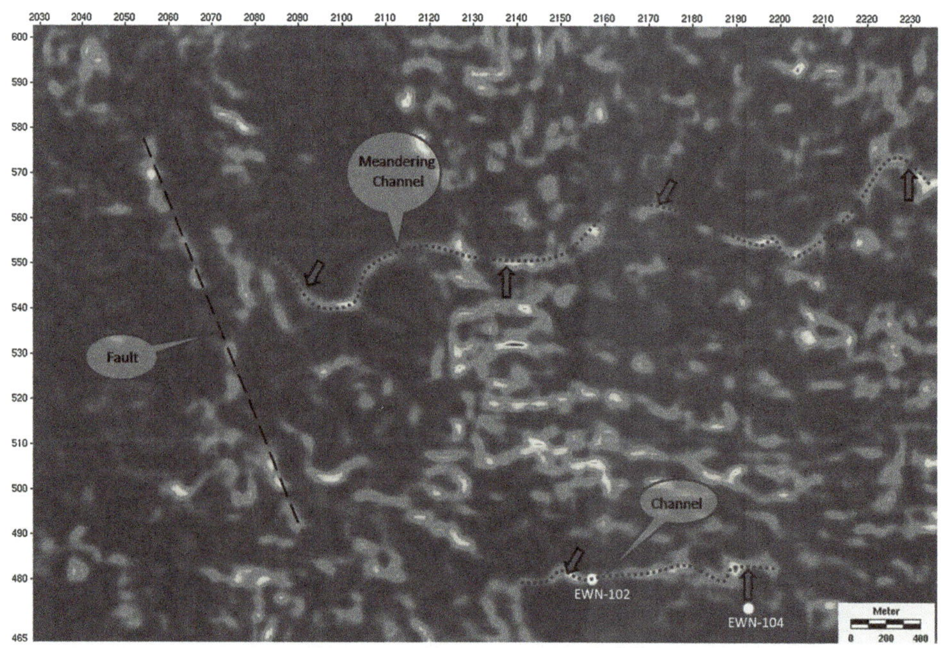

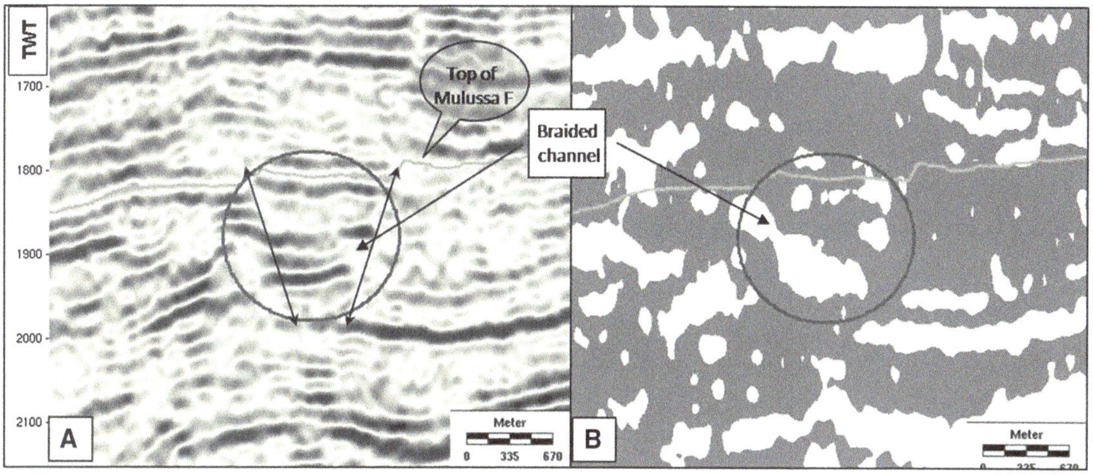

Fig. 8 **a**. Seismic section passing the braided channel. **b** Reflection strength attribute applied to the same section

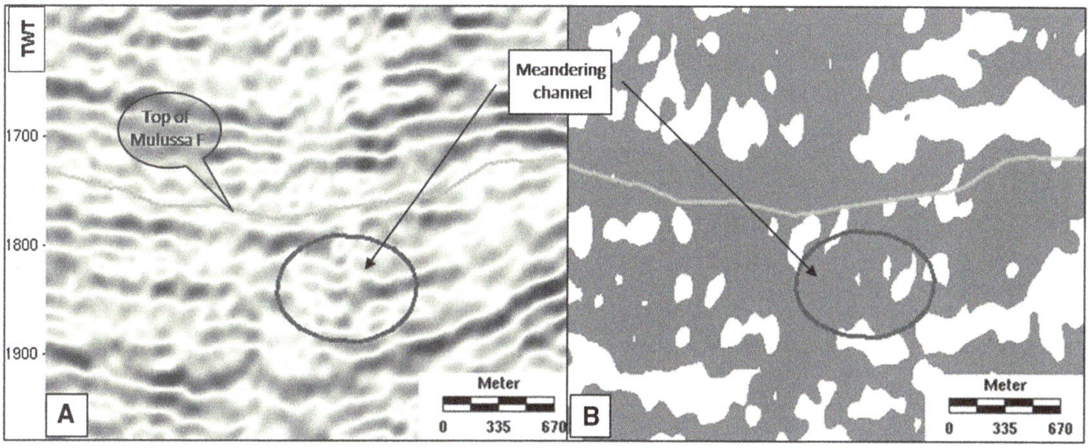

Fig. 9 **a** Seismic section passing the meandering channel. **b** Reflection strength attribute applied to the same section

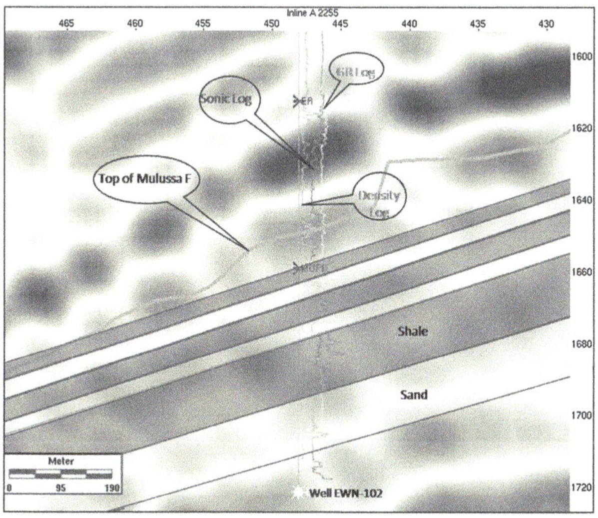

Fig. 10 Enlarged seismic section (*cross-line 477*) crossing the EWN-102 well and showing alternating shale and sandstone beds measured by the well logs in the Mulussa F formation

main meandering channel is estimated to be almost 24 km, with an average width of 80 m. The reflection strength attribute has been applied to a seismic reflection section (Fig. 9a) to infer the lithology of the channel fill. The figure shows that the channel fill mainly contains shales (grey colour) surrounding a little sandstone (light colour) (Fig. 9b).

The anastomosing channel is shown in Fig. 6. It forms three parallel straight channels. In general, it is defined as a branching, interlacing stream with a net-like appearance. It is traced on the attribute map, and it is approximately 2.5 km long.

The prediction of fluvial sandstone beds in the Mulussa F formation

Along the mapped meandering channel profile, there are plenty of point bar sandstone beds, some of which may contain stratigraphical hydrocarbon traps. Shale bed filled

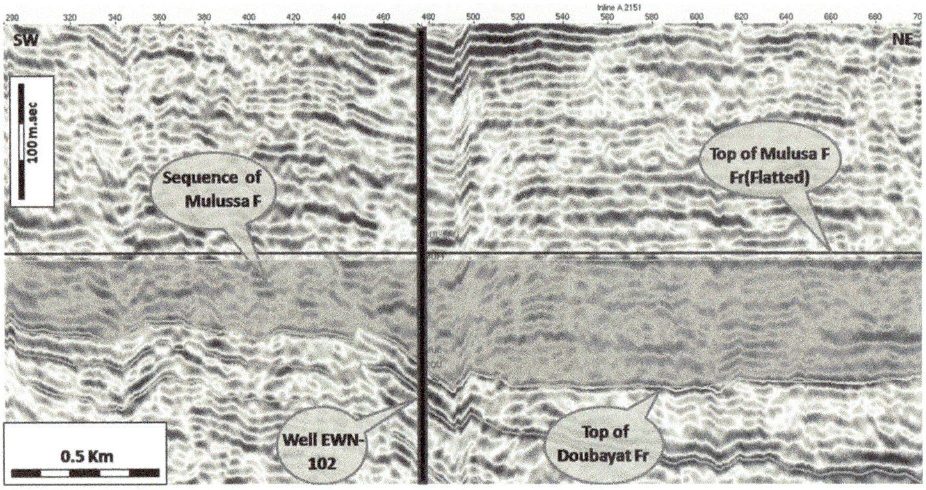

Fig. 11 A seismic section (part of *inline 2150*) flattened on the Mulussa F formation top (the *black* horizon) showing increasing formation thickness as the terrain dip gradient increases

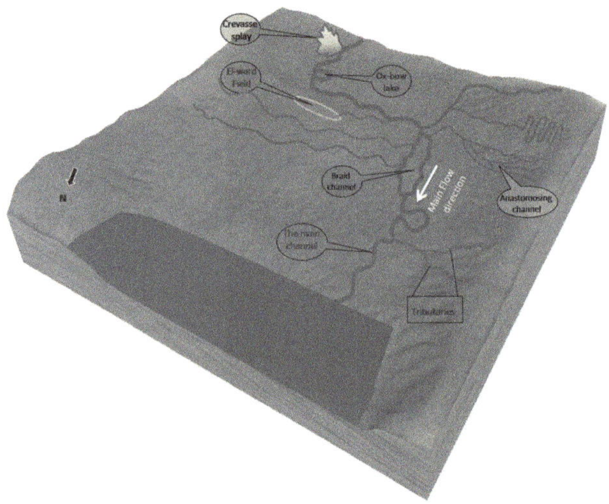

Fig. 12 Schematic depositional model of the Mulussa F formation in the Upper Triassic. The distribution of different channels system based on attribute maps is shown

channels spread widely across the floodplain within the Mulussa F and form seal rocks that fully surround the point bar, crevasse splay and ox-bow sandstone beds. Some shale-filled channels intersect in the Elward field wells. For example, the EWN-102 well is shown in Fig. 7. The central portions of the compacted shale-filled channels within the Mulussa F formation range from 20 to 25 metres, as measured on the well logs (Fig. 10).

Depositional model of the Mulussa F sequence

Integrating seismic data, log data, paleo-structural analysis and channels style information enables reconstruction of the depositional model of the Mulussa F formation Channels. The seismic section flattened on the Mulussa F formation top (inline 2150 in Fig. 4) shows an almost slight dip and uniform thickness until it reaches the EWN-102 well. Then, its thickness becomes suddenly larger due to an increasing terrain dip gradient at the braided channel (Fig. 11). The parallel reflection configurations indicate deposition over semi-flat surfaces during a tectonic serenity period. Palinspastic analysis shows paleo-highs along the eastern and western wings of the study area, from which a set of tributaries declines toward the main channels. A reconstructed depositional model of various channel styles is shown in Fig. 12.

Results and discussion

The coherence and relative acoustic impedance maps (Figs. 5, 6, 7) display better delineation of the channels at locations that correspond to crevasse splay, ox-bow lake facies and different fluvial systems such as the braided, anastomosing and meandering channels.

The coherence image clearly reveals several discrete meandering channels, as indicated by the dotted curve in Fig. 7 (low coherency white area). It also shows numerous subtle faults and other discontinuities inside the Mulussa F sequence.

The presence of various channels, as indicated in the attribute map (Fig. 6), provides significant information about the paleo-relief, the river discharge and the direction in the Upper Triassic period. The anastomosing channel occurs in the regions of low stream gradient. The relative acoustic impedance map indicates that the paleo-flow channel is directed to the southeast, near Rutba high, whereas the braided channel indicates that the river channel was not very stable during the deposition of the Mulussa F formation. This style is usually created in the areas of steep dips and large supply sediments and in areas where the discharge of the river fluctuates daily or annually (Walker 1984). However, based on the aforementioned attribute maps and palinspastic restoration analysis, the main paleo-

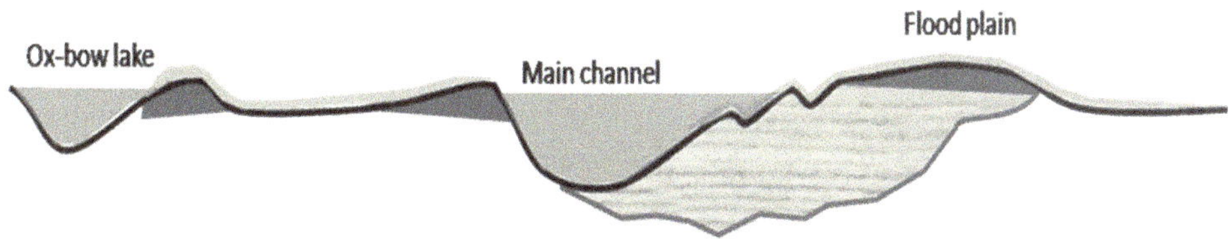

Fig. 13 A schematic diagram showing the water level during deposition of the Mulussa F formation

flow direction was toward the northeast, from Rutba high to the Central Euphrates, whereas the paleo-flow directions of all tributaries were toward the larger main channel.

Coarse sandy channel deposition reflects a high-energy environment, and a muddy flood plain reflects a low-energy environment. These features are obvious in the seismic sections. The first one is a strong reflection, and the other has good reflection continuity (Yuhua et al. 2005).

The lithology within the channels can be inferred from the reflection strength attribute (Figs. 8, 9), which is used as an effective indicator to show the acoustic impedance contrast. Consequently, the reflection strength attribute is useful for recognising the stratigraphic variations within the fluvial channels (Walls et al. 2002). As a result of integrating the reflection configurations, the relative acoustic impedance attribute and the nearby well data, it can be concluded that the mapped channels are mainly filled with shaly beds of relatively low impedance (dark colour), whereas the sandy beds have high impedance (light colour) and are surrounded by shale beds that were deposited on the channel banks, point bars and crevasse splays, as shown in Figs. 5, 6. All of these features may be formed by high-quality potential stratigraphic hydrocarbon reservoirs that are rich in sandstones. Hence, researchers try to concentrate on the exploration of undiscovered potential traps due to uncertainties in the prediction of their distributions. A lack of understanding about the reservoir spatial distribution limits the drilling of wells in the locations of stratigraphic traps within the Mulussa F formation.

Fluvial systems react to changes in the base level (Martin et al. 2011), which can be defined as the level at which a stream can no longer erode its bed. This level corresponds to sea level for rivers that flow into the sea. Changes in the base level cause changes in the shape, width, and depth of fluvial channels (Blum and Tornqvist 2000). In addition, other factors such as sea-level variations, tectonic events, and changes in fluvial discharge play a significant role in determining the stream shape via controlled erosion and deposition processing.

The Euphrates Graben area experienced a wet climate in the Upper Triassic period, which led to an abundant discharge of water loaded with sediments into the basin across many fluvial channels. At the same time, the tectonic serenity period prevailed in the study area, as indicated in the seismic data. These factors, along with sea-level rises, can be used to interpret the reason for the creation of unincised channels within the Mulussa F formation. It is generally known that incised channels are formed by at least three main mechanisms (Posamentier 2001): (1) as a result of sea-level fall, (2) by tectonic tilting of alluvial settings, or (3) during significant decreases in fluvial discharge. Based on the aforementioned information, it can be concluded that the Mulussa F formation was deposited on the continental fluvial deltaic depositional basin by unincised fluvial channels. The river water level was likely close to the flood plain level during deposition (Fig. 13).

Conclusion

3-D seismic data have been interpreted using powerful seismic volume attribute techniques. The braided, meandering and anastomosing channels, in addition to the associated fluvial facies, have been mapped. The attribute shows that the sandstones surrounded by shaly beds have higher impedance and were deposited on the channel bank, point bar and crevasse splay. These fluvial facies are considered to be high-quality potential hydrocarbon reservoirs that could be future exploratory targets. Many of the channels indicate that a wet climate prevailed in the Euphrates Basin region in the Upper Triassic period. Water discharge and sea-level rises may have contributed to creating the unincised channels. The presence of various types of fluvial channels provides important information about the paleo-relief during the deposition of the Mulussa F Formation.

Acknowledgments This article was prepared under the supervision and permission of the Syrian Petroleum Company (SPC). The authors would like to thank the staff of the Syrian Petroleum Company, especially Mr Wasef Al-Himed, Exploration Manager, for close cooperation and for the permission to publish this paper including the seismic profiles.

References

Bahorich M, Farmer S (1995) 3-D seismic discontinuity for faults and stratigraphic features, The coherence cube. Lead Edge 17:1053–1058

Becquey M, Lavergne M, Willm C (1979) Acoustic impedance logs computed from seismic traces. Geophysics 44(9):1485–1501

Best JA, Barazangi M, Al-Saad D, Sawaf T, Gebran A (1993) Continental margin evolution of the northern Arabian Platform in Syria. Am Assoc Pet Geol Bull 77(2):173–193

Blum MD, Tornqvist TE (2000) Fluvial responses to climate and sea level change, a review and look forward. Sedimentology 47:2–48

Brew G, Lupa J, Barazangi M, Sawaf T, Al-Imam A, Zaza T (2001a) Structure and tectonic development of the Dead Sea Fault System and Ghab Graben in Syria. J Geol Soc 158(4):665–674

Brew G, Barazangi M, Al-Maleh AK, Sawaf T (2001b) Tectonic and geologic evolution of Syria. GeoArabia 6(4):573–616

Chopra S, Marfurt K (2005) SEG75th Anniversary, Seismic attributes—a historical perspective. Geophysics 70:3–28

Chopra S, Misra S, Marfurt K (2011) Coherence and curvature attributes on preconditioned seismic data. Lead Edge 30(4):386–393

Litak RK, Barazangi M, Beauchamp W, Seber D, Brew G, Sawaf T, Al-Youssef W (1997) Mesozoic-Cenozoic evolution of the intraplate Euphrates fault system, Syria: implications for regional tectonics. J Geol Soc 154(4):653–666

Litak RK, Barazangi M, Brew G, Sawaf T, AL-Imam A, ALYoussef W (1998) Structure and evolution of the petroliferous Euphrates Graben System, Southeast Syria. Am Assoc Pet Geol Bull 82(6):1173–1190

Martin J, Cantelli A, Paola C, Blum M, Wolinsky M (2011) Quantitative modeling of the evolution and geometry of incised valleys. J Sediment Res 81:64–79

McClusky S, Balassanian S, Barka A, Demir C, Ergintav S, Georgiev I, Gurkan O, Hamburger M, Hurst K, Kahle H, Kastens K, Kekelidze G, King R, Kotzev V, Lenk O, Mahmoud S, Mishin A, Nadariya M, Ouzounis A, Paradissis D, Peter Y, Prilepin M, Reilinger R, Sanli I, Seeger H, Tealeb A, Toksöz MN, Veis G (2000) Global Positioning System constraints on plate kinematics and dynamics in the eastern Mediterranean and Caucasus. J Geohys Res 105(B3):5695–5719

Posamentier HW (2001) Low stand alluvial bypass systems, incised vs unincised. Am Assoc Pet Geol Bull 85:1771–1793

Radovich BJ, Oliveros RB (1998) 3D sequence interpretation of seismic instantaneous attributes from the Gorgon field. Lead Edge 17(9):1286–1293

Sangree JB, Widmier M (1978) Seismic stratigraphy and global changes of sea level, Part 9: seismic interpretation of clastic depositional facies. Am Assoc Pet Geol Bull 62:752–771

Schlumberger (2007a) Interpreter's guide to seismic attributes. http://www.rocksolidimages.com/pdf/attrib_revisited.htm. Accessed 22 Jun 2006

Taner MT (2000) Attributes revisited, rock solid images. http://www.rocksolidimages.com/pdf/attrib_revisited.htm. Accessed 10 Apr 2009

Taner MT, Koehler F, Sheriff RE (1979) Complex seismic trace analysis. Geophysics 44:1041–1063

Vail PR, Mitchum RM, Thompson S (1977) Seismic stratigraphy and global changes of sea level, part 4, global cycles of relative changes of sea level. Am Assoc Pet Geol Bull 26:83–97

Walker R (1984) Facies model, Geoscience Canada, Geological Association of Canada, Reprint series 1, Toronto, pp 71–89

Walls JD, Taner MT, Taylor G, Smith M, Carr M, Derzhi N, Drummond J, McGuire D, Bregar S, Lakings J (2002) Seismic reservoir characterization of a US Midcontinent fluvial system using rock physics, poststack seismic attributes, and neural networks. Lead Edge 21:428–436

Yuhua B, LiYuhai FZ, Bao L (2005) The study and practice of fluvial sandstone reservoir prediction in Gangdong Oilfield. SEG Annual Meeting, Houston, Texas

10

A rigorous method to calculate the rising speed of gas kick

Kegang Ling · Jun He · Jun Ge · Peng Pei · Zheng Shen

Abstract The rising speed of gas kick is an important parameter in well control operation. The position of the gas kick dictates the pressure at the casing shoe, which is usually the weakest point in the openhole section, and the wellhead pressure, which is one of the key factors affecting the blowout preventer and choke folder. In this research, we derived a rigorous model to estimate the rising speed of gas kick. Starting from the force analysis and mass conservation, we developed equations to calculate the forces exerting on the gas kick. With the mass of the gas kick, the rising speed of the gas kick is calculated. The effect of wellbore temperature profile on the rising of the gas kick is taken into account in the derivation. Before the development of this model, the estimation of gas kick position is commonly based on experience. In many cases, the experience alone is not good enough for well control. The proposed model provides a new approach with solid theoretical base to characterize the rising of gas kick in the hole. It makes the procedure of the well control simple and makes drilling engineers feel more comfortable to control the well. The new model can be combined with engineers experience to predict the downhole situation, shut-in casing pressure, and mud rate as a functions of position of gas kick. Any deviation from the forecast indicates accidents or downhole problems. Therefore, the proposed model is a valuable tool to diagnose the problems in well control.

K. Ling · J. He · J. Ge · P. Pei
University of North Dakota, Grand Forks, TX, USA

Z. Shen
Texas A&M University, College Station, ND, USA

Keywords Gas kick · Well control · Kick migration speed

Nomenclature

$A_{\text{cone base}}$	= Area of base of cone of gas column
$A_{\text{cone surface}}$	= Characteristic area of the cone of gas column
$A_{\text{gas column base}}$	= Area of base of gas column
$A_{\text{gas column flank}}$	= Surface area of flank of gas column
a	= Acceleration
D_1	= Drillpipe diameter
D_2	= Wellbore diameter
E_k	= Kinetic energy per unit volume
F_1	= Fluid force behind the gas column
F_2	= Fluid force in front of the gas column
F_{D1}	= Drag force on the flank of the gas column
F_{D2}	= Drag force on the cone surface of gas column
$F_{D2\text{-}Vert}$	= Vertical drag force resulting from drag force on the cone surface of gas column
F_G	= Gravitational force
F_{net}	= Net force
f	= Friction factor
f_i	= Gas-mud interfacial friction factor
g	= Gravitational acceleration
$h_{\text{gas base}}$	= Location of base of gas column
$h_{\text{gas center}}$	= Location of center of gas column
h_{well}	= Well depth
K	= Consistency index of mud
$L_{\text{bottomhole-gas base}}$	= Distance between bottomehole and base of gas column
$L_{\text{gas cone}}$	= Height of cone part of gas kick

$L_{\text{gas cylinder}}$ = Height of cylinder part of gas kick
$L_{\text{gas kick}}$ = Height of gas kick
L_{migrated} = Gas migrated distance
$L_{\text{wellhead-gas top}}$ = Distance between wellhead and top of gas column
M = Molecular weight of gas
m_g = Mass of gas column
n = Flow-behavior index of mud
p_1 = Pressure behind the gas column
p_2 = Pressure in front of the gas column
p_{casing} = Casing pressure
p_f = Frictional pressure drop
p_p = Pore pressure
p_{SIDP} = Shut-in drillpipe pressure
$\bar{p}_{\text{gas}}$ = Average pressure of gas column
R = Universal gas constant
T = Temperature at location
T_{BH} = Bottomhole temperature
u_g = Gas velocity
u_m = Mud velocity
$V_{\text{gas cone}}$ = Volume of cone part of gas kick
$V_{\text{gas cylinder}}$ = Volume of cylinder part of gas kick
$V_{\text{gas kick}}$ = volume of gas kick at any location
$V_{\text{gas kick,BH}}$ = Volume of gas kick at bottomhole
$V_{M,\text{out}}$ = Volume of mud flow out the hole
$V_{M,\text{in}}$ = Volume of mud flow into the hole
z = gas deviation factor
z_{BH} = Gas deviation factor at bottomhole
ρ_g = Gas density
$\rho_{g,\text{BH}}$ = Gas density at botomhole
ρ_m = Mud density
θ = Angle between F_{D2} and vertical direction
θ_{300} = The 300-rpm dial reading in mud viscometer
θ_{600} = The 600-rpm dial reading in mud viscometer
μ_g = Gas viscosity
τ_{gm} = Shear stress between gas and mud
Δt_l = 1st time step

Introduction

The rising speed of gas kick is an important parameter in well control operation. The position of the gas kick dictates the pressure at the casing shoe, which is usually the weakest point in the openhole section, and the wellhead pressure, which is one of the key factors affecting the blowout preventer and choke folder. In many gas kick well control operations, the estimations of gas kick position are commonly based on experience. In many cases, the experience alone is not good enough for well control. A model with theoretical base to predict the gas kick rising speed is highly desired.

Many studies have been focused on gas–liquid two-phase flow in wellbore. Some researchers developed model to analyze two-phase flow in annuli during drilling. LeBlanc and Lewis (1968) built a mathematical model to calculate the backpressure during circulating gas kick out of well. In their model, the frictional pressure drop was ignored. Hoberock and Stanbery (1981a, b) combined different models to analyzed pressure distribution in wells during gas kicks assuming constant temperature along the annulus. Santos and Bourgoyne (1989) estimated pressure profile in wellbore for two-phase flow basing on flow regime. Van Slyke and Huang (1990) used a dynamic wellbore model to predict gas kick behavior in oil-based drilling mud. The mass of free gas changes with the temperature and pressure because the solution gas in oil-base mud varies along the wellbore. Johnson and White (1991) conducted experiment to examine gas migration rate in drilling mud in a 49-ft long, 7.8-in ID inclinable flow loop. Skalle et al. (1991) studied gas rising velocity and its effect on bottomhole pressure (BHP) in a vertical well using experiment. Three empirical two-phase flow correlations were used to analyze the experimental data. Frank and Rolv (1991) ran full-scale kick experiments and studied the effect of different parameters on gas-rise velocity. Johnson and Steven (1993) investigated the gas migration velocities during gas kicks in deviated wells using the same facilities used by Johnson and White in 1991. Martins Lage et al. (1994) tested the gas kick migration in closed and open wells. Tarvin et al. (1994) analyzed data from test-well experiment and found that gas rises through drilling mud faster than the migration rates generally accepted in the drilling industry. Ashley et al. (1995) reviewed different gas migration velocity at different gas concentration. Choe (2001) developed a two-phase flow model to calculate pressure in annulus using flow regime. Nunes et al. (2002) used Beggs and Brill method to analyze gas kicks in deepwater well drilling. Yu et al. (2009) developed a mechanistic model for gas–liquid flow in upward vertical annuli. Flow regimes are applied in their model. Chirinos et al. (2011) proposed a simplified method to estimate peak casing pressure during managed pressure drilling well control.

Methods to detect a kick

It is crucial to detect a kick as the early beginning. Early detection can minimize the kick size and reduce the risk of blowout when controlling the well. Kick-detection equipment should be installed. The followings are important kick indications:

1. An abrupt increase in penetration rate or drilling break
2. An increase in pump rate and a decrease in pump pressure
3. An increase in the mud return flow rate
4. Pit gains due to the increase in the mud return flow rate
5. An increase in drillstring weight
6. Gas cutting or salinity changes in the drilling fluid
7. Mud flows when pumps are off.

Well control procedures to circulate out gas kick and kill the well

When there is a kick, two methods are usually applied to circulate the kick out of the wellbore and keep the well under control. They are driller's method and wait and weight method. A thoroughly understanding of procedures of these two well control methods helps the development of governing equation for gas kick rising speed calculation. The basic principle of both methods is to keep BHP constant at the formation pressure. The driller's method differs from wait and weight method in the circulation number. Driller's method needs two circulations to circulate the kick out of hole and kill the well. Following steps are used in driller's method:

1. Shut in the well and get casing pressure and drillpipe pressure
2. Calculate the kill mud weight
3. Start up the pump by holding casing pressure constant
4. Pump old mud and circulate the kick out of hole while keeping drillpipe pressure constant
5. After circulating kick out of hole, pump kill mud; start up the pump by holding casing pressure constant
6. Hold casing pressure constant and pump kill mud until kill mud flow to the bit
7. Switch to constant drillpipe pressure and circulate kill mud until it flows out of the choke
8. Shut down pumps by holding casing pressure constant
9. Check casing pressure and drillpipe pressure to make sure both pressures are zero psi
10. If both pressures are zero psi, complete well control.

The procedure of wait and weight method is as follows:

1. Shut in the well and get casing pressure and drillpipe pressure
2. Calculate the kill mud weight and mix the kill mud
3. Start up the pump by holding casing pressure constant
4. Pump kill mud and circulate the kick out of hole while keeping BHP constant, manipulate the choke to make sure drillpipe pressure, and follow the pressure reduction schedule
5. Circulate kill mud until it flows out of the choke
6. Reduce pump speed while closing the choke
7. Shut down the pump
8. Check casing pressure and drillpipe pressure to make sure both pressures are zero psi
9. If both pressures are zero psi, complete well control.

Gas kick rising speed in well control

According to the procedure of driller's method, the old mud is pumped into the drillpipe to circulate the kick out of the hole, which occurs in the first circulation. Therefore, the mud in drillpipe and annulus has same properties. To analyze the gas kick rising speed during the circulation, following assumptions are made:

1. A volume of gas kick, $V_{g,\mathrm{BH}}$, entered into the bottomhole when the well is shut in
2. The compressibility of mud is neglected comparing with gas compressibility
3. Gas kick rises from the bottomhole to surface as a single column
4. There are two mud annuli between gas column and walls of wellbore and drillpipe due to the wettability effect. The thicknesses of these two annuli are very small comparing with the radius of the gas column
5. The temperature of gas column follows the temperature gradient in mud
6. Water base mud with negligible gas solubility
7. Drilling mud follows power-law model.

For a volume of gas kick, $V_{g,\mathrm{BH}}$, enters into the bottomhole, the volume of the gas kick equals the difference between the mud flow out of hole and into the hole.

$$V_{\text{gas kick,BH}} = V_{M,\text{out}} - V_{M,\text{in}} \tag{1}$$

where $V_{\text{gas kick,BH}}$ = volume of gas kick at bottomhole, $V_{M,\text{out}}$ = volume of mud flow out the hole, $V_{M,\text{out}}$ = volume of mud flow out the hole, $V_{M,\text{in}}$ = volume of mud flow into the hole.

Since the time period between well shut in and starting pumping old mud to circulate the kick is very short, we can assume the migration of gas begin as the pump is started up. Now we analyze the rising speed of gas column at the beginning of first circulation. As old mud is pumped into the drillpipe, the gas kick migrates upward inside the annulus like a piston as shown in Fig. 1. At the time the gas kick begins to move upward, the height of gas kick is the sum of heights of cone part and cylinder part, which is expressed as

$$L_{\text{gas kick}} = L_{\text{gas cone}} + L_{\text{gas cylinder}} \tag{2}$$

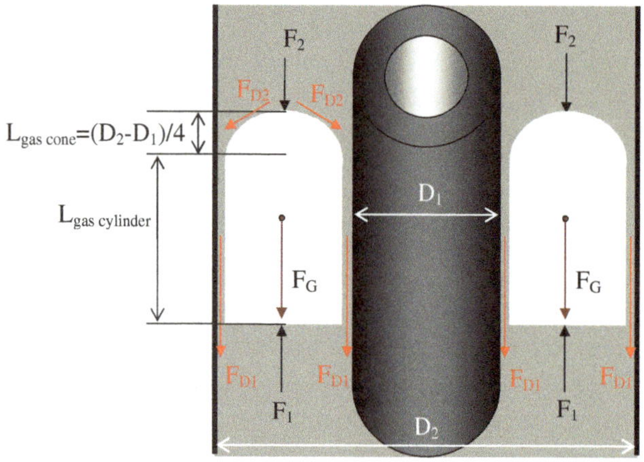

Fig. 1 Distribution of gas kick in the annuli

where $L_{\text{gas kick}}$ = height of gas kick, $L_{\text{gas cone}}$ = height of cone part of gas kick, $L_{\text{gas cylinder}}$ = height of cylinder part of gas kick.

Due to the effect of gas–mud interfacial tension, the height of cone should equal $\frac{D_2-D_1}{4}$assuming the annuli between gas column and walls of wellbore and drillpipe are small and can be neglected. Therefore, the volume of gas kick can be expressed as

$$\begin{aligned} V_{\text{gas kick}} &= V_{\text{gas cone}} + V_{\text{gas cylinder}} \\ &= 2\pi\left(\frac{D_1}{2}+\frac{D_2-D_1}{4}\right)\frac{\pi}{2}\frac{(D_2-D_1)^2}{16} \\ &\quad + \pi\frac{D_2^2-D_1^2}{4}L_{\text{gas cylinder}} \\ &= \pi^2\left(\frac{D_2^2-D_1^2}{4}\right)\left(\frac{D_2+D_1}{16}\right)+\pi\frac{D_2^2-D_1^2}{4}L_{\text{gas cylinder}} \end{aligned} \tag{3}$$

where $V_{\text{gas kick}}$ = volume of gas kick at any location, $V_{\text{gas cone}}$ = volume of cone part of gas kick, $V_{\text{gas cylinder}}$ = volume of cylinder part of gas kick, D_1 = drillpipe diameter, D_2 = wellbore diameter.

Therefore, the height of gas cylinder can be estimated from gas kick volume, which is

$$\begin{aligned} L_{\text{gas cylinder}} &= \frac{V_{\text{gas kick}} - \pi^2\left(\frac{D_2^2-D_1^2}{4}\right)\left(\frac{D_2+D_1}{16}\right)}{\pi\frac{D_2^2-D_1^2}{4}} \\ &= \frac{V_{\text{gas kick}}}{\pi\frac{D_2^2-D_1^2}{4}} - \pi\left(\frac{D_2+D_1}{16}\right) \end{aligned} \tag{4}$$

As the gas column moves upward, the gas expands. According to Eq. (4), the shape of gas cone remains constant while the height of gas cylinder becomes longer due to gas expansion.

To estimate the rising velocity of gas kick, force analysis is required. Forces on gas column can be analyzed in two dimensions, horizontal and vertical directions. For the purpose of this study, horizontal forces are not considered. According to the U-tube theory, the pressure inside the drillpipe should be balanced by pressure in the annulus. When the old mud is pumped into the drillpipe, the gas column will move upward along the annulus. The gas column is subjected to five forces, the gravitational force, the drag force on the cone surface of gas column, the drag force on the flank of the gas column, the fluid forces in front of and behind the gas column. Because the forces in the vertical direction control the upward movement of gas kick, they are analyzed here. The net force in vertical direction is calculated by

$$F_{\text{net}} = F_1 - F_2 - F_{D1} - F_{D2\text{-Vert}} - F_G$$

or

$$F_{\text{net}} = F_1 - F_2 - F_{D1} - \int_{0^\circ}^{90^\circ} F_{D2}\cos\theta\,\mathrm{d}\theta - F_G \tag{5}$$

where F_{net} = net force, F_G = gravitational force, F_{D2} = drag force on the cone surface of gas column, $F_{D2\text{-Vert}}$ = vertical drag force resulting from drag force on the cone surface of gas column, F_{D1} = drag force on the flank of the gas column, F_2 = fluid force in front of the gas column, F_1 = fluid force behind the gas column, θ = angle between F_{D2} and vertical direction.

The fluid force behind the gas column is

$$F_1 = p_1 A_{\text{gas column base}} = \pi\frac{D_2^2-D_1^2}{4}p_1 \tag{6}$$

where p_1 = pressure below the gas column, $A_{\text{gas column base}}$ = area of base of gas column.

According to the basic principle in the driller's and engineer's methods, BHP is kept constant during the circulation, which means BHP always equals pore pressure. Therefore, the pressure below the gas column can be expressed in terms of pore pressure, frictional pressure drop, and pressure change due to potential energy change, which is

$$p_1 = p_p - \rho_m g L_{\text{bottomhole-gas base}} - p_f \tag{7}$$

where p_p = pore pressure, ρ_m = mud density, g = gravitational acceleration, $L_{\text{bottomhole-gas base}}$ = distance between bottomhole and base of gas column, p_f = frictional pressure drop between bottomhole and base of gas column.

Pore pressure can be estimated from the shut-in drillpipe pressure and static hydraulic pressure due to drill mud, which is

$$p_p = p_{\text{SIDP}} + \rho_m g h_{\text{well}} \tag{8}$$

where p_{SIDP} = shut-in drillpipe pressure, h_{well} = well depth.

The frictional pressure drop depends on the flow regime. When the Reynolds number is less than 2,100, the flow is laminar; otherwise, it is turbulent flow. The Reynolds number is calculated through (Bourgoyne et al. 1986)

$$N_{\mathrm{Re}} = \frac{109000 \rho_m u_m^{2-n}}{K} \left[\frac{0.0208\,(D_2 - D_1)}{2 + 1/n} \right]^n \tag{9}$$

where

$$K = \frac{510 \theta_{300}}{511^n} \tag{10}$$

$$n = 3.32 \log \frac{\theta_{600}}{\theta_{300}} \tag{11}$$

where K = consistency index of mud, n = flow-behavior index of mud, u_m = mud velocity, θ_{300} = the 300-rpm dial reading in mud viscometer, θ_{600} = the 600-rpm dial reading in mud viscometer.

If the flow is laminar flow, the frictional pressure drop can be calculated by (Bourgoyne et al. 1986)

$$p_f = \frac{\mathrm{d}p_f}{\mathrm{d}h} L_{\text{bottomhole–gas base}} = \frac{K u_m^n \left(\frac{2+1/n}{0.0208} \right)^n}{144000 (D_2 - D_1)^{1+n}} L_{\text{bottomhole–gas base}} \tag{12}$$

If the flow is turbulent flow, the frictional pressure drop can be calculated by

$$p_f = \frac{\mathrm{d}p_f}{\mathrm{d}h} L_{\text{bottomhole–gas base}} = \frac{f \rho_m u_m^2}{21.1 (D_2 - D_1)} L_{\text{bottomhole–gas base}} \tag{13}$$

where f = friction factor.

The friction factor can be read from Fig. 2.

The fluid force in front of the gas column is

$$F_2 = p_2 A_{\text{cone base}} = p_2 A_{\text{gas column base}} = \pi \frac{D_2^2 - D_1^2}{4} p_2 \tag{14}$$

where p_2 = pressure in front of the gas column, $A_{\text{cone base}}$ = area of base of cone of gas column, which equals area of base of gas column.

The pressure in front of the gas column can be expressed in terms of casing pressure, frictional pressure drop, and pressure change due to potential energy change, which is

$$p_2 = p_{\text{casing}} + \rho_m g L_{\text{wellhead–gas top}} - p_f \tag{15}$$

where p_{casing} = casing pressure, $L_{\text{wellhead–gas top}}$ = distance between wellhead and top of gas column.

Casing pressure is readily available during circulation. The calculation of frictional pressure drop is akin to the frictional pressure drop between bottomhole and base of gas column. In case of laminar flow, the frictional pressure drop is

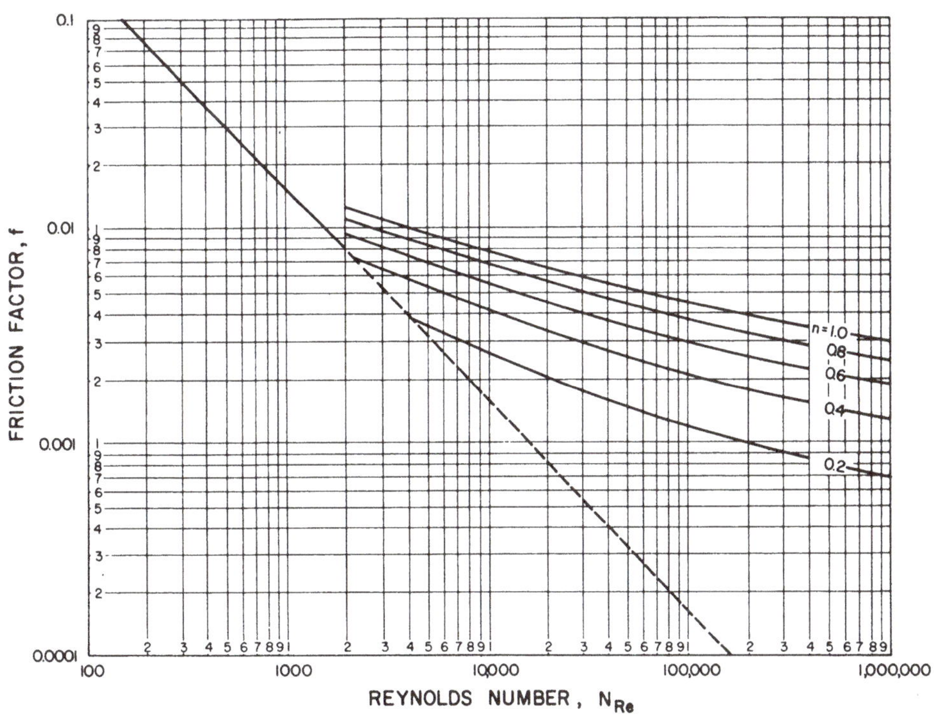

Fig. 2 Friction factor for power-law fluid model, after Bourgoyne et al. (1986)

$$p_f = \frac{dp_f}{dh} L_{\text{wellhead-gas top}} = \frac{K u_m^n \left(\frac{2+1/n}{0.0208}\right)^n}{144000 (D_2 - D_1)^{1+n}} L_{\text{wellhead-gas top}} \quad (16)$$

If the flow is turbulent flow, the frictional pressure drop can be calculated by

$$p_f = \frac{dp_f}{dh} L_{\text{wellhead-gas top}} = \frac{f \rho_m u_m^2}{21.1 (D_2 - D_1)} L_{\text{wellhead-gas top}} \quad (17)$$

Again, Reynolds number is calculated from Eq. (9) and friction factor is read from Fig. 2.

The drag force on the flank of the gas column can be derived according to the flow regime: laminar and turbulent flow conditions. When gas column moves upward along the annulus, the drag force on the flank of the gas column is

$$F_{D1} = A_{\text{gas column flank}} \tau_{\text{gm}} = \pi (D_1 + D_2) L_{\text{gas cylinder}} \tau_{\text{gm}} \quad (18)$$

where $A_{\text{gas column flank}}$ = surface area of flank of gas column, τ_{gm} = shear stress between gas and mud.

Shear stress between gas and mud is calculated through (Taitel 1995)

$$\tau_{\text{gm}} = \frac{f_i \rho_g u_g^2}{2} \quad (19)$$

where f_i = gas–mud interfacial friction factor, u_g = gas velocity, ρ_g = gas density.

Gas–mud interfacial friction factor is calculated by

$$f_i = C N_{\text{Re}}^{-n} \quad (20)$$

where $C = 16$ and $n = 1.0$ for laminar flow, and $C = 0.046$ and $n = 0.2$ for turbulent flow. If f_i from Eq. (20) is larger than 0.014, $f_i = 0.014$ should be used.

Reynolds number is calculated by

$$N_{\text{Re}} = \frac{\rho_g u_g (D_2 - D_1)}{\mu_g} \quad (21)$$

where μ_g = gas viscosity.

The drag force on the cone surface, F_{D2}, can be estimated by Ling's (2010) method. In Fig. 1, the cone of gas column experiences a drag force resulting from the viscous mud flow around the cone surface. The magnitude of drag force depends on the flow regime, laminar, or turbulent flow. For laminar flow, the drag force is calculated from Stokes law. Stokes law has shown that for creeping flow (Castleman 1926), the drag force is related to the gas cone velocity through the fluid by:

$$F_{D2} = 3\pi \left(\frac{D_2 - D_1}{2}\right) \frac{\pi \left(\frac{D_2 - D_1}{4}\right) 2\pi \left(\frac{D_1}{2} + \frac{D_2 - D_1}{4}\right)}{4\pi \left(\frac{D_2 - D_1}{4}\right)^2} \mu_g u_g = \frac{3}{4} \pi^2 (D_2 + D_1) \mu_g u_g. \quad (22)$$

Decomposing the drag force on the cone surface, we obtain vertical direction force, which is

$$F_{D2\text{-Vert}} = \int_{0^\circ}^{90^\circ} F_{D2} \cos\theta \, d\theta = F_{D2} [\sin(90^\circ) - \sin(0^\circ)] = F_{D2} = \frac{3}{4} \pi^2 (D_2 + D_1) \mu_g u_g. \quad (23)$$

Equation (23) is found to give acceptable accuracy for Reynolds numbers below 0.1. For Reynolds numbers greater than 0.1, the drag force needs to be estimated using friction factor. The friction factor is defined by:

$$f = \frac{F_{D2}}{A_{\text{cone surface}} E_k} \quad (24)$$

where $A_{\text{cone surface}}$ = characteristic area of the cone of gas column, E_k = kinetic energy per unit volume.

Then, the drag force can be expressed as:

$$F_{D2} = f A_{\text{cone surface}} E_k \quad (25)$$

The characteristic area of the cone of gas column is given by:

$$A_{\text{cone surface}} = \frac{1}{4} \pi \left(\frac{D_2 - D_1}{4}\right) 2\pi \left(\frac{D_2 + D_1}{4}\right) = \frac{\pi^2}{32} \left(D_2^2 - D_1^2\right) \quad (26)$$

The kinetic energy per unit volume is given by:

$$E_k = \frac{\rho_g u_g^2}{2} \quad (27)$$

The friction factor f can be calculated by Eq. (20). If f_i from Eq. (20) is larger than 0.014, $f_i = 0.014$ should be used.

The gravity force resulting from the gas column is expressed as

$$F_G = \rho_{g,\text{BH}} g V_{\text{gas kick,BH}} \quad (28)$$

where

$$\rho_{g,\text{BH}} = \frac{p_p M}{z R T_{\text{BH}}} \quad (29)$$

M = molecular weight of gas, R = universal gas constant, z = gas deviation factor, T_{BH} = bottomhole temperature, $\rho_{g,\text{BH}}$ = gas density at bottomhole.

The molecular weight can be calculated from gas-specific gravity. Gas-specific gravity can be calculated from shut-in drillpipe pressure, shut-in casing pressure, mud density, and pit gain, or from offset well gas property. Bottomhole temperature and temperature at any depth can be estimated using regional temperature gradient. Gravity force of gas column is constant during the gas migration.

Substituting Eqs. (6), (14), (18), (23), and (28) into (5), we have

$$F_{\text{net}} = \pi\frac{D_2^2 - D_1^2}{4}p_1 - \pi\frac{D_2^2 - D_1^2}{4}p_2 - \pi(D_1 + D_2)L_{\text{gas cylinder}}\tau_{\text{gm}} - \frac{3}{4}\pi^2(D_2 + D_1)\mu_g u_g - \rho_{g,\text{BH}}gV_{\text{gas kick,BH}} = \pi\left(\frac{D_2^2 - D_1^2}{4}\right)(p_1 - p_2) - \pi(D_2 + D_1)u_g\left[L_{\text{gas cylinder}}\frac{8}{N_{\text{Re}}}\rho_g u_g + \frac{3}{4}\pi\mu_g\right] - \rho_{g,\text{BH}}gV_{\text{gas kick,BH}} \tag{30}$$

for laminar flow.

Substituting Eqs. (6), (14), (18), (25), and (28) into (5), we have

$$F_{\text{net}} = \pi\frac{D_2^2 - D_1^2}{4}p_1 - \pi\frac{D_2^2 - D_1^2}{4}p_2 - \pi(D_1 + D_2)L_{\text{gas cylinder}}\tau_{\text{gm}} - fA_{\text{cone surface}}E_k - \rho_{g,\text{BH}}gV_{\text{gas kick,BH}} = \pi\left(\frac{D_2^2 - D_1^2}{4}\right)(p_1 - p_2) - \pi(D_2 + D_1)L_{\text{gas cylinder}}\frac{\frac{0.046}{N_{\text{re}}^{0.2}}\rho_g u_g^2}{2} - f\frac{\pi^2}{32}(D_2^2 - D_1^2)\frac{\rho_m u_g^2}{2} - \rho_{g,\text{BH}}gV_{\text{gas kick,BH}} \tag{31}$$

for turbulent flow.

With the calculated net force, the acceleration of gas column can be calculated through

$$a = \frac{F_{\text{net}}}{m_g} = \frac{F_{\text{net}}}{\rho_{g,\text{BH}}V_{\text{gas kick,BH}}} \tag{32}$$

where a = acceleration, m_g = mass of gas column.

Therefore, substituting Eqs. (30) and (31) into (32) gives the governing equations for accelerations of gas column under laminar and turbulent flows, respectively.

$$a = \frac{\pi\left(\frac{D_2^2 - D_1^2}{4}\right)(p_1 - p_2) - \pi(D_2 + D_1)u_g\left[L_{\text{gas cylinder}}\frac{8}{N_{\text{Re}}}\rho_g u_g + \frac{3}{4}\pi\mu_g\right]}{\rho_{g,\text{BH}}V_{\text{gas kick,BH}}} - g \tag{33}$$

and

$$a = \frac{\pi\left(\frac{D_2^2 - D_1^2}{4}\right)(p_1 - p_2) - \pi(D_2 + D_1)L_{\text{gas cylinder}}\frac{\frac{0.046}{N_{\text{re}}^{0.2}}\rho_g u_g^2}{2} - f\frac{\pi^2}{32}(D_2^2 - D_1^2)\frac{\rho_m u_g^2}{2}}{\rho_{g,\text{BH}}V_{\text{gas kick,BH}}} - g \tag{34}$$

Calculation procedure

The calculation of gas column migrating up the annuli can be broken into the following steps:

1) Calculate the acceleration of gas column when it starts to migrate; at this moment, there is no drag force, so Eqs. (33) and (34) reduce to

$$a = \frac{\pi\left(\frac{D_2^2 - D_1^2}{4}\right)(p_1 - p_2)}{\rho_{g,\text{BH}}V_{\text{gas kick,BH}}} - g \tag{35}$$

2) Select a small 1st time step, Δt_1, and calculate the velocity of gas column at the end of 1st time step by

$$u_g = a\Delta t_1 \tag{36}$$

3) Calculate the migrated distance, L_{migrated}, by

$$L_{\text{migrated}} = \frac{1}{2}a\Delta t_1^2 \tag{37}$$

4) Calculate the location of base of gas column, $h_{\text{gas base}}$, by

$$h_{\text{gas base}} = h_{\text{well}} - L_{\text{migrated}} \tag{38}$$

Table 1 Input data for gas kick migration calculation

Molecular weight of gas	20	lb/lb-mole
Gas specific gravity	0.69	air = 1.0
Well depth at kick occur	9,975	ft
Wellbore diameter	8.5	in.
Drillpipe length	9,445	ft
Drillpipe diameter (OD)	4.5	in.
Drill collar length	530	ft
Drill collar diameter (OD)	6.25	in.
Shut-in drillpipe pressure	275	psi
Shut-in casing pressure	448	psi
Bottomhole temperature	180	°F
Temperature gradient	1.1	°F/100 ft
Pit gains or kick volume at bottomhole	13.6	bbl
Mud density	11.7	lbm/gal
The 300-rpm dial reading in mud viscometer	29	lbf/100 ft^2
The 600-rpm dial reading in mud viscometer	46	lbf/100 ft^2
Consistency index of mud	234	cp
Flow-behavior index of mud	0.67	
Mud pressure gradient	0.608	psi/ft
Mass of kick	1,304	lbm
Gravitational acceleration	31.174	ft/s^2
Bottomhole pressure	6,338	psi
Kick density at bottomhole	17.08	lbm/ft^3

5) Assuming a new gas volume, which is larger than gas volume at bottomhole; calculate the height of gas column, $h_{\text{gas kick}}$ through Eq. (3)
6) Calculate the location of center of gas column $h_{\text{gas center}}$, by

$$h_{\text{gas center}} = h_{\text{gas base}} - \frac{L_{\text{gas kick}}}{2} \quad (39)$$

7) Calculate the average pressure of gas column, $\bar{p}_{\text{gas}}$
8) Using real gas law, calculate the volume of gas kick at new location, which is

$$V_{\text{gas kick}} = \frac{z_{\text{BH}} T_{\text{BH}} \bar{p}_{\text{gas}}}{z T p_p V_{\text{gas kick,BH}}} \quad (40)$$

where T = temperature at location, z_{BH} = gas deviation factor at bottomhole, Δt_1 = 1st time step, L_{migrated} = gas migrated distance, $h_{\text{gas base}}$ = location of base of gas column, $h_{\text{gas center}}$ = location of center of gas column, $\bar{p}_{\text{gas}}$ = average pressure of gas column.

9) If calculated gas volume is different from assumed volume in Step 5, repeat Steps 5 through 8 until a converged volume is obtained
10) With the gas column location after 1st time step, we can calculate the acceleration of gas column using Eq. (33) or (34)
11) Select 2nd time step and calculate the velocity and migrated distance at the end of 2nd time step
12) Repeat Steps 4 through 11 until base of gas column migrates to the surface

Case study to illustrate the validation and application of model

Field data from a gas kick detection and control in a well in Southeast Asia were used to verify the model. A kick was detected when the well was drilled to a depth of 9,975 ft. The well was shut in; the influx was contained and further entry of formation fluid was prevented. The pit gains were 13.6 bbl when the well was shut in. Shut-in casing and drillpipe pressures were recorded. The driller's method was used to circulate the kick out the hole and control the well. Table 1 shows the key parameters used in the calculations.

The calculated time for gas migrating to wellhead is 123.1 min. Field observed that it takes 129.3 min for gas migration. The absolute error is −6.2 min and the relative error is −4.8 %. The errors can be results of irregular borehole, inaccurate temperature profile in wellbore, inaccurate measurement of pit gains, variation of mud properties along the wellbore, inaccurate kick properties, and any deviation from the aforementioned assumptions. Therefore, the model gives reasonable results.

A computer program is coded to calculate the rising of gas kick. Thus, the calculation can be done within acceptable time period after the well is shut in. Then, the calculation can provide a forecast of gas kick location versus circulating time for circulating kick out of hole operation. The real-time data during well control can be compared with the predicted values. Any deviation from forecast could be an indication of downhole problem. The proposed method can be combined with engineers experience to predict the downhole situation, shut-in casing pressure, and mud rate as functions of position of gas kick. Therefore, the new model is a valuable tool in well control.

Conclusions

Following conclusions can be drawn from this study:

The governing equation to estimate the gas kick migration velocity in the annuli has been developed.

The procedure to calculate the migration of gas kick from bottomhole to surface has been proposed.

Differences between forecast values and real-time data in well control could be signs of downhole problems.

Acknowledgments This paper was firstly presented at the SPE Middle East Unconventional Gas Conference & Exhibition held in Muscat, Sultanate of Oman on January 28–30, 2013. The authors thank Society of Petroleum Engineers (SPE) in allowing us to publish this paper with Journal of Petroleum Exploration and Production Technology. The authors are grateful to The Petroleum Engineering Department in University of North Dakota. This research is supported in part by the U.S. Department of Energy (DOE) under award number DE-FC26-08NT0005643.

References

Ashley J, Ian RC, Tim B, Dominic M (1995) Gas migration: fast, slow or stopped, SPE 29342, SPE/IADC drilling conference, Amsterdam, Netherlands, 28 Feb–2 March 1995

Bourgoyne TA, Chenevert EM, Millhein KK, Young SF (1986) Applied drilling engineering. Textbook Series, SPE, Richardson

Castleman RA (1926) The resistance to the steady motion of small spheres in fluids. In: National advisory committee for aeronautics, Technical note no. 231, Washington, DC, February

Chirinos JE, Smith JR, Bourgoyne D (2011) A simplified method to estimate peak casing pressure during mpd well control, SPE

147496, SPE annual technical conference and exhibition, Denver, Colorado, USA, 30 Oct–2 Nov 2011

Choe J (2001) Advanced two-phase well control analysis. J Canad Petrol Technol 40(11):39–47

Frank H, Rolv R (1991) Analysis of gas-rise velocities from full-scale kick experiments, SPE 24580, SPE annual technical conference and exhibition. Washington, DC, 4–7 Oct 1992

Hoberock LL, Stanbery SR (1981a) Pressure dynamics in wells during gas kicks: part 1—component models and results. J Petrol Technol 33(8):1357–1366

Hoberock LL, Stanbery SR (1981b) Pressure dynamics in wells during gas kicks: part 2—component models and results. J Petrol Technol 33(8):1367–1378

Johnson AB, Steven C (1993) Gas migration velocities during gas kicks in deviated wells, SPE 26331, SPE annual technical conference and exhibition, Houston, Texas, 3–6 Oct 1993

Johnson AB, White DB (1991) Gas-rise velocities during kicks. SPE Drill Eng 6(4):257–263

LeBlanc JL, Lewis RL (1968) A mathematical model of a gas kick. J Petrol Technol 20(8):888–898

Ling K (2010) Gas viscosity at high pressures and high temperatures, Ph.D. dissertation, Texas A&M University, College Station

Martins Lage ACV, Nakagawa EY, Cordovil AGDP (1994) Experimental tests for gas kick migration analysis, SPE 26953, SPE Latin America/Caribbean petroleum engineering conference, Buenos Aires, Argentina, 27–29 April 1994

Nunes JOL, Bannwart AC, Ribeiro PR (2002) Mathematical modeling of gas kicks in deep water scenario, SPE 77253, IADC/SPE Asia Pacific Drilling Technology, Jakarta, Indonesia, 8–11 Sept 2002

Santos OL, Bourgoyne Jr AT (1989) Estimation of pressure peaks occurring when diverting shallow gas, SPE 19559, SPE annual technical conference and exhibition, San Antonio, Texas, 8–11 Oct 1989

Skalle P, Podio AL, Tronvoll J (1991) Experimental study of gas rise velocity and its effect on bottomhole pressure in a vertical well, SPE 23160, Offshore Europe, Aberdeen, United Kingdom, 3–6 Sept 1991

Taitel Y, Barnea D, Brill JP (1995) Stratified three phase flow in pipes. Int J Multi Flow 21(1):53–60

Tarvin JA, Hamilton AP, Gaynord PJ, Lindsay GD (1994) Gas rises rapidly through drilling mud, SPE 27499, SPE/IADC drilling conference, Dallas, Texas, 15–18 Feb 1994

Van Slyke DC, Huang ETS (1990) Predicting gas kick behavior in oil-based drilling fluids using a pc-based dynamic wellbore model, SPE 19972, SPE/IADC drilling conference, Houston, Texas, 27 Feb–2 March 1990

Yu TT, Zhang HQ, Li MX, Sarica C (2009) A mechanistic model for gas/liquid flow in upward vertical annuli, SPE 124181, SPE annual technical conference and exhibition, New Orleans, Louisiana, 4–7 Oct 2009

11

Dynamic asphaltene deposition control in pipe flow through the application of DC potential

Muhammad Ihtsham Hashmi · B. Ghosh

Abstract Asphaltene–resin moieties in crude oil are found to carry residual surface electric charge, which is characteristics to their colloidal structure, asphaltene–resin ratio and system pH. This research investigated the possibilities of controlling asphaltene deposition in oil well by applying static electrical potential taking advantage of their residual surface charge. Laboratory experiments were conducted at static and also in dynamic condition constructing a dual flow loop set-up, equipped with precision pumping, pressure recording and regulated DC power supply. Neat and heptane-diluted crude oil having inherent asphaltene deposition tendencies is used to investigate the influence of DC electrical potential on asphaltene deposition tendencies. Real time deposition trend is interpreted through differential pressure build-up across the flow loops and also through quantitative estimation of deposited mass. The results were encouraging, showing up to 180 % reduction in asphaltene deposition in the cathode loop and about 140 % increase on the anode loop at an optimised potential of 60 V DC. Further, it was observed that higher the *n*-heptane dilution, higher is the effect of static potential in terms of arresting deposition. Based on these optimistic results, further studies and upscaling are planned (considering the changes of pressure, temperature and fluid composition in an oil well) and looking at the possibilities of controlling asphaltene deposition, by converting the well into a cathode, along with a nearby sacrificing anode well, applying optimum electrical potential.

M. Ihtsham Hashmi · B. Ghosh
Petroleum Engineering Department,
The Petroleum Institute, 2533, Abu Dhabi, UAE

Keywords Asphaltene · Oil · Flow assurance · Electrodeposition · Resin

Introduction

Asphaltenes in petroleum flow streams are often compared with low-density lipoproteins in blood stream, both being responsible for flow restriction, pressure surge and finally complete blockage, resulting system shut down owing to phase separation and deposition on the flow lines. In case of a petroleum production system, this may lead to permeability damage of the reservoir pore matrix, blocking production tubing, surface flow lines and fouling of surface equipment.

Asphaltene, the major constitution of the residue fraction of crude petroleum is commonly known as poly-condensed aromatic molecules of 4–10 rings (Groenzin and Mullins 2000) with multiple alkyl side chains. Hetero-atomic (NSO) functional groups are the source of their polarity and amphoteric characteristics (Goual and Firoozabadi 2002), and the coordination metals (Fe, Ni, V) are confined within porphyrin type moieties (Speight 1998).

In general crude petroleum is a physical mixture of three major groups of compounds; asphaltenes are the largest in molecular size, resins fall in the mid range and the rest are the lighter aromatics and paraffins (maltenes). Asphaltenes and resins are polar, while the maltene compounds are either non-polar or mildly polar. The interaction among these species strongly affects asphaltene precipitation from petroleum fluids (Goual and Firoozabadi 2004).

Due to extremely wide molecular weight distribution and the ability to form molecular aggregates at different intensity at different chemical and physical condition, only average chemical structures and properties of asphaltenes and resins could be established so far. Stability of asphaltene in crude oil depends upon a variety of factors (temperature, pressure, composition, pH, etc.) of which resin content in oil is attributed to have the greatest influence (Andersen and Speight 2001). It is an established fact that petroleum resins with lesser aromaticity but higher polarity compared with asphaltene act as peptizing agents and help stabilizing asphaltene–resin colloids. The mechanism behind this molecular association is believed to be due to opposite surface charge of resins and asphaltene moieties (Taylor 1998). It is observed that destabilization of colloidal asphaltene in oil production systems may occur if the surface charge and the resulting attraction forces between adsorbed resins and asphaltenes are disturbed (Leontaritis 1989; Al-kafeef 2001). Later works of Fotland and Anfindsen (1996) and Behar et al. (1998) on electrical conductivity of asphaltene helped in understanding their molecular association phenomena in solution phase, while works of Igor and Aleksander (2010) on electrical conductivity of solid asphaltenes showed that asphaltenes in solid state behave like semiconductors, conductivity increasing at higher temperature, possibly due to structural phase transition.

Electrokinetic, electrophoretic and adsorption properties of asphaltenes are studied through potentiometric and electrokinetic measurements in great details on various crude oils which firmly established its charge-bearing properties and streaming potential (Kokal et al. 1995; Leon et al. 2000; Gonzalez et al. 2003; Parra-Barraza et al. 2003). Asphaltenes in aqueous solutions are found to possess surface charge as a result of protonation and dissociation reactions of functional groups enabling electrophoretic mobility (Janusz and Jablonski 2000). Solvent-induced polarity is seen to reverse the electrophoretic mobility of destabilized asphaltene particles. Dispersed in polar (water) media, the particles presented a negative electrophoretic mobility, whereas in toluene their mobility was positive. When resins were present in the precipitating medium, co-precipitation of resins occurred, indicating a binding process of the resins on the nascent asphaltene particles (Parra-Barraza et al. 2003).

Due to the dependency of several factors (solvent, temperature and influence of resins), the nature of native surface charge on asphaltene is still not fully established. Whereas some researchers claimed that asphaltene possess a predominantly positive charge in most non-aqueous dispersants (James and Richard 1963; Lichaa and Herrera 1975; Leon et al. 2000; Gonzalez et al. 2003), some reported their net negative surface charges based on electro-deposition experiments with neat crude oil samples (Idem and Ibrahim 2002; Nasser and Belhaj 2010; Belhaj and Khalifeh 2013). Also noteworthy is the work of Leontaritis and Mansoori (1987) who showed that the net negative charge on asphaltene colloidal surface is the prime cause of deposition on metal pipes due to the potential difference across them. This resulted with the hypothesis that, any effort in neutralising this charge may result in controlling asphaltene precipitation and deposition (Yen 1994). Leon et al. (2000) from their electrophoretic mobility studies with 27 crude oils established that the surface charges on asphaltene micelles were predominantly negative in crude oils and at low heptanes dilution but the surface charges were reversed at higher dilution. The charge reversal was attributed to removal of positively charged peptizing resins. Khvostichenko and Andersen (2009) investigated the charge reversal phenomenon under various conditions of dilution and found that application of a DC electric field can arrest asphaltenes aggregation or precipitation when diluted in aromatic solvent (toluene).

Taylor (1998) studied the quantum of electro-deposition of asphaltene and resins as a function of the electric field strength, residence time, dilution ratio and oil type. It was observed that stable asphaltene 'micelles' in crude oils possess little net charge and are apparently not affected by the DC electric field. However, upon partial or total destabilization of micelles with *n*-alkanes, electro-deposition occurred, depending on the crude oil characteristics, alkane used and their relative dilution. At lower dilutions (partial destabilization), the deposited material possesses a small net negative charge; higher dilutions resulted in a positively charged deposit. Although these findings are in contradiction with some of the observations described above, they are consistent with the structural model for asphaltene micelles comprising negatively charged 'peptizing' resins surrounding a positively charged central asphaltene core. The ambiguities existing on the surface charge of asphaltene and resins have been clarified by the work of Hashmi and Firoozabadi (2012) in which they have concluded that asphaltene colloids may possess heterogeneous surface charge. Some particles may carry a small net positive charge that can deposit on cathode, while others may carry a small net negative charge that can deposit on anode surface. Metallic components, pi-electron clouds and acid–base functional groups on the periphery contribute to the surface charge which may vary from crude to crude depending on source, maturation and migration history of the oil.

From the above discussion, it is evident that the subject needs extensive and systematic research to understand the mechanisms that generate the asphaltene surface charge and the physical conditions which may have impact on them. In this study, a new approach is adopted in which the rate of asphaltene deposition is measured in static and dynamic (flowing) condition under variable DC electric potential, using dead crude oil samples from selected Abu Dhabi reservoirs where frequent asphaltene deposition in the downhole production tubing and surface flow lines is a potential flow assurance problem. Oil samples are diluted with *n*-heptane at different ratios to mimic oil swelling at different production nodes. The work is focused on understanding the behaviour of surface charge potential of stabilised and destabilised (due to dilution) asphaltene as well as investigating the possibilities of controlling the rate of asphaltene deposition in flow lines by the application of DC electric potential. To achieve these objectives, a dynamic dual loop flow system was constructed, equipped with precision syringe pump, differential pressure transducers with data acquisition system and a DC power supply system with voltage regulator. Oil samples were flown for a long time till equilibrium is reached between deposition rate inside the tube surface and dislodgement due to pressure surge and corresponding differential pressure across the individual loops remained nearly constant. Further experiments were conducted to understand the effect of solvent stripping mechanism of asphaltene–resin moieties and the shielding effect of resin, using oil samples diluted with various ratios of *n*-heptane.

Materials and methods

Crude oil

Crude oil sample form Arab-X field was collected in sealed container under nitrogen environment, vacuum degassed, dewatered through centrifugation and finally filtered through 0.2-μm membrane filter (Millipore) to remove suspended materials if any. Three oil samples were prepared mixing crude oil and *n*-heptane (Sigma–Aldrich) in 100:0, 80:20 and 50:50 ratios to represent oil in its original state, maximum saturate expansion at bubble point and a severe dilution scenario under natural gas injection. These samples will be identified as neat oil, 80:20 oil and 50:50 oil. All experiments were conducted at 20 ± 1 °C.

The asphaltene content of the crude oil was measured by mixing the petroleum fluid with heptane at 40 mL/g. The mixtures are sonicated for 1 min and filtered through 0.2-μm membrane filter. The filtrate is collected, dried and weighed to determine the fraction of asphaltenes in the oil. SARA (saturates, aromatics, resins and asphaltenes) analysis of the crude oil is conducted using standard method (IP-469). Anton Paar Elementar is used for elemental analysis of the crude oil. Detail properties of the crude oil are given in Table 1.

Static or base experiments

In order to establish that the asphaltene moieties in the sample crude oil bear excess surface charge and electrical attractive forces can have impact on their electrophoresis and deposition behaviour, a series of static electro-deposition experiments were conducted on a specially designed double half-cell apparatus (Fig. 1). Oil was placed in two separate temperature controlled plastic cells of 250 mL capacity, each having a graphite electrode (12 cm in length and 0.8 cm in diameter), one connected to the positive terminal and the other to the negative terminal of a DC power source (maximum capacity of 300 V). Static potentials of 20, 50, 100, 150 and 200 V were applied across these electrodes. Asphaltene deposition was studied up to 30 h. Finally, the electrodes were removed from the cells and inspected visually for asphaltene deposition, and quantitative measurements were conducted following ASTM D2007-80 method, i.e. dissolving in toluene and re-precipitating with excess *n*-heptane. In addition to understanding the nature of asphaltene and resin characteristics, these experiments helped in optimising the voltage and deposition time for next phase of experiments.

Dynamic flow studies

Schematic diagram of the experimental set-up used for the dynamic flow studies is given in Fig. 2. The main section of the set-up consists of two spiral loops (designated as test loop A and B, made of stainless steel tubes (length 10 ft, OD $\frac{1}{4}$ and ID $\frac{1}{8}$ in.). Both the loops were fitted with highly sensitive pressure transducers along with data acquisition system. One of the loops is connected to the negative terminal and the other to the positive terminal of a table top DC power source. Two floating piston cylinders A and B, each having a capacity of 2,000 mL were used to store and flow crude oil through the test loops. While one cylinder is put in pumping mode, the other cylinder is in receiving mode that way nearly continuous flow could be maintained. A dual (continuous mode) ISCO 260D positive displacement pump was used to inject the test fluid into the test loops through the piston cylinders. The set-up used several valves, so that continuous pumping of oil through the test loops could be possible with minimum flow interruption. Stainless steel tubes were substituted by high-pressure

Table 1 Properties of the test crude oil

Specific gravity at 60 °F	Kinematic viscosity at 70 °F (Cst)	Total acid number (mg KOH/gm)	Sulphur content (wt%)	N_2 content (wt%)	O_2 content (wt%)	Saturates (wt%)	Aromatics (wt%)	Resins (wt%)	Asphaltenes (wt%)
0.826	4.32	0.021	1.32	0.03	4.28	70.56	25.08	3.93	0.42

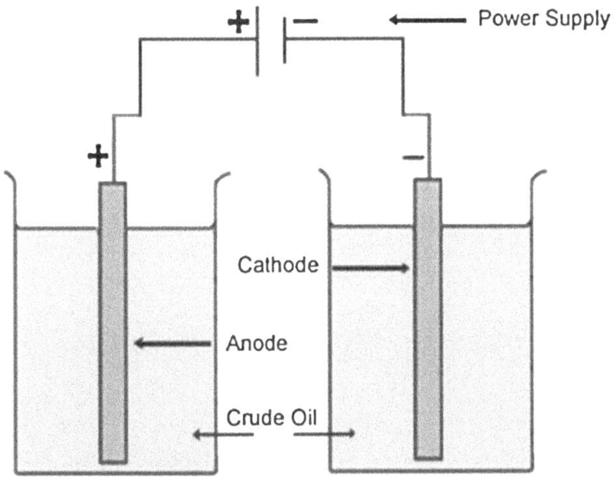

Fig. 1 Static experiment for the characterisation of charge-bearing properties of asphaltene

plastic tubes at different locations to make sure that the loops and pumping sections are electrically isolated.

Using this flow set-up, a series of dynamic electro-deposition tests were conducted using neat crude oil as well as crude oil diluted with *n*-heptane at 80:20 and 50:50 ratios. For each run, 30 L of test fluid was flown through the loops at a constant flow rate of 5 mL/min (laminar flow regime). In addition to blank runs, tests were conducted applying DC potential until the optimum voltage requirement is reached. All tests were conducted in duplicate interchanging the polarity of the loops and average readings were taken. This measure nullified error due to positional advantage of the loops. During the flow period, change of differential pressure across the loops was measured and plotted. On completion of the flow period, the coils were cleaned thoroughly with hot toluene, and total asphaltene content is measured following standard methodology (ASTM D2007-80). Three sets of tests were conducted as follows: (A) with neat oil, (B) crude oil partially destabilised by 25 % dilution with heptane and (C) crude oil fully destabilised with 100 % *n*-heptane.

Results and discussion

Elemental analysis

Analytical results given in Table 1 show that the sulphur and oxygen contents of the sample crude oil are much higher compared with nitrogen. It is a common knowledge that NSO functional groups are generally concentrated in asphaltene and resin fraction of the crude oil. In asphaltenes, they are found mostly as polar functional groups on the molecule itself, inducing polarity. The resin fraction being a complex mixture, they may, however, be present as long-chain carboxylic or sulphur-containing acids and nitrogen-containing basic compounds among numerous others. From the elemental analysis, it is apparent that the presence of N-containing basic functional groups (which give rise to positive surface charge) is expected to be significantly less than the S- and O-containing acidic functional groups (which contribute to negative surface charge).

Fig. 2 Schematic of experimental set-up for asphaltene deposition study

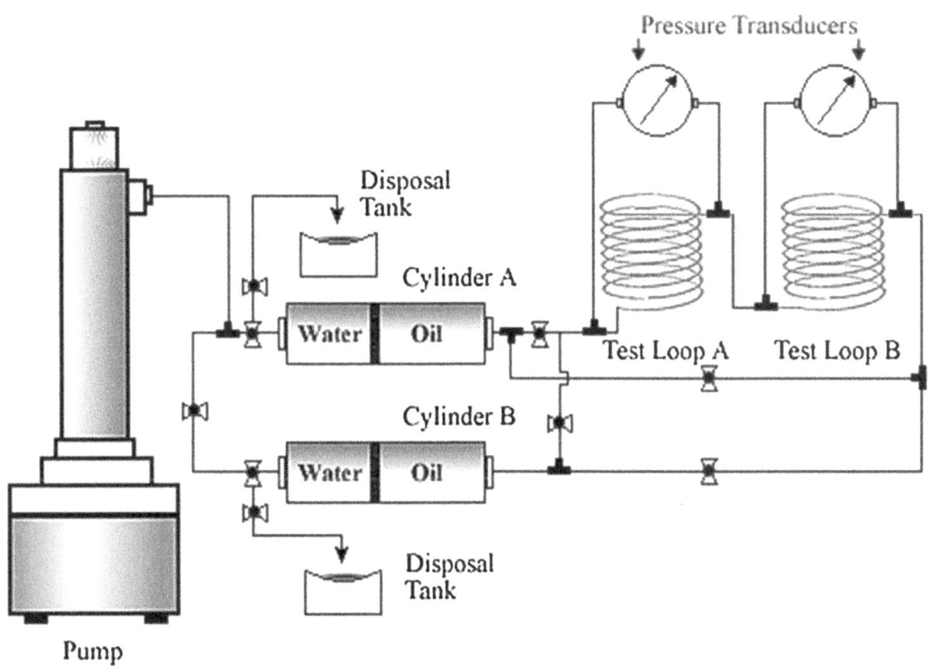

SARA and colloidal instability index

From the SARA (saturates, aromatics, resins and asphaltenes) analysis results, we can calculate the colloidal instability index or CII of the oil, which is an empirical measure of stability of asphaltene in a crude oil system.

$$CII = (Saturates + Asphaltenes)/(Aromatics + Resin)$$

It is observed that when CII < 1.0, the combined presence of resins and aromatics in oil is sufficient to stabilise the asphaltenes in crude. Presence of higher ratio or addition of saturates or removal of aromatics enhances CII value and reduces the asphaltene stability (Asomaning and Watkinson 2000).

It can be seen that CII value of the crude oil used in this study is 2.44 indicating that the crude oil is colloidally unstable and the possibility of asphaltene precipitation and deposition exists if suitable physical condition is available.

Static electro-deposition studies

Static electro-deposition experiments were conducted with three oils in a truly static condition (in terms of mechanical agitation and flow of current). DC potential was increased progressively to understand the threshold and optimum voltage requirement. Results of the 12 static jar tests are given in Table 2. In all the experiments, deposition of asphaltene like solids was observed on anode only, indicating that the deposited materials bear a net negative surface charge. The threshold voltage required to initiate precipitation was found to reduce with increasing oil dilution. Increase in DC potential resulted in an increase in the deposition rate on anode, no deposition was observed on the cathode electrode. Optimum voltage is seen to be between 80 and 120 V for all the oil samples. These observations lead us to believe that the asphaltene fraction in the crude oil possess excess negative charge which increases progressively with heptane dilution. Since asphaltene micelles are attracted by the anode, it would be reasonable to assume that cathode will repel the asphaltenes and inhibit their deposition. These results were used as a guiding factor in conducting further studies in dynamic condition.

Table 2 Test data on the electro-deposition of asphaltene at static condition on anode and cathode

Test no.	Oil–heptane ratio	DC potential	Deposit at anode (mg)	Deposit at cathode
SA1	Neat	40	Not detectable	Not detectable
SA2	Neat	80	13	Not detectable
SA3	Neat	120	47	Not detectable
SA4	Neat	160	39	Not detectable
SB1	80:20	40	29	Not detectable
SB2	80:20	80	57	Not detectable
SB3	80:20	120	63	Not detectable
SB4	80:20	160	54	Not detectable
SC1	50:50	40	42	Not detectable
SC2	50:50	80	93	Not detectable
SC3	50:50	120	113	Not detectable
SC4	50:50	160	127	Not detectable

Table 3 Test data on the dynamic electro-deposition on anode and cathode coils

Test no.	Oil to heptane ratio	Static potential (v)	ΔP across anode (psid)	ΔP across cathode (psid)	Deposit at anode (mg)	Deposit at cathode (mg)
DA-1	Nil	0	0.024	0.012	142	118
DA-2	Nil	20	0.025	0.021	150	119
DA-3	Nil	40	0.023	0.030	173	116
DA-4	Nil	60	0.079	0.022	176	108
DA-5	Nil	80	0.094	0.031	178	122
DA-6	Nil	100	0.091	0.033	181	117
DB-1	80:20	0	0.136	0.068	328	298
DB-2	80:20	20	0.176	0.026	362	204
DB-3	80:20	40	0.195	0.023	392	184
DB-4	80:20	60	0.170	0.03	415	169
DB-5	80:20	80	0.162	0.012	399	151
DC-1	50:50	0	0.25	0.20	654	553
DC-2	50:50	20	0.345	0.07	819	238
DC-3	50:50	40	0.405	0.09	896	162
DC-4	50:50	60	0.51	0.05	923	127
DC-5	50:50	80	0.55	0.047	929	127

Electro-deposition behaviour in dynamic or flowing condition

During electro-deposition studies at dynamic condition, we were careful in maintaining oil flow rate in the laminar region to avoid erosion of the expected deposits. Each study was conducted for about 10 h, circulating nearly 30 L of oil through each flow loop. DC voltage was applied, progressively stepping up in each subsequent test till no further enhancement of either deposition rate or deposited quantity was observed. As mentioned earlier, measurement of differential pressure across the flow loops was the means of monitoring real time deposition rate, while quantitative assessment of net deposition was conducted at the end of each experiment through tube cleaning and re-precipitation method. Results of the asphaltene deposition behaviour of the three oils across the anode and cathode loops under various DC potential are given in

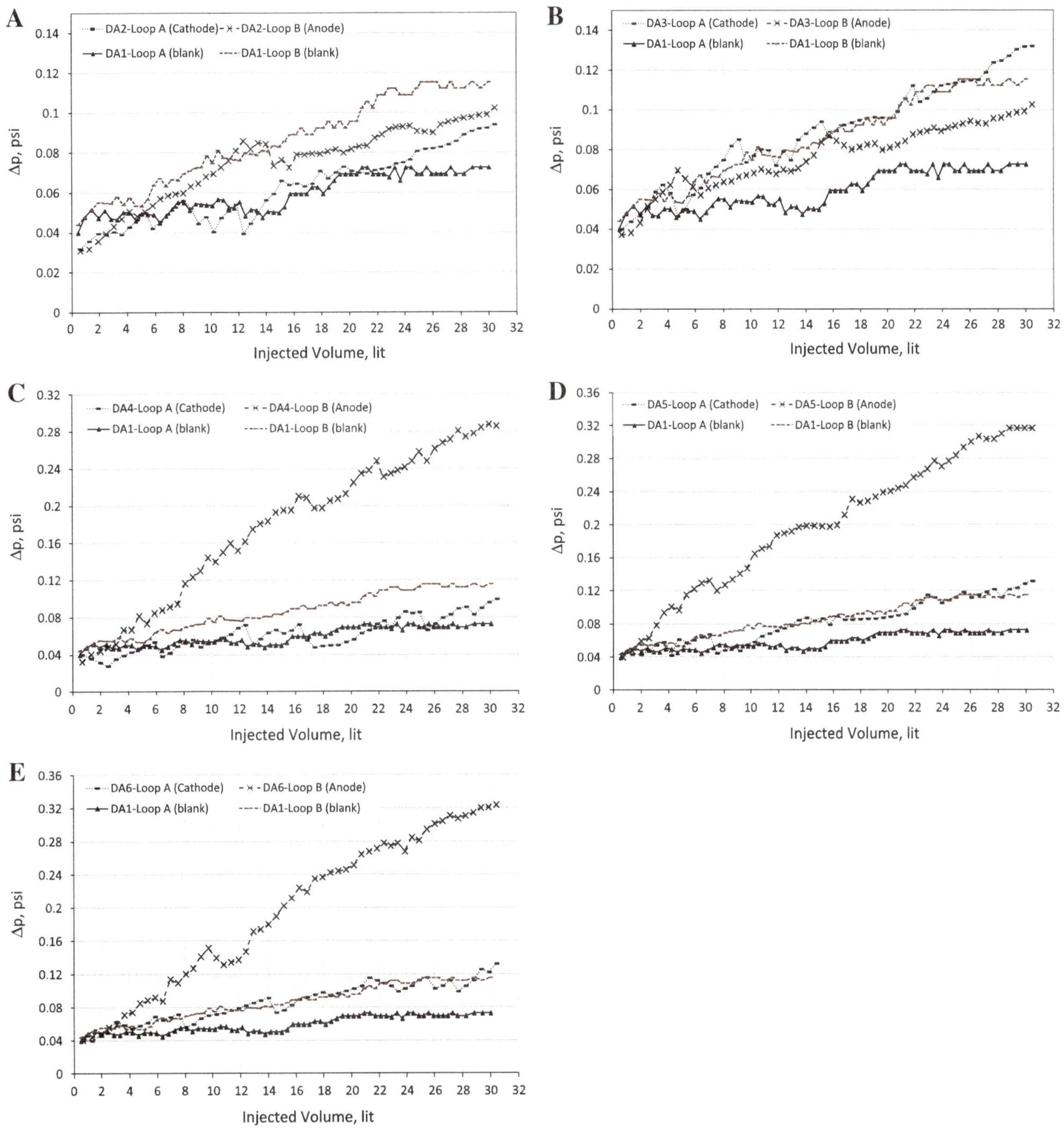

Fig. 3 **a** Pressure build-up during blank DA-1 and test DA-2 (20 V, neat oil). **b** Pressure build-up during blank DA-1 and test DA-3 (40 V, neat oil). **c** Pressure build-up during blank DA-1 and test DA-3 (60 V, neat oil). **d** Pressure build-up during blank DA-1 and test DA-4 (80 V, neat oil). **e** Pressure build-up during blank DA-1 and test DA-5 (100 V, neat oil)

Table 3. It must be mentioned at this stage that the current flow in these experiments was bare minimum (1–3 mA); thus, the electric field applied on the flow tubes can be considered as static potential.

Figure 3a through 3e depicts the dynamic electro-deposition of the neat oil under DC potential of 0, 20, 40, 60, 80 and 100 V, respectively. Similarly, the figure in 4a through 4d and Fig. 5a through 5d are graphical representation of the experimental results, when 80:20 and 50:50 oils were used for investigation.

It appears from the blank test with neat oil (Exp. DA-1) that precipitation and deposition of asphaltene on its flow path are influenced not only due to solvent dilution but also due to sedimentation potential and/or streaming potential arising due to the flow through the metal tube. (*Sedimentation potential* is an electrical potential arises due to the

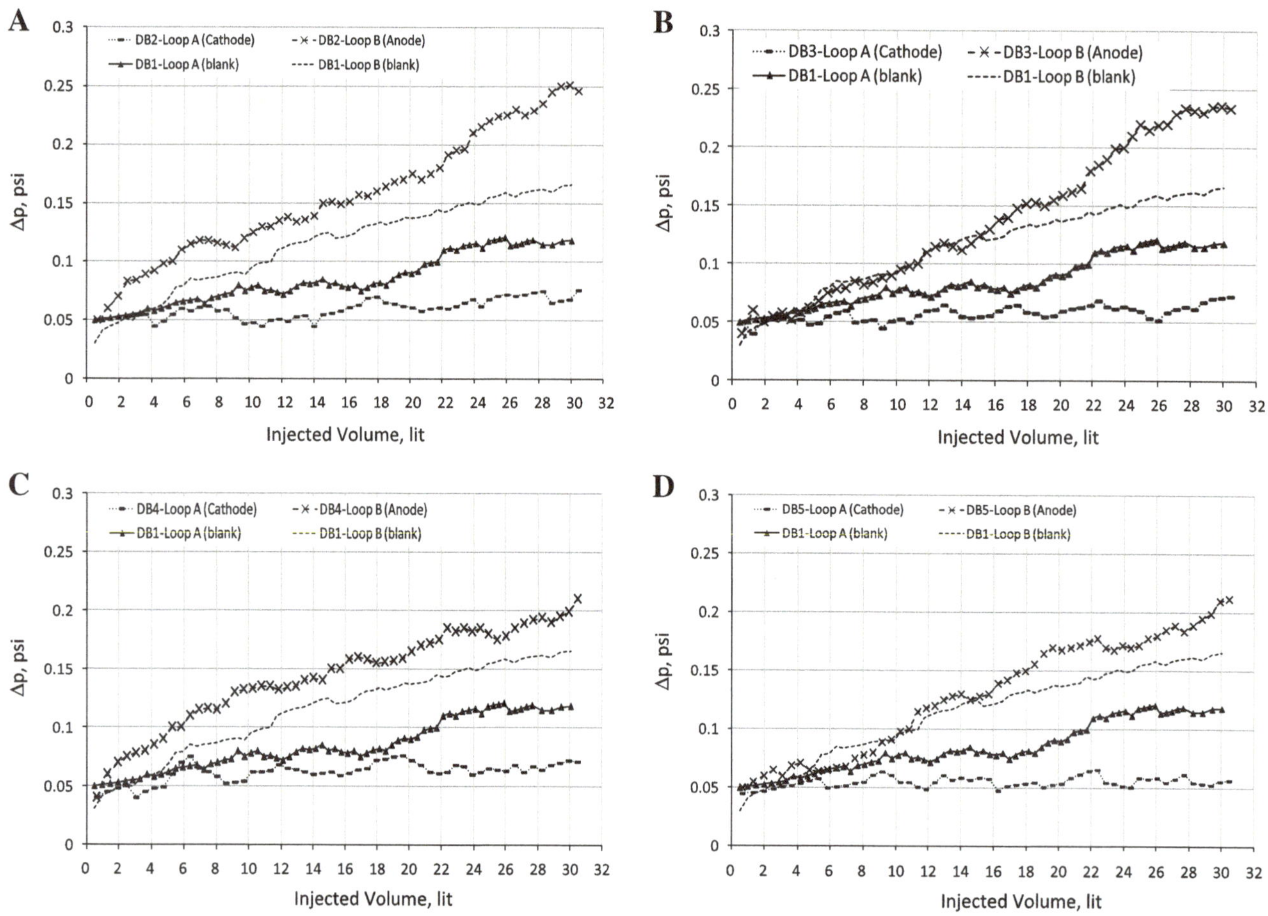

Fig. 4 **a** Pressure build-up during blank DB-1 and test DB-2 (20 V, 80:20 oil). **b** Pressure build-up during blank DB-1 and test DB-3 (40 V, 80:20 oil). **c** Pressure build-up during blank DB-1 and test DB-4 (60 V, 80:20 oil). **d** Pressure build-up during blank DB-1 and test DB-4 (80 V, 80:20 oil)

movement of charged particles through a liquid by gravity or pressure differential, and *Streaming potential* is the electric potential generated when a liquid is forced to move along a stationary charged surface). From CII value, we have established that the crude oil is supposed to be colloidally unstable. However, on long-term storage, we did not find sedimentation at the bottom of the bottle, indicating that at static condition the asphaltene particles are either small enough to enable sedimentation or due to similar surface charges they generate sufficient repelling force to prevent colloidal association and sedimentation. This observation explains the reason of blockage of production tubing even though the oil looks stable at laboratory condition. In subsequent flow studies, when the oil was flown through a charged surface (Exp. DA-2–DA-6, Fig. 3a through 3e), the asphaltene colloids were subjected to an increasingly stronger streaming potential in addition to its inherent sedimentation potential, resulting larger amount of asphaltene deposition. This is evident from the increase in differential pressure across the anode loop (as well as larger quantities of deposited mass). Important and notable facts observed form these comparative plots are prominence of asphaltene deposition on the anode loop only and almost no deposition on the cathode loop, strongly suggesting negative residual charge of the asphaltene colloids. Electro-deposition behaviour of the heptane-diluted samples was found to be similar but more pronounced than the neat oil. The blank experiments (DB-1 and DC-1) yielded faster and more asphaltene deposition compared with neat oil (Table 3), and upon application of DC potential, the rate of deposition and quantity of deposits (on anode) increased proportionately with little or no deposition on the cathode loops (Fig. 4a through 5d). To understand the combined effect of heptane dilution and application of static potential, we refer to Fig. 6, which represents comparative analysis of all the flow studies in terms of deposited mass compared with blank experiments with same oil. It is evident from this figure that as the deposition on the anode is progressively increasing with higher static potential, higher level of deposition control can be achieved on the cathode loop. Also, it is evident that control of asphaltene deposition is more prominent when

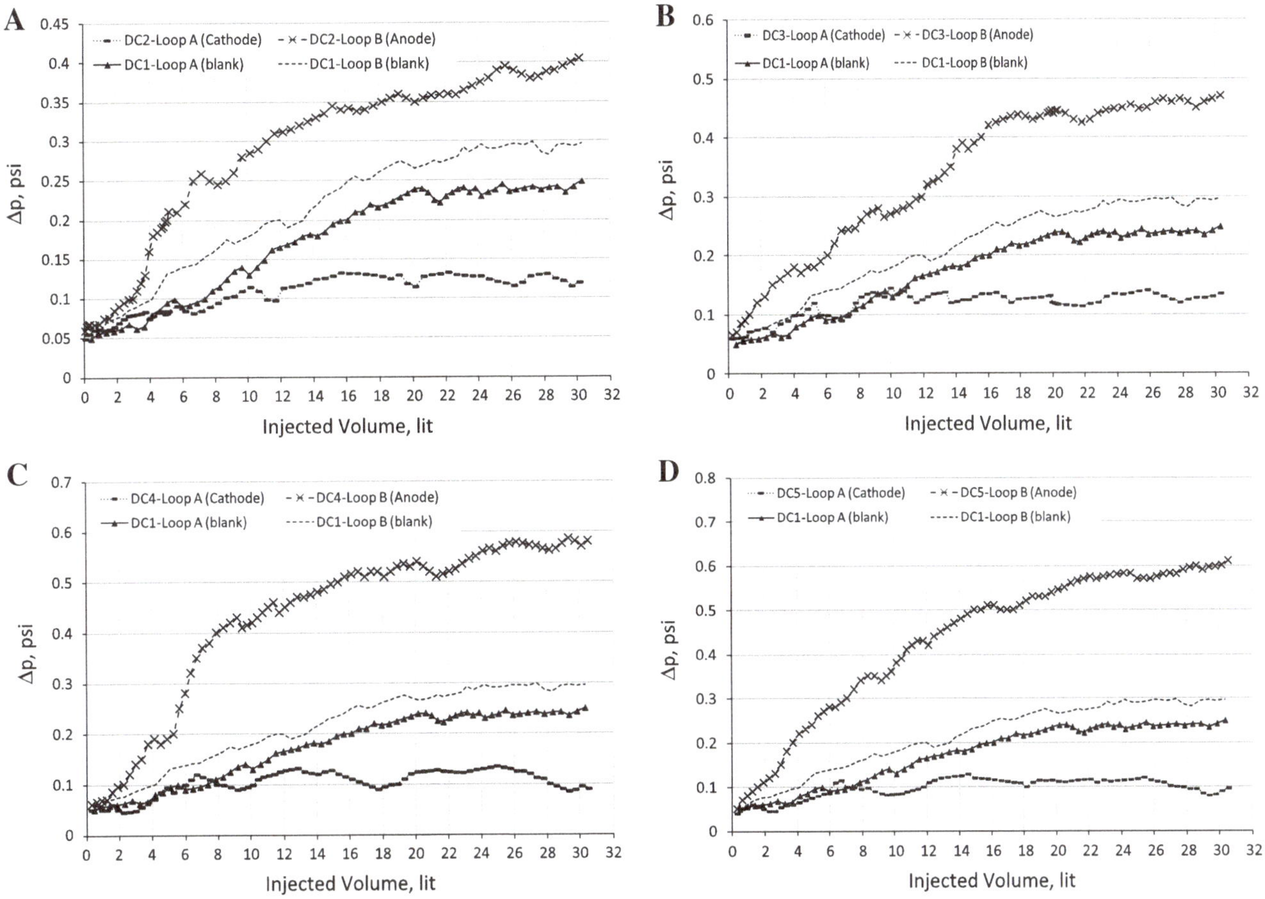

Fig. 5 **a** Pressure build-up during blank DC-1 and test DC-2 (20 V, 50:50 oil). **b** Pressure build-up during blank DC-1 and test DC-3 (40 V, 50:50 oil). **c** Pressure build-up during blank DC-1 and test DC-4 (60 V, 50:50 oil). **d** Pressure build-up during blank DC-1 and test DC-4 (80 V, 50:50 oil)

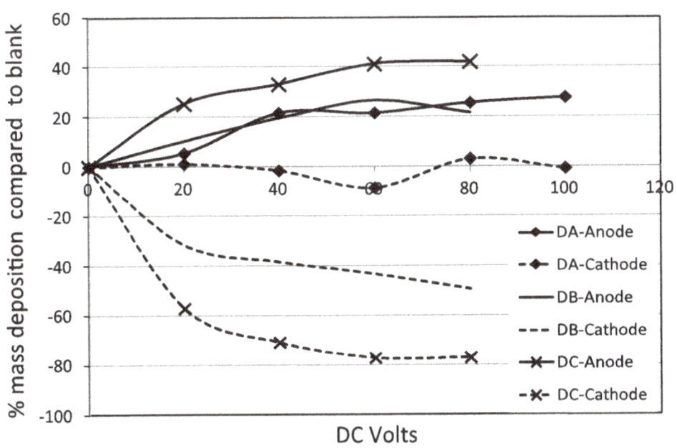

Fig. 6 Incremental and reduced asphaltene deposition in anode and cathode

the oil is diluted. Considering the worst case scenario, when the oil is diluted 100 % (50:50), 180 % deposition control is achieved with the application of 60 V compared with almost no control when neat oil is used.

The above observations clearly suggest the colloidal nature of asphaltene with excess negative surface charge. Negative surface charge on asphaltenes arises upon dissociation of acidic peripheral functional groups such as carboxylic acids and sulphonic acids. (Jada and Salou 2002; Vega et al. 2009). The positive surface charge arises upon protonation of basic nitrogen-containing functional groups such as pyridine. Parra-Barraza et al. (2003) measured zeta potential of heptane-diluted crude oils and found that at acidic pH asphaltene goes below isoelectric point and becomes positively charged whereas at neutral and higher pH de-protonation occurs resulting negative surface charge. Since the crude oil under present investigation originated in a carbonate reservoir with basic pH environment, it can be assumed that the asphaltenes have exceeded isoelectric point and acquired negative charges. Two possible scenarios that may be encountered in asphaltene–resin system are (A) asphaltene is negatively charged and resin is positively charged and (B) asphaltene is positively charged and resin is negatively charged. These scenarios and the effect of heptane dilutions are depicted through Fig. 7. Careful analysis of these two scenarios show that in case of scenario A, higher the dilution of crude

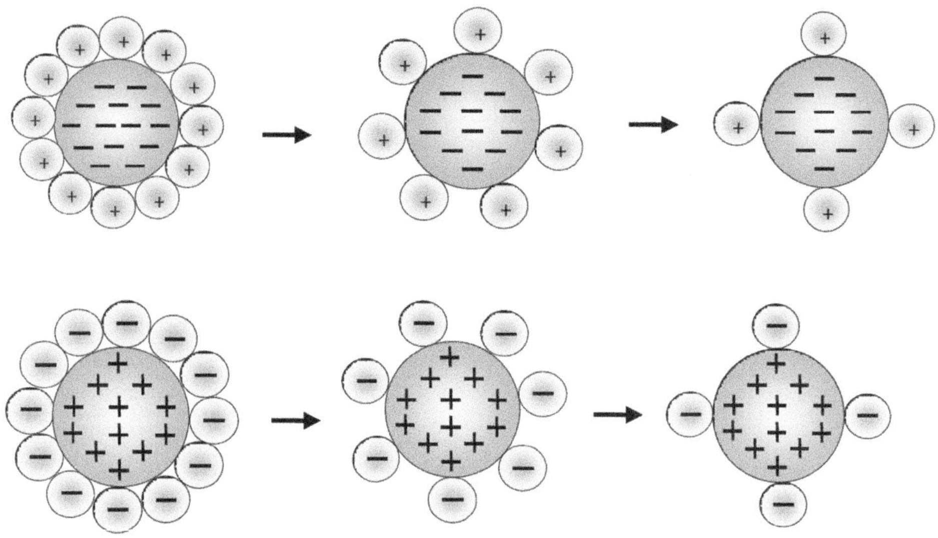

Fig. 7 Possible depletion of surface charge of colloidal micelle due to heptane dilution. Case 1 (*top*) Asphaltene contains excess negative charge. Due to heptane dilution, positively charged resins go away resulting increasing negative charge on the micelle. Case 2 (*bottom*) opposite phenomenon would occur if charges on asphaltene and resins are reversed

oil, larger will be negative residual charge on the asphaltene–resin colloid, resulting higher attraction towards anode, whereas in case scenario B, the loss of resin due to dilution would result in increasing positive charge and larger deposition of asphaltene mass on cathode. Since the experimental results show asphaltene deposition on anode only and very little or no deposition on the cathode, it can be safely assumed that the asphaltenes in the crude oil possess net negative charge and resins are positively charged.

Based on the experimental results, it is evident that it may be possible to control asphaltene deposition in a flowing system by applying optimum DC potential and converting the pipe into a cathode (with an anode pair nearby) similar to the cathodic protection technology commonly used to prevent corrosion. However, careful investigation and upscaling would be required with consideration for system temperature, pressure, salinity, pH, crude oil composition, fluid velocity and gas–oil ratio. In case of downhole application, the experiments must be conducted considering asphaltene onset pressure, system bubble point pressure and the changes of fluid composition during the upward flow.

Conclusions

Based on the results of the experimental works conducted in this study, the following conclusions may be drawn:

1. The asphaltene–resin colloids in the crude oil have excess negative charge which is the cause of asphaltene deposition while in dynamic state.
2. Application of DC potential resulted in rapid and excess asphaltene deposition on anode surface, while little or no deposition occurred on cathode surface.
3. The amount of *n*-heptane used to dilute the crude oil determines the surface charge of asphaltene–resin colloids which in turn affects the deposition rate and quantity. Higher the dilution, higher is the effect of DC potential in terms of deposition rate and quantity.

Acknowledgments The authors are grateful to the Petroleum Institute, Abu Dhabi, UAE for providing the laboratory facilities to conduct this research work and other financial assistance.

References

Alkafeef SF (2001) An investigation of the stability of colloidal asphaltene in petroleum reservoirs. Presented at the SPE international symposium on oilfield chemistry, Houston, 13–16 Feb 2001 (SPE 65018)

Andersen SI, Speight JG (2001) Petroleum resins: separation, character, and role in petroleum. Pet Sci Technol 19:1–34

Asomaning S, Watkinson AP (2000) Petroleum stability and heteroatom species effects in fouling of heat exchangers by asphaltenes. Heat Transf Eng 21(3):10–16

Behar E, Hasnaoui N, Achard C, Rogalski M (1998) Study of asphaltene solutions by electrical conductivity measurements. Oil Gas Sci Technol 53(1):41–50

Belhaj H, Khalifeh HA (2013) Asphaltene stability in crude oil during production process. J Pet Environ Biotechnol 4(3):1–5

Fotland P, Anfindsen H (1996) Electrical conductivity of asphaltenes in organic solvents. Fuel Sci Technol Int 14(1/2):101–115

Gonzalez G, Neves GBM, Saraiva SM, Lucas EF, De Sousa MA (2003) Electro-kinetic characterization of asphaltenes and the asphaltenes-resins interaction. Energy Fuels 17:879–886

Goual L, Firoozabadi A (2002) Measuring asphaltenes and resins, and dipole moment in petroleum fluids. AIChE J 48(11): 2646–2662

Goual L, Firoozabadi A (2004) Effect of resins and DBSA on asphaltene precipitation from petroleum fluids. AIChE J 50(2):470–479

Groenzin H, Mullins OC (2000) Molecular size and structure of asphaltenes from various sources. Energy Fuels 14:677–684

Hashmi SM, Firoozabadi A (2012) Controlling nonpolar colloidal asphaltene aggregation by electrostatic repulsion. Energy Fuel 26:4438–4444

Idem RO, Ibrahim HH (2002) Kinetics of CO_2-induced asphaltene precipitation from various Saskatchewan crude oils during CO_2 miscible flooding. J Pet Sci Eng 35(3):233–246

Igor NE, Aleksander PL (2010) Electrical conductivity and dielectric properties of solid asphaltenes. Energy Fuels 24(7):3959–3969

Jada A, Salou M (2002) Effects of the asphaltene and resin contents of the bitumens on the water–bitumen interface properties. J Pet Sci Eng 33:185–193

James W, Richard RM (1963) The electrophoretic mobility of asphaltenes in nitromethane. J Colloid Sci 18:223–236

Janusz M, Jablonski W (2000) Electrochemical properties of asphaltene particles in aqueous solutions. J Dispers Sci Technol 21(6):785–802

Khvostichenko DS, Andersen SI (2009) Electrodeposition of asphaltenes preliminary studies on electrodeposition from oil-heptane mixtures. Energy Fuels 23:811–819

Kokal S, Thompson T, Laurier S, Selim S (1995) Electrokinetic and adsorption properties of asphaltenes. Coll Surf A Physicochem Eng Asp 94:253–265

Leon O, Rogel E, Torres G, Lucas A (2000) Electrophoretic mobility and stabilization of asphaltenes in low conductivity media. Pet Sci Technol 18:913–927

Leontaritis KJ (1989) Asphaltene deposition: a comprehensive description of problem manifestations and modeling approaches. Presented at SPE production operations symposium, Oklahoma City, 13–14 Mar 1989 (SPE 18892)

Leontaritis KJ, Mansoori GA (1987) Asphaltene flocculation during oil production and processing: a thermodynamic-colloidal model. Presented at SPE international symposium on oilfield chemistry, San Antonio, 4–6 Feb 1987 (SPE 16258)

Lichaa PM, Herrera L (1975) Electrical and other effects related to the formation and prevention of asphaltene deposition problem in Venezuelan crudes. Presented at the SPE oilfield chemistry symposium, Dallas, 13–14 Jan 1975 (SPE 5304)

Nasser A, Belhaj H (2010) Modeling of crude oil asphaltenes deposition: in-depth colloidal prospective. . Presented at SPE production and operations conference and exhibition, Tunis, 8–10 Jun 2010 (SPE 136087)

Parra-Barraza H, Hernandez-Montiel D, Lizardi J, Hernandez ZJ, Urbina RH, Valdez M (2003) The zeta potential and surface properties of asphaltenes obtained with different crude oil/*n*-heptane proportions. Fuel 82:869–874

Speight JG (1998) The chemistry and technology of petroleum, 3rd edn. Marcel Dekker, New York

Taylor SE (1998) The electro-deposition of asphaltenes and implications for asphaltene structure and stability in crude and residual oils. Fuel 77:821–828

Vega SS, Herrera UR, Valdez CM, Galeana LC (2009) The zeta potential of solid asphaltene in aqueous solutions and in 50:50 water + ethylene glycol (v/v) mixtures containing ionic surfactants. J Pet Sci Eng 69:174–180

Yen TF, Chilingarian GV (1994) Asphaltenes and asphalts-1. Elsevier, 1st edn. ISBN: 978-0-444-88291-2

12

Potential of Welan Gum as mud thickener

Changhong Gao

Abstract Welan Gum (WLG) is a biopolymer employed in food industry as thickener. This paper evaluates its performance as mud thickener and filtrate loss additive. Tests revealed that WLG, even at low concentrations, effectively boosted mud viscosity and yield point. Tests were also conducted on carboxymethyl cellulose (CMC), a commonly used mud thickener. Compared with CMC, WLG is more effective as mud thickener, but WLG has no effect on mud filtrate loss. WLG mud was also tested at elevated temperature. The viscosity of WLG mud was stable under short-term and long-term heating. Sodium chloride and calcium chloride were added into WLG mud. Test results showed WLG mud was resilient to sodium and calcium contamination. WLG has good potential as mud thickener under high temperature and high salinity conditions.

Keywords Welan Gum · Drilling mud · Thickener · Salinity

Introduction

Drilling mud plays a critical role in the success of oil and gas drilling operations. It cools and lubricates drill bit, controls well pressure, transports and suspends cuttings. Mud must have a relatively high viscosity to effectively suspend cuttings. Therefore, mud thickeners are often added into mud to boost mud viscosity. The most widely used mud thickeners are carboxy methyl cellulose (CMC) and hydroxy ethyl cellulose (HEC) (Rabia 1985).

C. Gao
The Petroleum Institute, SINOPEC SLOF, Dongying, China

Welan Gum (WLG) is produced by fermentation of sugar by bacteria of the genus Alcaligenes. The molecule consists of repeating tetrasaccharide units with single branches of L-mannose or L-rhamnose (Rakitsky and Richey 1992). The molecular structure is presented in Fig. 1. It was reported that WLG exhibits good viscosity retention at elevated temperature, as well as in the presence of sodium chloride (Chen 2007). WLG was patented to be used as cement additive (Allen et al. 1991), as well as water shutoff chemical (Hoskin et al. 1991). However, limited research has been conducted to investigate its effect on mud properties. The objective of this study was to unveil potential of WLG as a mud additive.

Material and method

Welan Gum was provided by BDL Chemical Company in China. CMC (medium viscosity) was produced by Gumpro Chemical Company in India. Other materials included Bentonite clay, NaCl, $CaCl_2$, and tap water. Mud viscosity was measured with OFITE 8-speed viscometer. It measures the torque under 8 different rotational speeds, ranging from 600 to 3 rpm. The readings can be easily translated into mud plastic viscosity, mud yield point, and mud apparent viscosity. Mud filtrate was measured with OFITE Model 14030 filter press. Mud passes through a filter paper under 100 psi of pressure, and the filtrate volume after 7.5 min of filtration is recorded. Pressure is provided by a cartridge containing pressurized carbon dioxide. Mud was prepared with high-speed mixer. The mud viscometer, filter press and mud mixer are all industry-standard devices.

Totally, 23 tests were conducted. The detailed test matrix is given in Tables 1 and 2. First, a base mud was

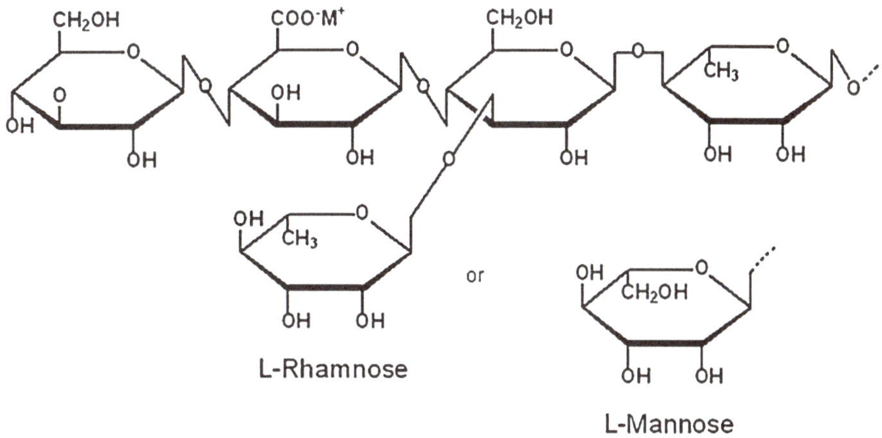

Fig. 1 Molecular structure of Welan Gum

Table 1 Summary of test matrix for WLG mud

Test no.	Tested mud	NaCl (g/L)	$CaCl_2$ (g/L)	WLG (g/L)	Temperature (°C)
1	Base mud	0	0	0	25
2	WLG mud	0	0	0.25	25
3	WLG mud	0	0	0.50	25
4	WLG mud	0	0	0.75	25
5	WLG mud	0	0	1	25
6	WLG mud	0	0	1	45
7	WLG mud	0	0	1	65
8	WLG mud	50	0	1	25
9	WLG mud	100	0	1	25
10	WLG mud	150	0	1	25
11	WLG mud	150	10	1	25
12	WLG mud	150	30	1	25
13	WLG mud	150	50	1	25
14	WLG mud	0	0	1	85 °C for 6 h

Table 2 Summary of test matrix for CMC mud

Test no.	Tested mud	NaCl (g/L)	$CaCl_2$ (g/L)	CMC (g/L)	Temperature (°C)
15	CMC mud	0	0	0.5	25
16	CMC mud	0	0	1	25
17	CMC mud	50	0	1	25
18	CMC mud	100	0	1	25
19	CMC mud	150	0	1	25
20	CMC mud	150	10	1	25
21	CMC mud	150	30	1	25
22	CMC mud	150	50	1	25
23	CMC mud	0	0	1	85 °C for 6 h

prepared by mixing 25 g of clay into 500 g of fresh water. Then mud viscosity and filtrate were measured. The measurements of mud plastic viscosity (PV), yield point (YP), apparent viscosity (AV) and mud filtrate loss followed textbook and industry standards (Bourgoyne et al. 1986). Second, the effect of WLG concentration on mud viscosity and mud filtrate was investigated by adding different dosages of WLG into the base mud.

The third group of tests was designed to test the performance of WLG mud under the influence of NaCl. Afterward, different dosages of $CaCl_2$ were added to WLG mud and the mud viscosity was measured. Finally, WLG mud was held at 85 °C for 6 h to test the effect of aging on mud properties. For the purpose of comparison, CMC was also tested for its stability under the influences of sodium ion, calcium ion, and long-term heating, as given in Table 2.

Results and analysis

Figure 2 presents the mud viscosity trends under different shear rates for base mud, as well as mud with WLG. The tests were conducted at room temperature of 25 °C. On this log–log plot, it can be seen that mud viscosity declines almost linearly at higher shear rates. This reveals that mud is a non-Newtonian fluid that abides by the power law (Gao 2013). It is also obvious that even a small dosage of WLG resulted in significant increase in mud viscosity.

Figure 3 presents the effect of WLG on mud YP, PV and AV at ambient temperature. It can be seen that WLG has a significant impact on mud YP and AV. Mud YP and AV steadily increased with increasing concentration of WLG. The mud PV also increased with addition of WLG. It was observed that when WLG concentration increased from 0.75 to 1 g/L, the mud YP and AV increased more sharply. This is possibly due to the formation of gel at relatively high WLG concentration.

Figure 4 compares the effects of WLG and CMC on mud properties. The addition of CMC resulted in mild increases in mud viscosity and yield point, while WLG is much more effective in thickening mud. Mud filtration test

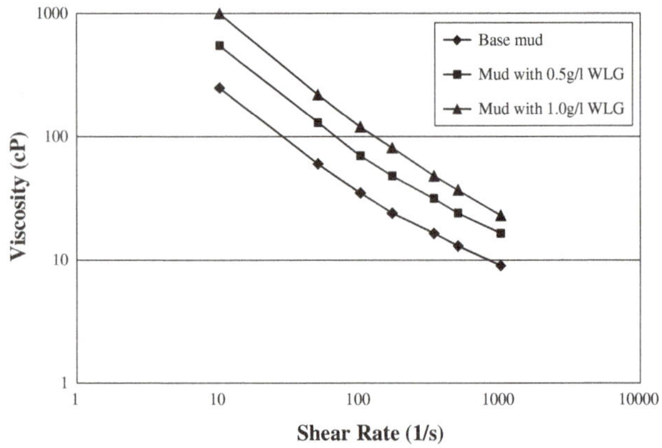

Fig. 2 Effect of shear rate and WLG concentration on mud viscosity

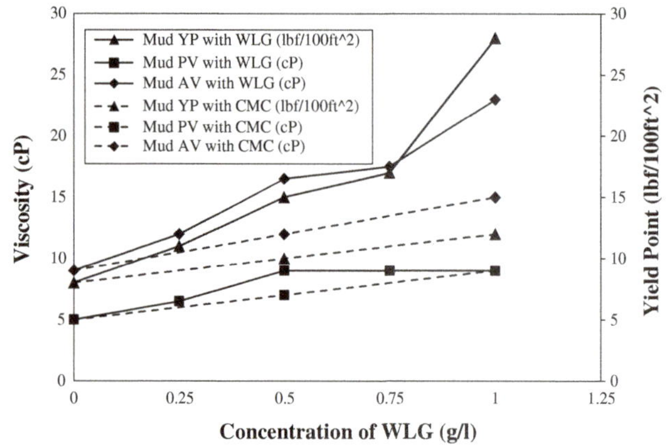

Fig. 4 Comparison of WLG mud and CMC mud

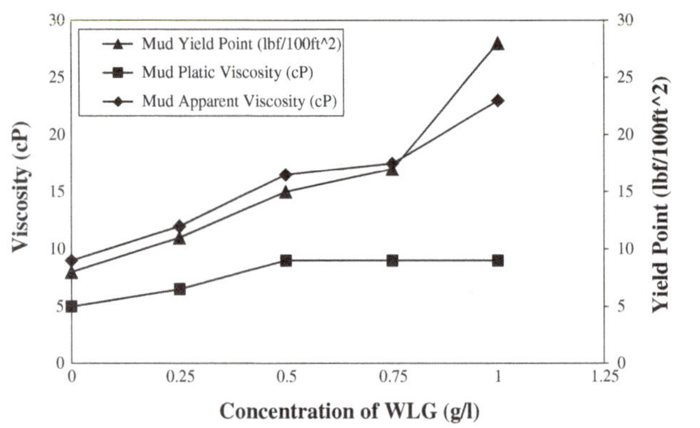

Fig. 3 Effect of WLG concentration on mud properties

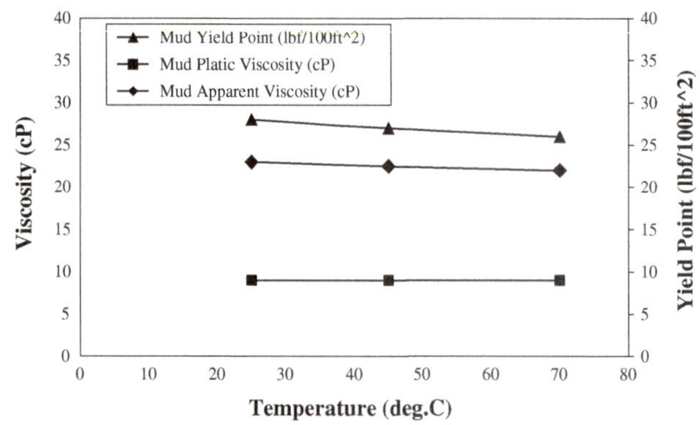

Fig. 5 Behavior of WLG mud at elevated temperature

was also conducted on WLG mud. Excessive filtrate loss during drilling can result in wellbore instability. Mud filtrate is often incompatible with formation fluid and can cause formation damage. The base mud without WLG had a filtrate of 10 mL after 7.5 min, while the mud containing 1 g/L WLG had a filtrate of 9 mL. WLG has no significant effect on filtrate loss.

The mud containing 1 g/L WLG was heated to 70 °C for 30 min, to test the stability of WLG at elevated temperature. The test data in Fig. 5 show good stability of WLG at high temperature. The mud properties after heating remained almost unchanged.

The mud samples were also tested for long-term stability under high temperature. The mud samples aged at 85 °C for 6 h, then cooled down to room temperature and mud properties were measured. The test data are presented in Table 3. For the mud containing WLG, both mud YP and AV increased after aging. This is probably due to the formation of weak gel at high temperature. On the other hand, the mud containing CMC showed reduction in YP and AV after aging. Comparison shows WLG is resilient to long-term heating.

Table 3 Properties of mud after aging tests

Tested fluid	Test condition	PV (cP)	YP (lb/ 100 ft^2)	AV (cP)
Base mud + 0.5 g WLG	25 °C	9	28	23
Base mud + 0.5 g WLG	Aged at 85 °C for 6 h then cool to 25 °C	9	39	29
Base mud + 0.5 g CMC	25 °C	9	12	15
Base mud + 0.5 g CMC	Aged at 85 °C for 6 h then cool to 25 °C	8	8	12

A good mud thickener must maintain stable viscosity under the attack of sodium and calcium ions. NaCl and $CaCl_2$ were mixed into WLG mud and the mud properties were measured. The test results are shown in Figs. 6 and 7. It can be seen that WLG is resilient to both sodium and calcium ions. The key parameters (YP, AV and PV) all remained quite stable after addition of salts.

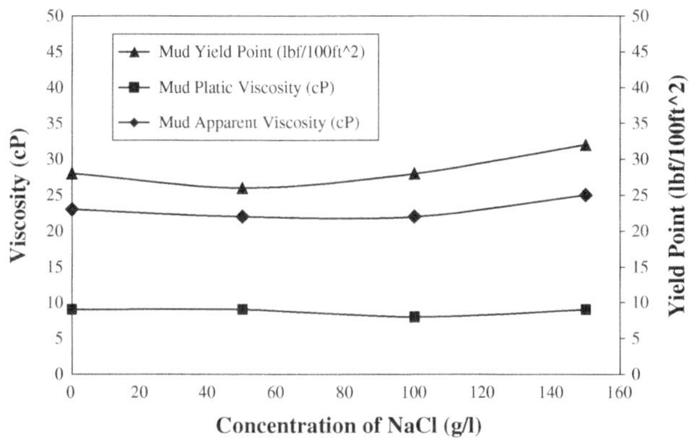

Fig. 6 Stability of WLG mud at varied concentrations of NaCl

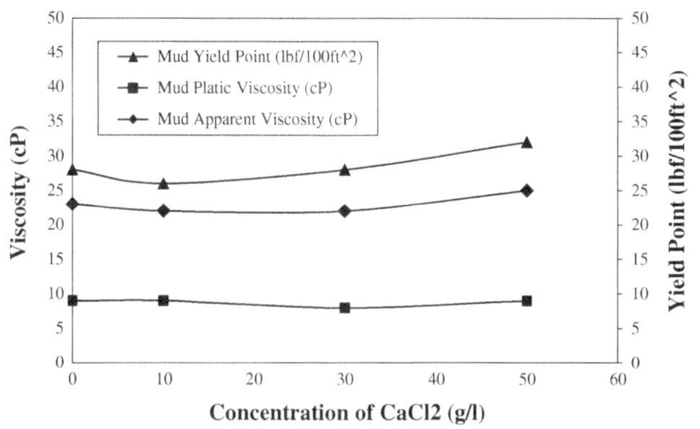

Fig. 7 Stability of WLG mud under varied concentrations of $CaCl_2$

Conclusions

Welan Gum is excellent as mud thicker, but not effective in reducing mud filtrate loss. Compared with CMC, WLG is more effective in boosting mud viscosity. WLG mud maintains stable viscosity under high temperature and long-term heating. Sodium and calcium ions have minor influence on WLG mud viscosity.

References

Allen F, Best G, Lindroth T (1991) Welan gum in cement compositions. United States Patent 5004506

Bourgoyne A, Chenevert M, Millheim K, Yang F (1986) Applied drilling engineering. SPE Textbook Series, Richardson, pp 40–50

Chen F (2007) Study of factors affecting property of welan gum solution. Food Sci 28(9):49–52

Gao C (2013) Viscosity of HPAM under shearing and heat. J Petrol Explor Prod Technol 3(3):203–206

Hoskin D, Mitchell T, Shu P (1991) Oil reservoir permeability profile control with cross linked welan gum biopolymers. United States Patent 4991652

Rabia H (1985) Oilwell drilling engineering. Graham and Trotman, London, pp 105–130

Rakitsky W, Richey D (1992) Rapidly hydrating welan gum. United States Patent 5175277

13

Numerical modeling of robust production scenarios from shared oil reservoirs

Sara Shokrolahzadeh-Behbahani ·
Mahdi Zeinali-Hasanvand · Mohammad-Ali Ahmadi

Abstract A reservoir that has formed through millions of years of sedimentation almost never conforms to the established borderlines that define political entities or legal jurisdictions of neighboring countries. The term shared reservoir is given to such hydrocarbon deposits. After the discovery of shared reservoirs, the neighboring countries could jointly develop a cross-border hydrocarbon deposit or disagree on any framework for unitizing or jointly developing the reservoir. Furthermore, in order for any negotiation to take place, the contending sides need to settle on an unbiased agreement. This can only be done after the term "unbiased" is defined quantitatively in technical terms. Although defining a reservoir before thorough exploration is challenging and rather unreliable, understanding the behavior of shared reservoirs is highly valued since many reservoirs around the globe, including the world's largest condensate gas reservoir, in under such collaborative ownership. It is evident that optimized production from such reservoirs will benefit their rightful possessors and the rest of the world. This study investigates all main factors that influence production from a shared oil reservoir and demonstrates how each neighboring country is affected by variations of each of these factors. Displays of simulated reservoir at potential production conditions complement our quantitative analysis throughout this study.

S. Shokrolahzadeh-Behbahani · M.-A. Ahmadi
Department of Petroleum Engineering, Ahwaz Faculty of Petroleum Engineering, Petroleum University of Technology, Ahwaz, Iran

M. Zeinali-Hasanvand
Research Institute of Petroleum Industry (RIPI), Tehran, Iran

Keywords Quantitative analysis · Shared oil reservoir · Optimized production · Crude migration · Influential factors · Numerical modeling

Introduction

A hydrocarbon reservoir is a porous medium that has formed as a result of sedimentation over long periods of time. The organic matter buried beneath the sediments is pressurized and introduced to higher temperatures; this induces chemical changes to the organic matter and alters it to valuable substances such as coal, tar, oil and gas. Due to the heterogeneous nature of sedimentation layers and hydrocarbon migration, hydrocarbon reservoirs can exist at any geographical location (Hou et al. 2011). Therefore any hydrocarbon deposit may be found to intersect the boundary dividing two separate territories. Such reservoirs that are known as shared reservoirs will demonstrate indefinite hydraulic fluid connection and heterogeneity of the reservoir rock which can lead to crude migration across the boundary. This complicates matters when there are two countries contending for maximized production. The shared reservoirs are distributed all over the world. In the Middle East, the main Energy production place of world there are many numbers. In Table 1 list of some Iranian-Neighbors shared oil and gas reservoirs and their original hydrocarbon reserves are presented.

Maximum production from the shared reservoir, pressure maintenance, enhanced oil recovery (EOR) processes, improved oil recovery (IOR) processes and optimized production are controversial subjects to be discussed when considering dual ownership of a shared reservoir as there are no laws or regulations that could enforce conditions to prohibit fluid migration or cross-boundary production

Table 1 List of shared hydrocarbon reservoirs of Iran and their original fluid and rock properties considered for simulated reservoir

Reservoir name	Type	Original oil in place (million barrel)	Original gas in place (billion Scf)	Countries
Naft Shahr	Oil	692	–	Iran–Iraq
Dehloran	Oil	4,212	–	Iran–Iraq
Azadegan	Oil	33,000	–	Iran–Iraq
Yadavaran	Oil	12,200		Iran–Iraq
Arash	Gas	83.6	22.5	Iran–Koweit
South Pars	Gas	18,000	131.3	Iran–Qatar
Sasan	Oil/gas	4,073	0.183	Iran–Abo Dubi
Hengam	Gas	–	22.4	Iran–Oman

(Weems and Fallon 2012). Only a quantitative analysis on the factors that influence production and fluid migration within a shared reservoir can determine whether or not the rights of each country have been served in the dual production process. A more consequential matter that should be studied is whether joint management of the reservoir is sufficiently economical to encourage a cooperative reservoir development.

Due to the diverse physical characteristics of oil and gas, separate investigations need to be carried out on gas and oil reservoirs and extending results of one fluid to the other only allows a qualitative analysis between the two. To assess this goal, current research focuses on shared oil reservoirs while through this research, the fluid hydraulics within a shared reservoir by studying the equations that can be carried out to the fluid in a porous medium to figure out the dominating factors that impose the migration of hydrocarbons across the borderline illustrated along with details. Characteristics that tend to assist the migration of oil or affect it in some way are the rock properties (such as permeability and porosity), fluid properties (such as viscosity, compressibility and formation volume factor) and other characteristics like arrangement of producing wells, reservoir's tilt, the effectiveness of the aquifer and time. After study on each factor and the extent to which it contributes to oil migration, we analyze if and how co-managing the shared reservoir can promote the ultimate recovery. It is worth bearing in mind that Eclipse's 2009 Black oil model was carried out to achieve addressed goal of this research due to high computational time of simulation runs in compositional approach and low computational time in black oil approach.

Fluid flow in porous media

Through petroleum reservoirs, the relationship commonly carried out to figure out fluid flow in porous media is the equation proposed by Henry Darcy in 1856 while he assumed that steady state isothermal one-dimensional flow of incompressible fluid met:

$$q = -\frac{kA}{\mu} \times \frac{dP}{dx} \tag{1}$$

where, q, k, A, μ, P and x represent fluid flow rate (m^3/s), rock permeability (m^2), rock cross sectional area (m^2), viscosity (Pa $\times$ s), pressure (Pa) and rock length (m), correspondingly (Ayala and Kouassi 2007). For the cylindrical flow of semi-compressible fluids, the Darcy equation develops into Eq. 2:

$$Q_o = \left[\frac{0.00708kh}{\mu_o B_o c_o \ln\left(\frac{r_e}{r_w}\right)}\right] \ln[1 + c_o(p_e - p_{wf})] \tag{2}$$

where, B_o, C_o, h, r_e, r_w, p_e, p_w, k, μ_o and Q_o are assigned to the oil formation volume factor (bbl/STB), oil compressibility (psi^{-1}), net formation thickness (ft), external (drainage) radius, wellbore radius, external pressure (psi), wellbore pressure (psi), permeability (millidarcy (mD)), viscosity (cP) and oil flow rate (STBD), correspondingly (Ahmed and McKinney 2005). Furthermore, if a compressible fluid like as gas is under study, the Darcy equation for cylindrical flow geometry will be:

$$Q_g = \frac{kh(\psi_e - \psi_w)}{1422T \ln\left(\frac{r_e}{r_w}\right)} \tag{3}$$

where, Q_g and T are assigned to the gas flow rate in Mscf/day and temperature in °R units, respectively. In the relationship given by Eq. 3, ψ_e is called the real gas potential and is defined as following expression:

$$m(p) = \psi = \int_0^p \left(\frac{2p}{\mu_g z}\right) dp \quad psi^2/cP \tag{4}$$

As these equations suggest, rock properties (such as porosity and permeability) and fluid properties (such as viscosity, compressibility and oil formation volume factor) impact the produced amount of oil and gas. In addition to these properties, the pressure decline trend and time play significant roles.

Moreover, by combining continuity equations for a cylindrical reservoir with the Darcy equation, we derive the diffusivity equation which is a second degree differential equation that correlates pressure as a function of time and radius of investigation.

$$\frac{\partial^2 p}{\partial r^2} + \frac{1}{r}\frac{\partial p}{\partial r} = \frac{1}{\eta}\frac{\partial p}{\partial t} \tag{5}$$

where:

$$\eta = \frac{0.000264k}{\phi \mu c_t} \tag{6}$$

where ϕ and c_t are rock porosity and total compressibility factor (psi^{-1}), respectively.

According to the diffusivity equation, pressure decline in the reservoir is not associated with the production flow rate and depends solely on physical characteristics of reservoir rock and fluid. The diffusivity equation for gases is derived from Eq. 5 by replacing the real gas potential (ψ) by pressure as follows (Al-Hussainy et al. 1966):

$$\frac{\partial^2 \psi}{\partial r^2} + \frac{1}{r}\frac{\partial \psi}{\partial r} = \frac{\phi \mu c_t}{0.000264k}\frac{\partial \psi}{\partial t} \tag{7}$$

Another matter of concern in oil reservoirs is the effect of boundary. We cannot determine production conditions until the radius of investigation has reached the drainage radius. After which, the production can continue in any of the three following states:

1. The reservoir will continue to produce with stable pressure and flow rate.
2. The pressure will decrease linearly at the wellbore.
3. The pressure and flow rate will begin to reduce non-linearly with time.

The Darcy equation states that if the pressure decline remains constant throughout the fluid's path (for example for having an active aquifer), fluid velocity will also be constant ($v = Q/A$) and water will replace produced oil. If the aquifer is ineffective or non-existent, fluid velocity is high in the vicinity of wellbore and low in more distant areas. Eventually, the fluid velocity will decline all throughout the reservoir.

Considering velocity and the presence of an aquifer in the reservoir boundary, the configuration of production wells is of importance, especially when we cannot drill wells further than the borderline. Alternatively, if the aquifer is active, well configuration does not have much influence on fluid flow because the velocity of flow is linear and equal all throughout the reservoir.

Thus, based on equations and discussions provided, the following are identified as factors that influence fluid flow in a shared reservoir:

1. Well configuration
2. Production time
3. Slope of reservoir
4. Aquifer performance
5. Rock properties in each territory
6. Fluid properties in each territory

Analyzing production from a shared oil reservoir

To analyze the production from a shared reservoir, we will simulate various conditions of production and investigate the role of each identified variable that may influence production.

In Eqs. 1 through 7, variables such as pressure, permeability, compressibility and etc. have been considered in their simplest forms. The fluid compressibility is a theoretically determined sum of fluid compressibility (water, oil and gas) multiplied by saturation of each phase. Permeability also varies with fluid saturation. Furthermore, gravity's role should be incorporated into the equations if the reservoir is tilted. To deal with complications of this kind of reservoir, a simulator, namely ECLIPSE, was used.

Eclipse's Tracer software was carried out to introduce a reservoir with 20 km in length, 100 m in thickness and 8 km in width. The porosity and permeability of homogenous reservoir rock are originally considered 13 % and 100 (mD), correspondingly. Implemented fluid properties and other relevant information such oil/water and gas/oil relative permeabilities which required for graph evaluation are demonstrated in Table 2; Figs. 1 and 2. The oil on both sides of the supposed border line is initially considered identical eliminate its influence. Figure 3 exhibits the reservoir before production has begun. Two proposed producing wells, the border line separating neighboring countries and the reservoir bulk itself are demonstrated.

The program's data file for the reservoir is available in the appendix for implementation for future augmentation. To figure out the effect of each indentified production variables, we modified the input data fed to the simulator. The results are illustrated in details as below sections.

Table 2 Original fluid and rock properties considered for simulated reservoir

Property	Quantity	Unit
Irreducible water saturation	0.2	–
Irreducible oil saturation	0.19	–
Irreducible gas saturation	0.1	–
Bubble pressure	2,800	Psia
Gas in solution	1,130	SCF/STB
Total compressibility	1×10^{-6}	Psia^{-1}
Formation water density	63.01	lb$_m$/Cuft
Dead oil density	44.98	lb$_m$/Cuft
Gas density	0.0702	lb$_m$/Cuft
Water viscosity	0.96	cP
Oil viscosity	0.16	cP
Gas viscosity	1	cP

Fig. 1 Relative permeability curves of gas and oil phases

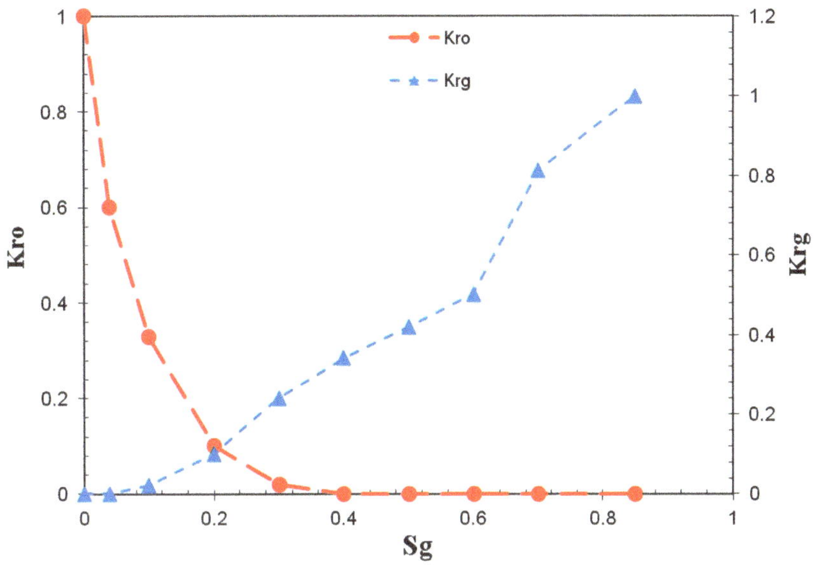

Fig. 2 Relative permeability curves of oil and water phases

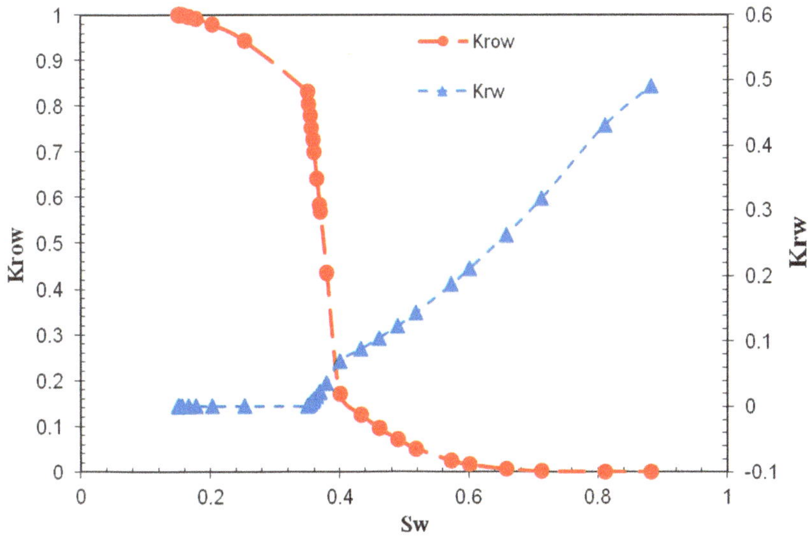

The effect of well configuration

Symmetrical well configuration across borderline

In the modeled reservoir we assume all fluid properties alike with only two producing wells, one well in each territory equidistant from the border line. According to the superposition principle, and acknowledging that pressure decline disseminates as production begins, we expect to see identical trends of pressure decline in both parts of the homogenous isotropic reservoir when well configuration is symmetrical across the assumed bisecting borderline. This means that oil within each territory is produced by its own producing well and no hydrocarbon migrates across the border. Figure 4 demonstrates oil saturation across the reservoir at various instances in time. The identicalness of production circumstances can be distinguished by similar saturations around producing wells. Analogous comparisons could be made via other parameters such as produced gas, pressures, and fluid movement by the border line as displayed in Figs. 5 and 6.

Based on results, we conclude that simultaneous production from an isotropic homogenous reservoir with similar well configuration on both sides of the borderline can lead to an evenhanded sharing of a shared reservoir. Needless to say, the reservoir may not be divided equally between neighboring countries and the borderline may not bisect the reservoir but other assumptions could be made to assure correspondence between production and share of reservoir.

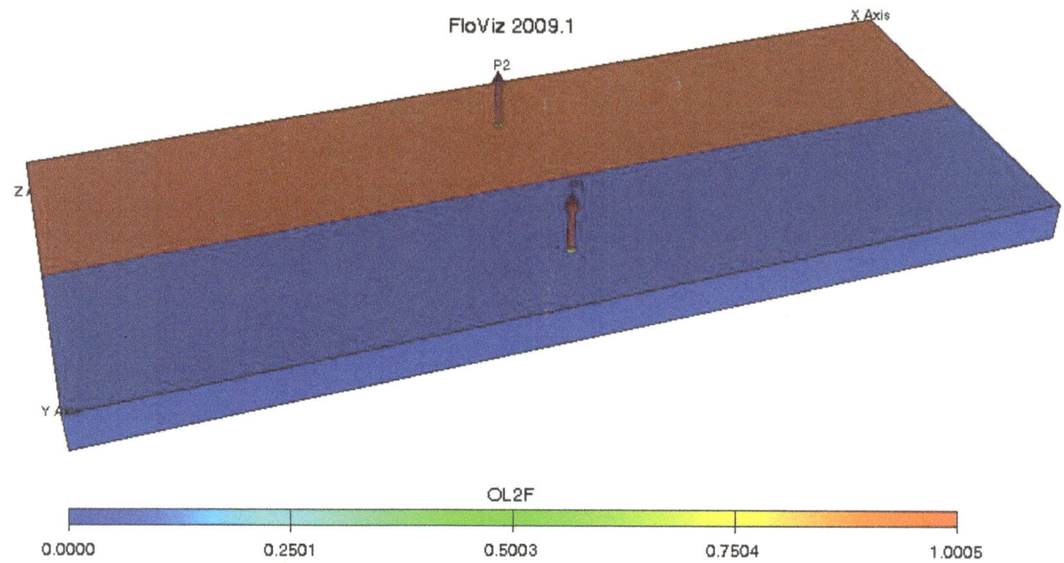

Fig. 3 Model of a shared hydraulically interconnected oil reservoir with one producing well in each of the two defined territories

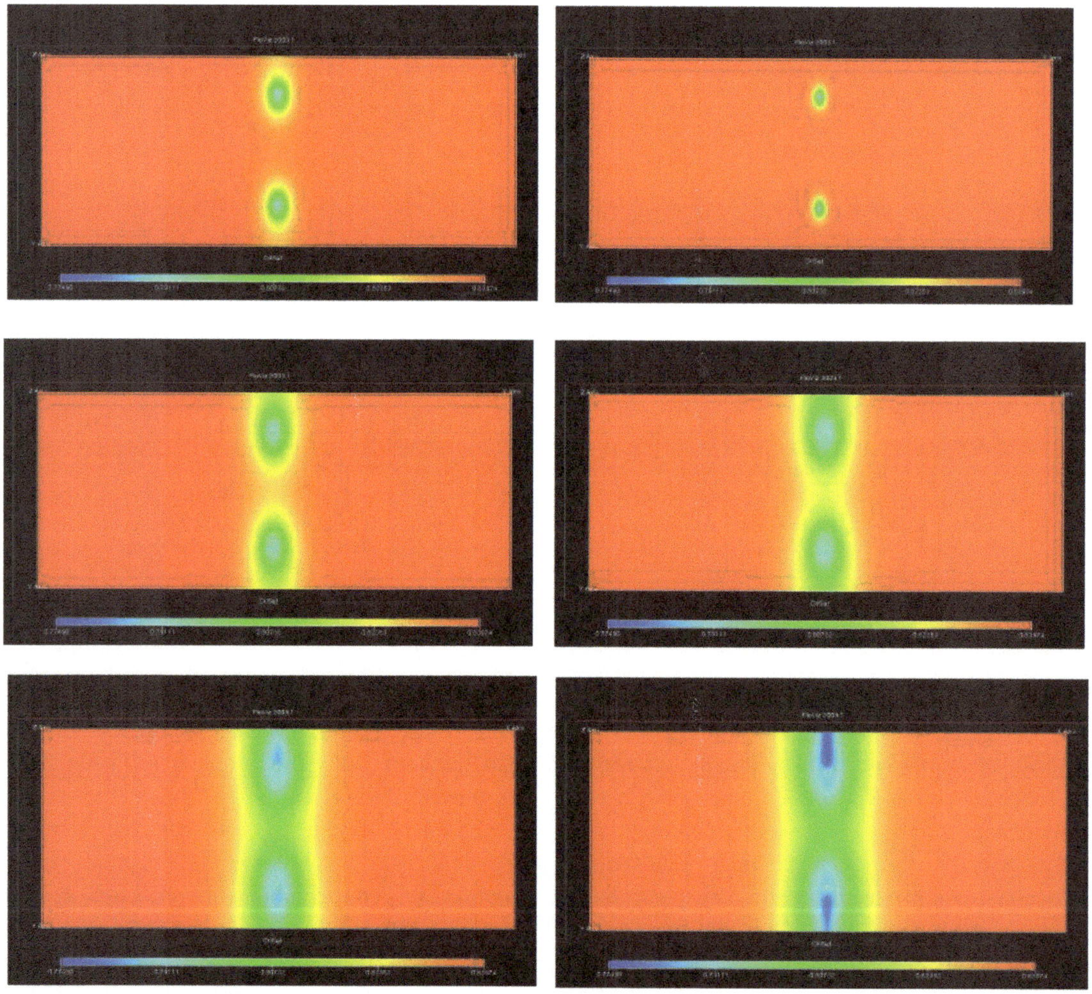

Fig. 4 Simulation results for symmetrical well configuration

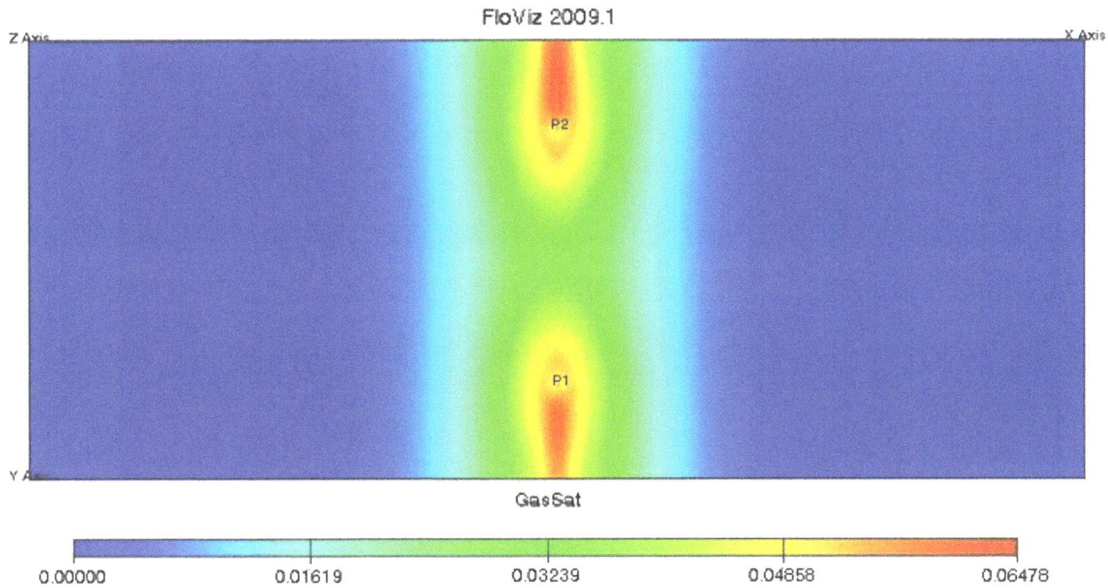

Fig. 5 Gas saturation in the environs of the two producing wells after production

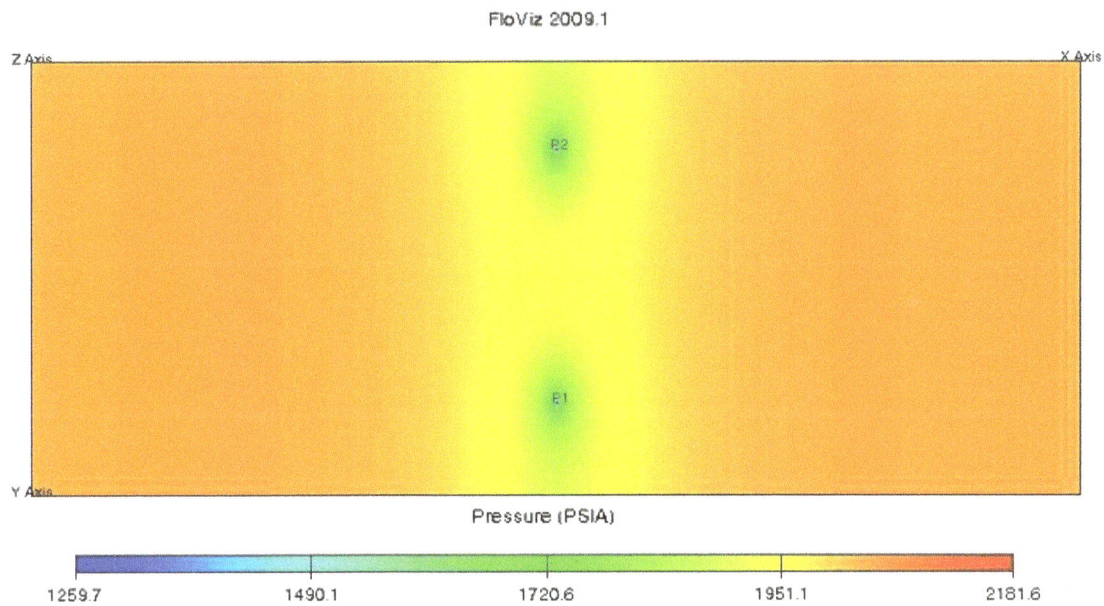

Fig. 6 Reservoir pressure in the environs of the two producing wells after production

Asymmetrical well configuration across borderline

To better investigate the role of well configuration on production from a shared oil reservoir, we define 10 producing wells within both territories producing at 160,000 STBD. The ten wells in the southern and northern countries are drilled 7,360 and 2,112 ft away from the borderline, respectively. As demonstrated by Fig. 7 the wells closer to the boundary produce some oil from their neighboring territory while the more distant wells are unable to do the same. The cross-boundary production of the northern country is 6,000 STBD while the southern country is unable to produce any oil from its neighbor's share of the reservoir fluid.

The origin of this migration is explained through fluid flow equations and the diffusivity equation in particular. The radius of investigation of wells closer to the borderline will reach the borderline and begin production from the area sooner. Even after farther wells begin production from the area near the border, the pressure required for production will not stronger than the opposing pressures of nearer wells. To reduce the harm, the suffering country

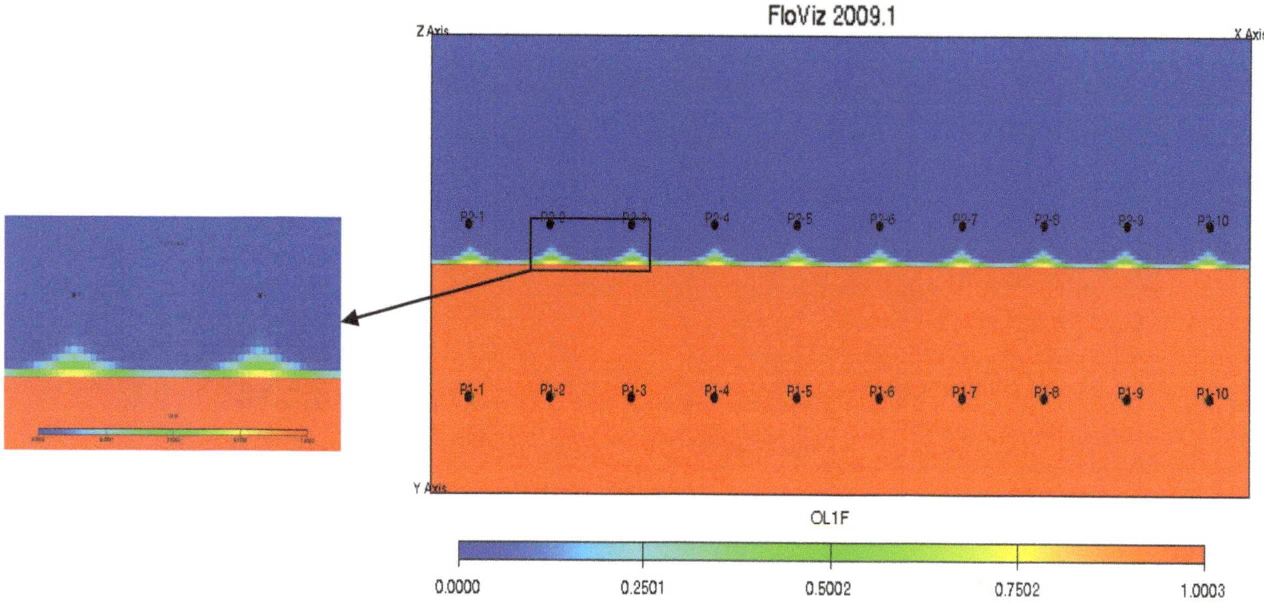

Fig. 7 The effect of asymmetrical well configuration on the oil saturation

may increase production flow rate to intensify the pressure decline or drill other wells at locations closer to the boundary.

Figure 8 demonstrates the reservoir in 40 years of production. Simulation results suggest that the country with wells farther away from the boundary has lower oil saturation and higher gas saturations in its side of the reservoir. Consequently, the country with farther wells will ultimately produce more gas, encounter more pressure decline and will have to abandon the reservoir sooner than the neighboring producer. Therefore, this country will not only lose migrating oil, it will be powerless to preserve optimum production conditions. Our investigations show that if the wells are located exactly on the boundary, as high as 50 % of production consists of fluid originating from the neighbor's share of the shared reservoir. Figures 9 and 10 demonstrate a graphical illustration of our results.

The effect of time of production

Time is a variable in the diffusivity and Darcy equations and will therefore influence fluid flow. To determine the extent to which production time affects hydrocarbon migration, all variables other than time will be held constant. Two parallel rows of ten wells spaced 2 km away from the border line are assumed to produce 160,000 STBD from an isotropic homogenous shared reservoir while the southern country begins production 3 years earlier than its northern neighbor.

Figure 11 demonstrates the reservoir at initial conditions and oil migration 3, 9 and 18 years after beginning production from a shared reservoir. As revealed by the figure, oil migrates toward the country that has begun production sooner. But the consequences caused by the time lag are mostly noticeable when the second country begins production with similar rates. The diffusivity equation suggests that fluid on borderline is produced only after the drainage radius has reached the borderline. The time it takes for a point at radius r to sense the pressure decline is (Ahmed and McKinney 2005):

$$t \propto \frac{r^2}{\eta} \quad \text{where} \quad \eta = \frac{0.000264k}{\phi \mu c_t} \tag{8}$$

According to Eq. 8, the country that has delayed production can minimize losses for by drilling wells that are closer to the borderline.

Simulation results exhibit that even after both countries have begun producing, oil migration continues to occur. It will take almost 15 years for the effects of beginning production after various times to fade away, for our reservoir of study. After 15 years, both countries will be able to produce at equal rates, but the leading country will have ultimately produced more oil. Despite the fact that the second country begins production only 3 years after and eventually produces with higher rates for some time, it will still not be able to produce as much oil as its neighboring country. The hydraulic connection of fluid in reservoir and reservoir pressure will only allow the late producer to save its share from what's left in the reservoir. Therefore, the late producer can never make up for the lost time and continues to be

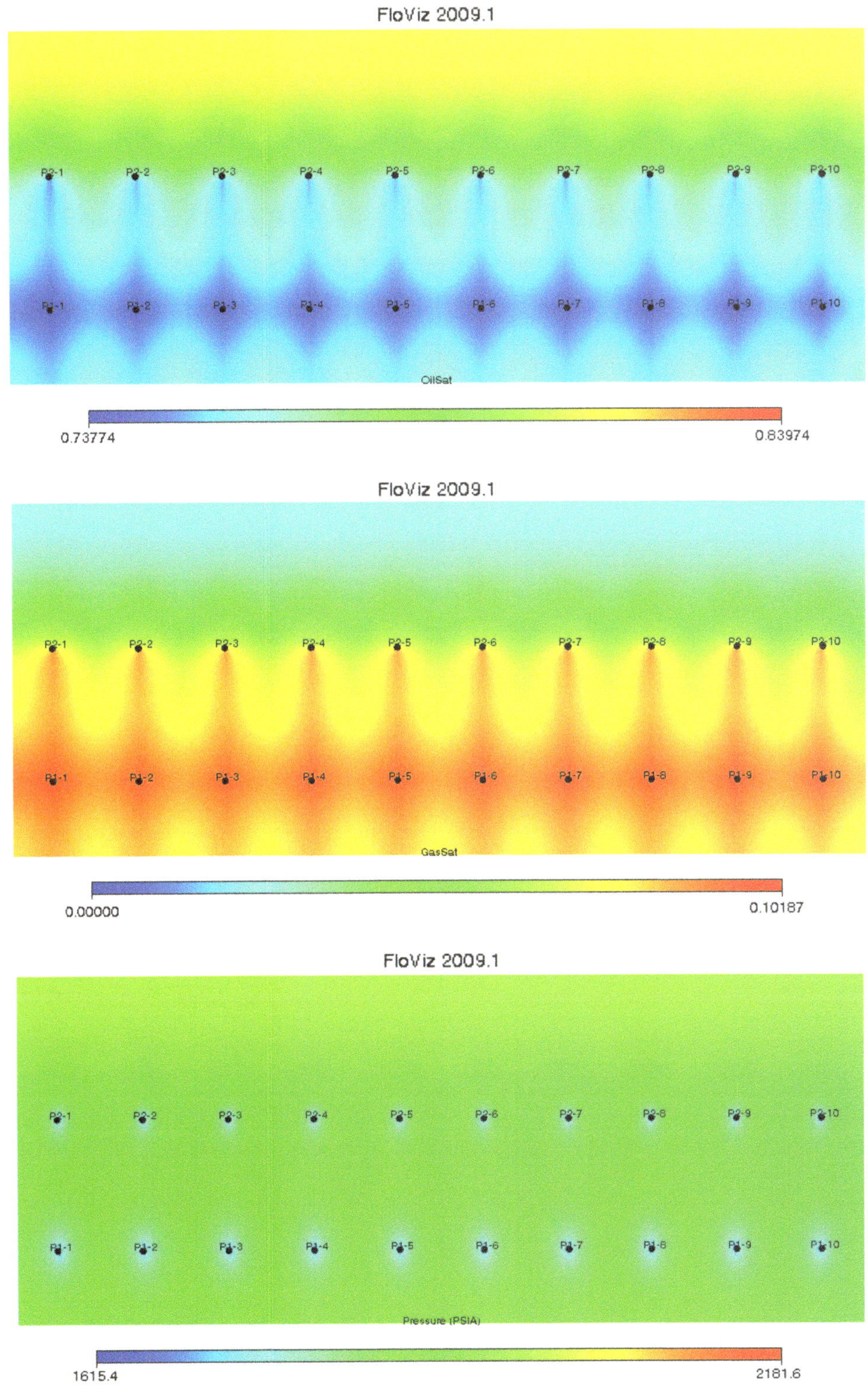

Fig. 8 Effect of 20 wells configured asymmetrically in oil saturation (*top*), gas saturation (*middle*) and pressure (*bottom*) after 13 years of production

bereaved from their share of oil until they begin producing it. Figure 12 illustrates the simulation results.

To compare the effect of time, we increased the time lag from 3 to 6 years. The results were similar. By doubling the time lag, migration across boundary increased twice as much and the final production doubled for the leading country.

The effect of reservoir slope

If a reservoir is tilted, the Darcy and diffusivity equation for pressure will have to be corrected for gravitational forces. To investigate the effect of a reservoir slope, we simulated a reservoir with previous initial data and a dip

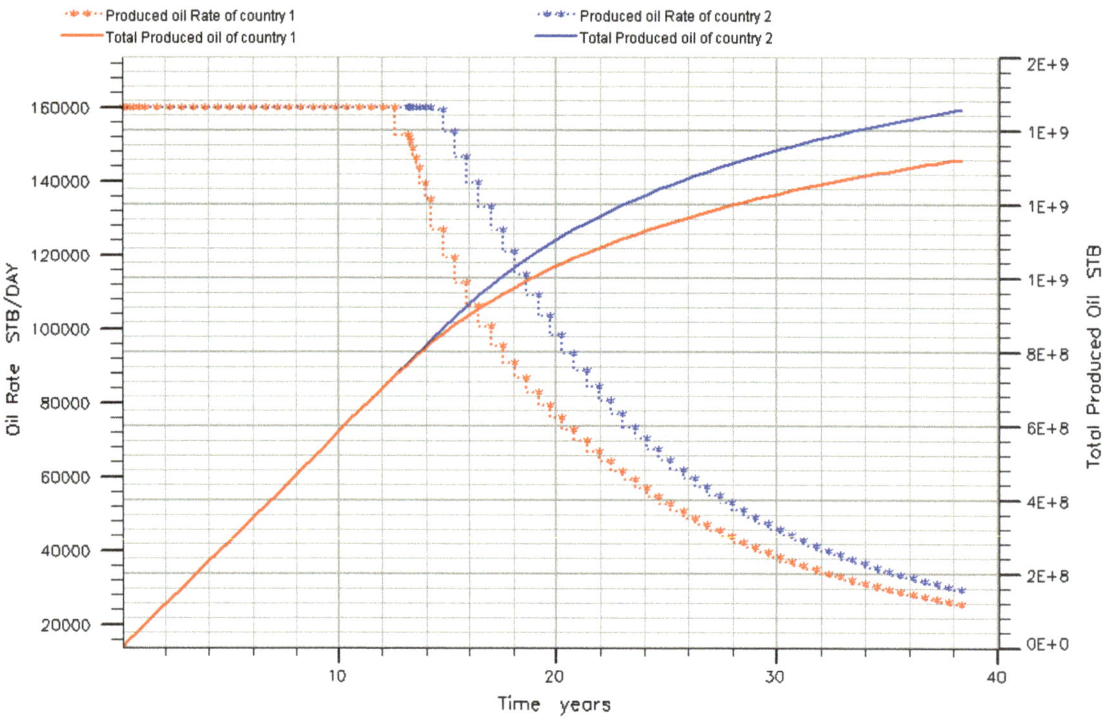

Fig. 9 Comparison of daily oil production and total oil production from asymmetrically located wells in the countries 1 and 2

Fig. 10 Comparison of oil production of each country from neighboring country's part of reservoir

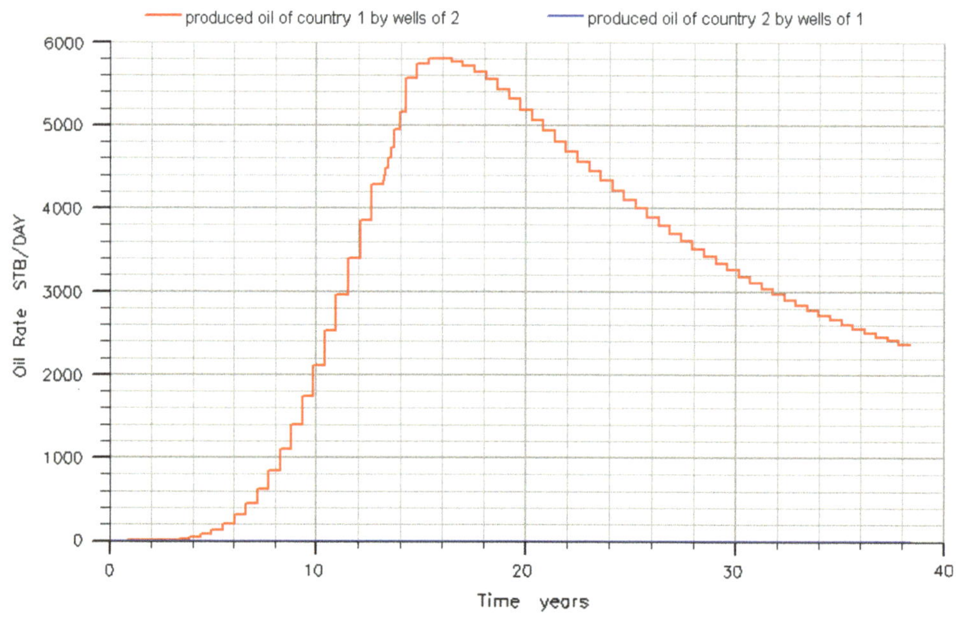

angle of 4.7°. Thus the part of reservoir under country 1 was located deeper down than the other half of reservoir.

Simulation results displayed that after 40 years of production with similar production conditions, the reservoir fluid migrates to the part of reservoir that is situated deeper down. Figure 13 demonstrates the results. The gravitational effects for our simulated reservoir were witnessed after 15 years of production and were measurable due to consistent conditions assumed throughout production time.

Furthermore, when the reservoir dip was increased from 4.7° to 10°, the effects of reservoir slope becomes more severe. Figure 14 illustrates the influence of gravity on two tilted reservoirs having dip angles of 4.7° and 10°.

Conclusively, the reservoir dip has an assessable influence on production. It will promote hydrocarbon fluid migration toward part of reservoir lying lower down and increases production in the same area.

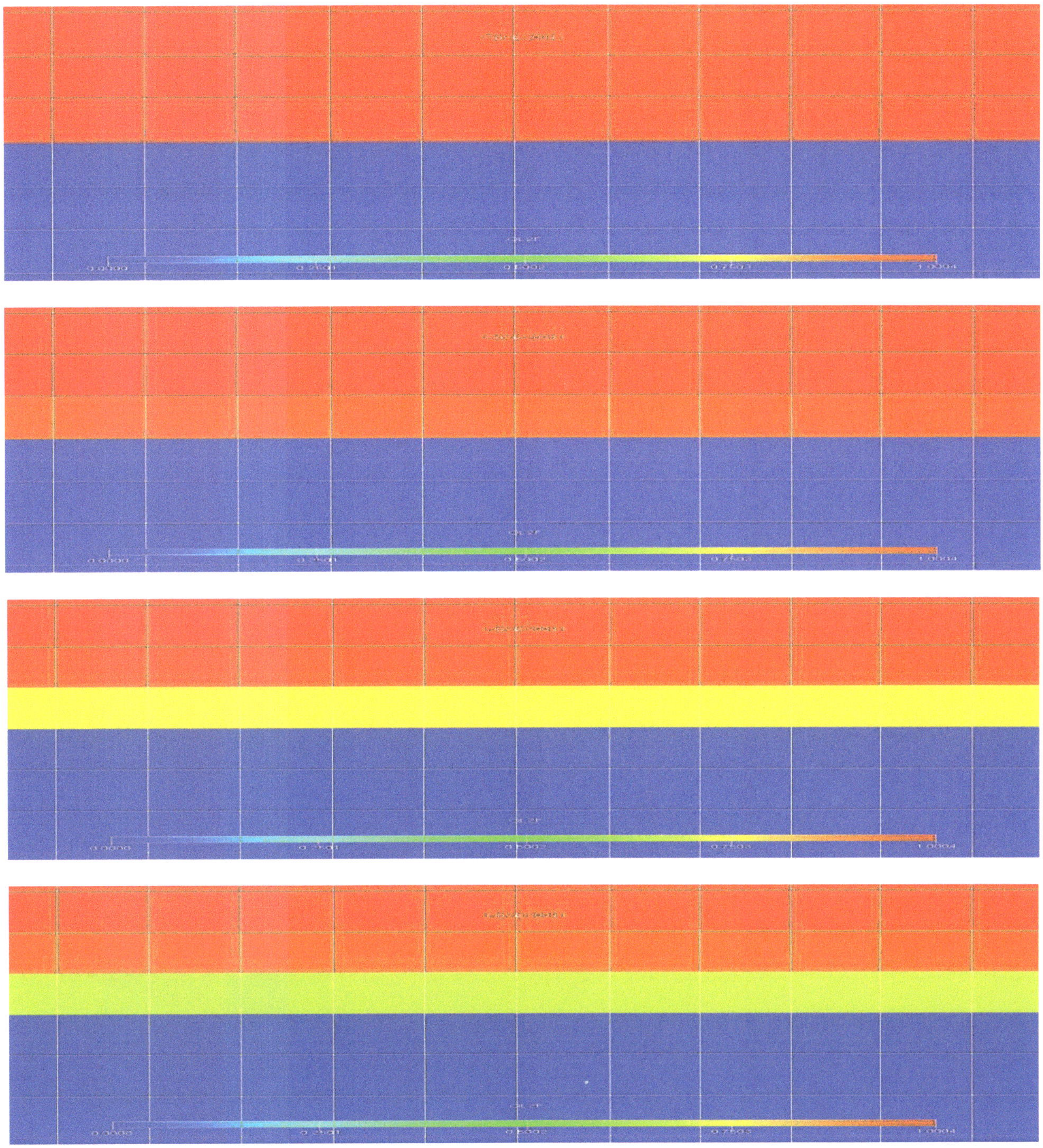

Fig. 11 Comparison of oil saturation at initial conditions (*top*) and migration across geographical border 3 years after (*second down*), 9 years after (*third down*) and 18 years after (*bottom*) the beginning of production

The effect of rock properties

To investigate the role of rock properties in production from a shared hydrocarbon deposit, we assume heterogeneity in the reservoir rock.

Permeability is a constant in the diffusivity equation and therefore we expect variations in permeability to affect fluid flow. To study the effects, permeabilities of 10 and 100 (mD) are presumed for reservoir rock while all other influential factors are similar.

Simulation results demonstrate oil residing in part of reservoir with lower permeability migrates toward where permeability is higher. According to the diffusivity equation, the drainage radius develops more rapidly where

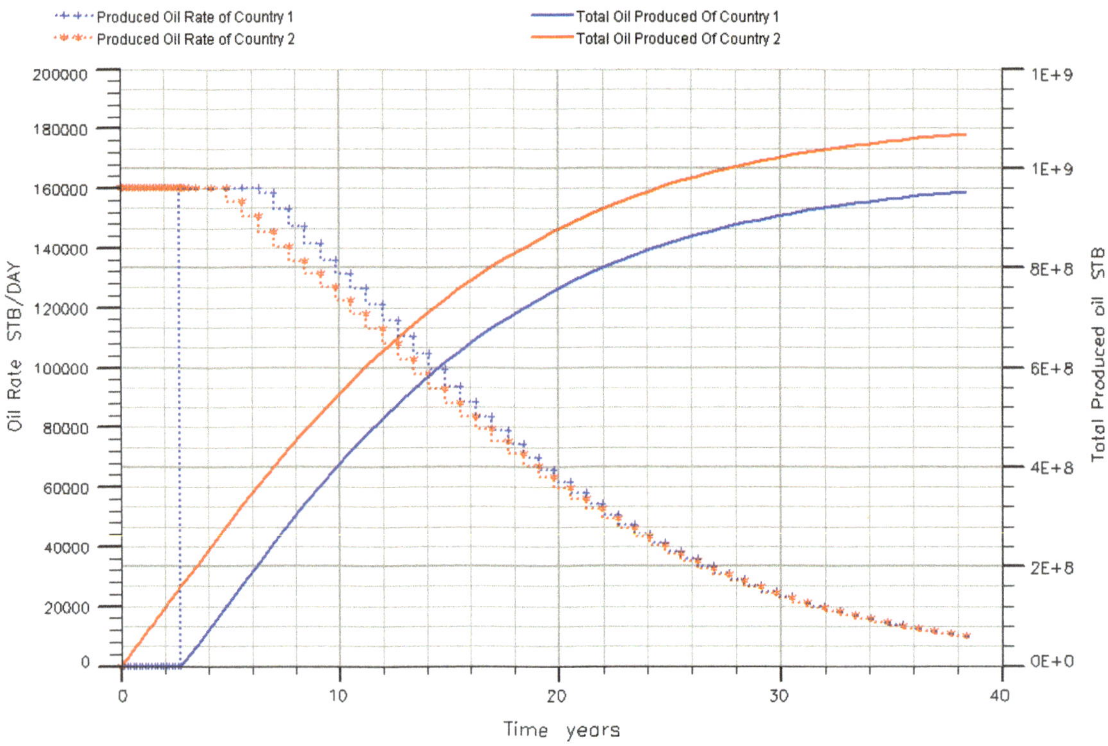

Fig. 12 Comparison of daily production flow rate and total production from a shared reservoir when country 1 begins production three years after country 2

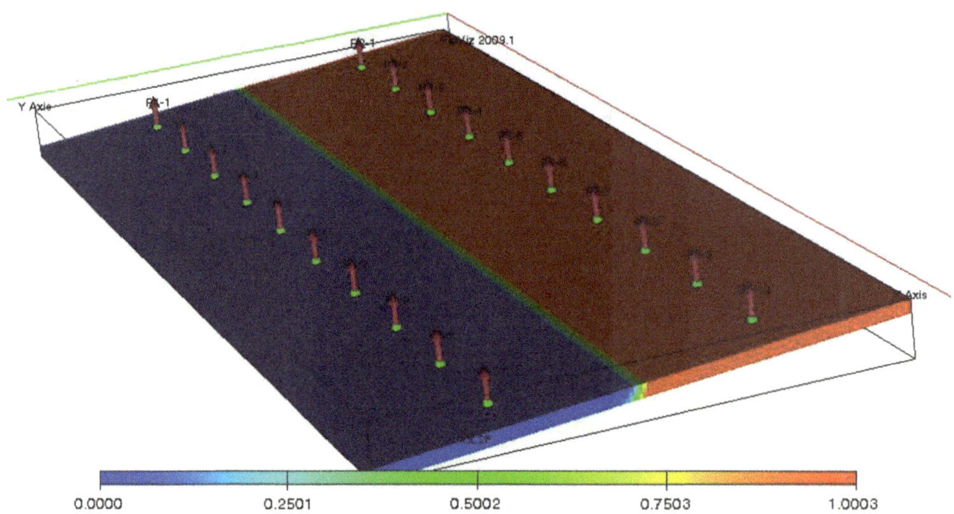

Fig. 13 Oil migration across borderline in a tilted reservoir with a dip angle of 4.7° after almost 40 years of production

permeability is higher. Therefore production from the borderline will occur sooner where permeability is greater (Ahmed and McKinney 2005). Figure 15 illustrates total production and production flow rate of each country from such a reservoir. The graphs demonstrate that although territory with higher reservoir rock permeability will begin production at a higher rate, but production potential will decline faster that neighboring country with reservoir rock of lower permeability. Thus eventually, production within part of reservoir with lower permeability exceeds production rates encountered in neighboring country. Furthermore, since lower mobility ratios ensues lower permeability, the drainage radius will take longer to develop and more oil will be in place at any time, as demonstrated in Fig. 16.

Other than permeability, porosity is another rock property that influences fluid flow. Porosity affects production in two ways. First, higher porosities indicate presence of more oil in place and second, the diffusivity equation demonstrates that porosity is directly related to the time it takes for drainage radius to develop. Therefore, the lower

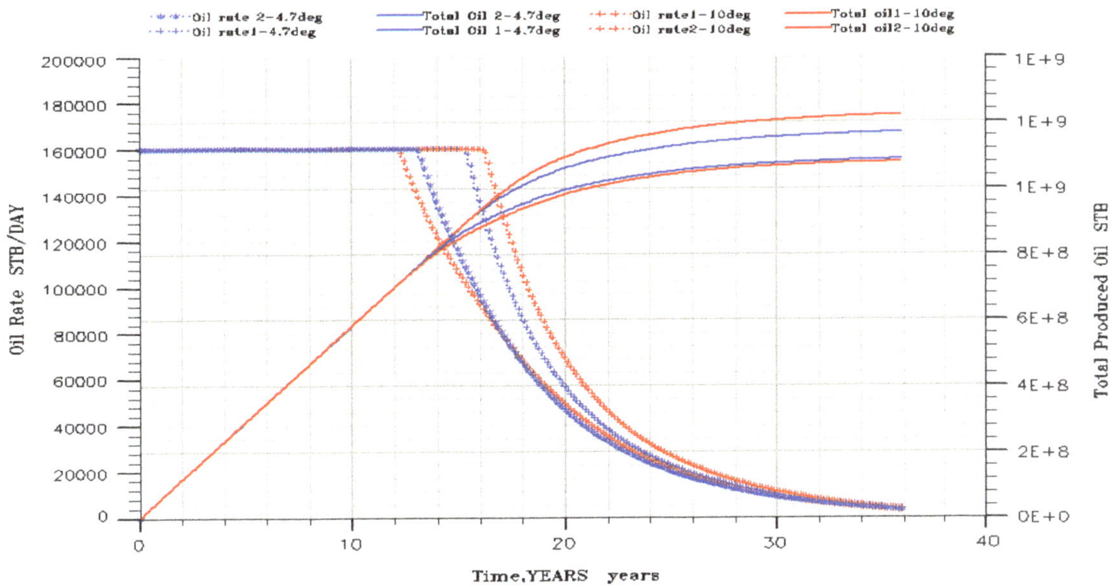

Fig. 14 Effect of dip angle on the oil flow rate and total oil production from a shared reservoir

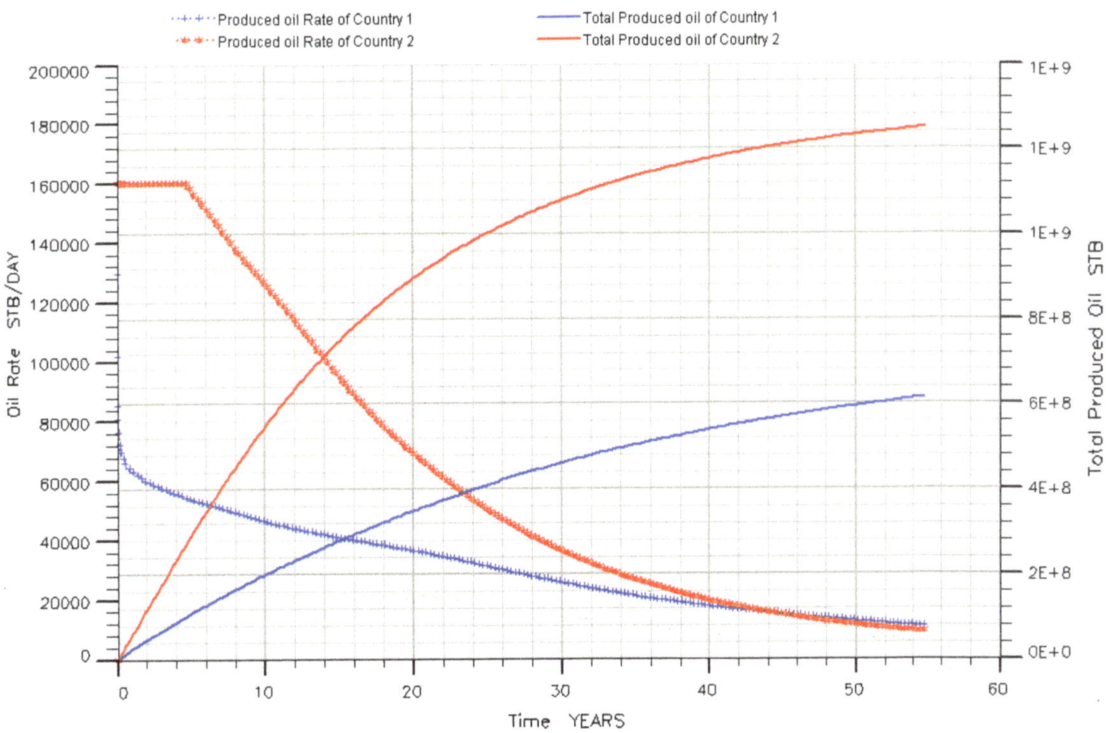

Fig. 15 Effect of permeability on the oil production flow rate and total production from a shared reservoir

the porosity is, the more rapidly the well drainage radius will reach the borderline.

Porosities of 10 and 30 % have been assigned to reservoir within country 2 and 1, respectively. Results demonstrate that reservoir with lower porosity produces less oil. But the oil produced, although less than the neighboring country was more than what was initially present within the pores of less porous reservoir rock, considering the lower available oil in the less porous rock. Simulation results exhibited that even though the porosity of country 2 to 1 was 1/3, the ratio of total produced oil of each country was 13/17. Since 13/17 is larger than 1/3, country with less porous reservoir rock produced more than its initial oil in place of 1/3. Figure 17 depicts the simulation results.

Although rock properties such as permeability and porosity are not elective properties and can hardly be

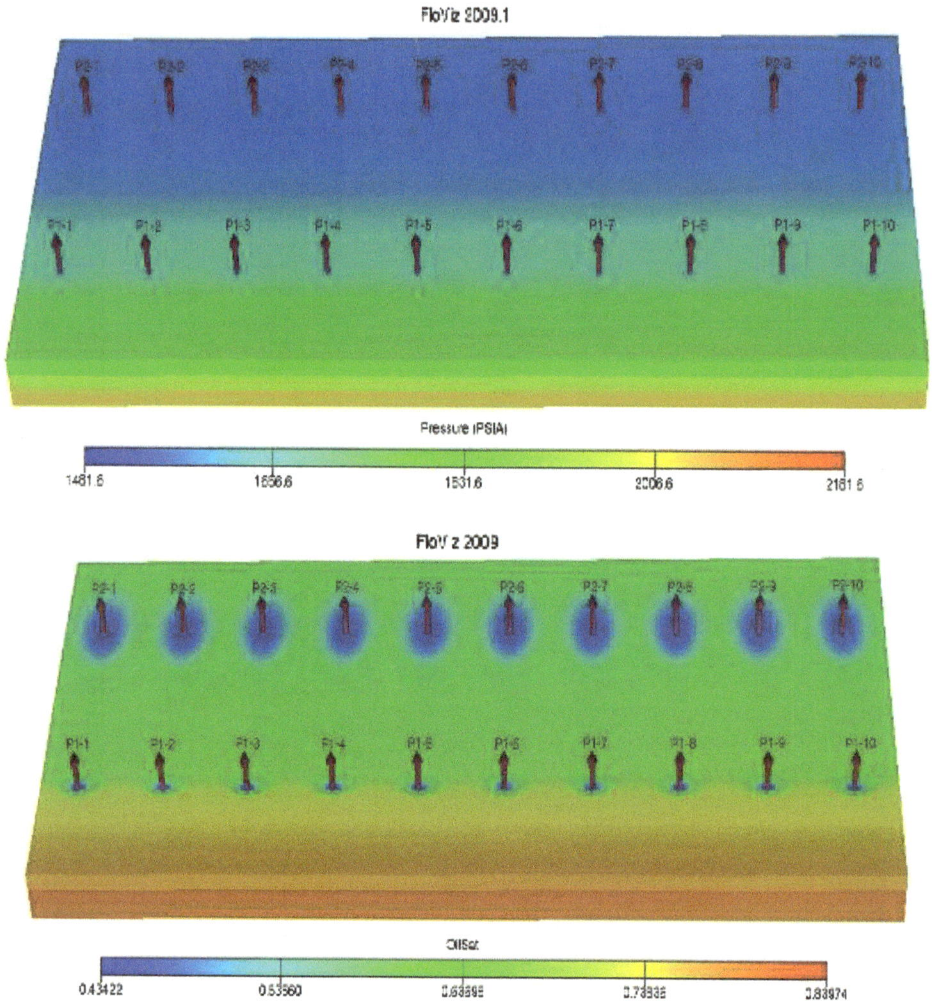

Fig. 16 Comparison between pressure (*top*) and oil saturation (*bottom*) in a shared reservoir with varied permeabilities of 10 md (northern country) and 100 md (southern country)

altered in large scales, but understanding their role and selecting means to manage their influences is vital when producing from a shared reservoir. Accordingly if a reservoir rock demonstrates heterogeneity, the long term effects of the rock properties has to be taken into account when planning reservoir production scenarios.

The effect of fluid properties

Fluid properties, similar to rock properties, affect production via two means. First, fluid properties directly affect production and second, according to the Darcy and diffusivity equations, fluid characteristics such as oil formation volume factor (B_o), viscosity (μ) and compressibility factor (c) influence fluid flow.

According to the Darcy equation, oil with higher formation volume factor (B_o) will be produced less than oil with smaller B_o when other influential factors are identical. The diffusivity equation constant, given by Eq. 6, suggests that viscosity and oil compressibility factor play roles that are analogous to the role of porosity.

Due to the resemblance of results to findings derived from analyses on rock properties, we exempt simulation results for this section.

The role of an aquifer

The presence of an aquifer, its ability to maintain pressure, outer boundary conditions, flow regimes and flow geometries directly influence production rates. For optimizing production, each country needs to study characteristics of aquifer in connection with their side of the reservoir. For this reason we've studied the role that an aquifer plays in a shared reservoir. We study the influence of an edge water drive on a reservoir-aquifer system.

A strong aquifer can impede pressure drop and maintain flow rates for longer periods of time by sweeping oil toward production wells of each country. The aquifer,

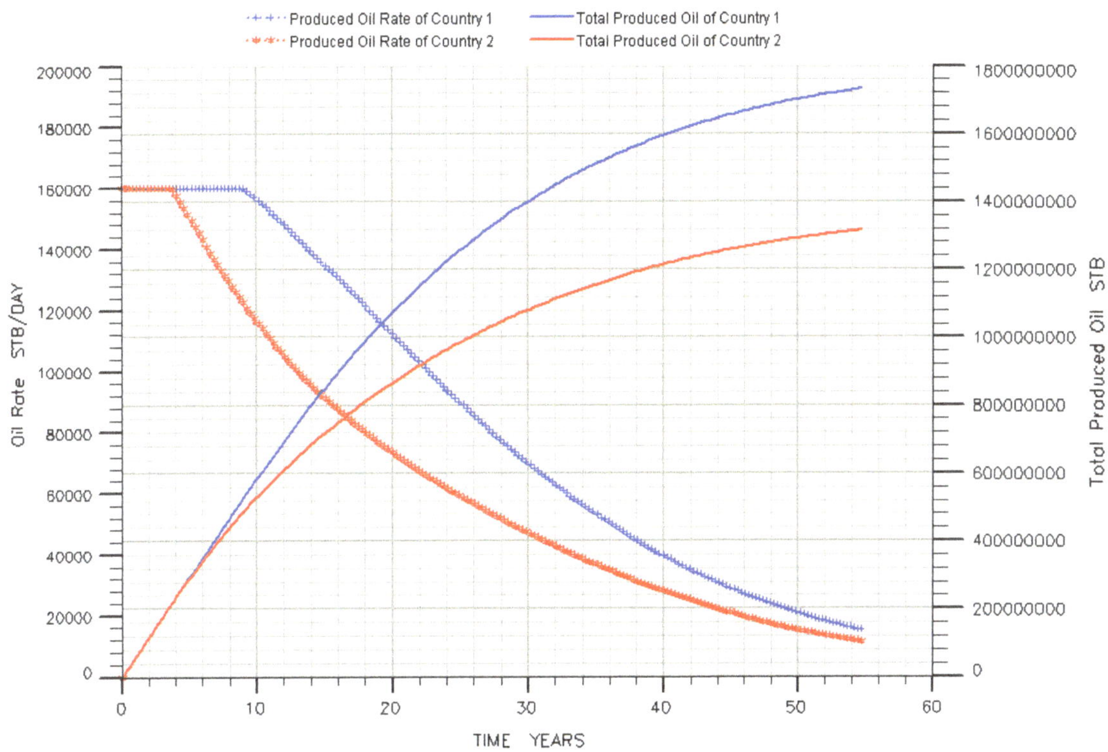

Fig. 17 Effect of porosity on the oil production flow rate and total production from a shared reservoir

Fig. 18 Oil saturation distribution in a producing shared oil reservoir with an edge aquifer

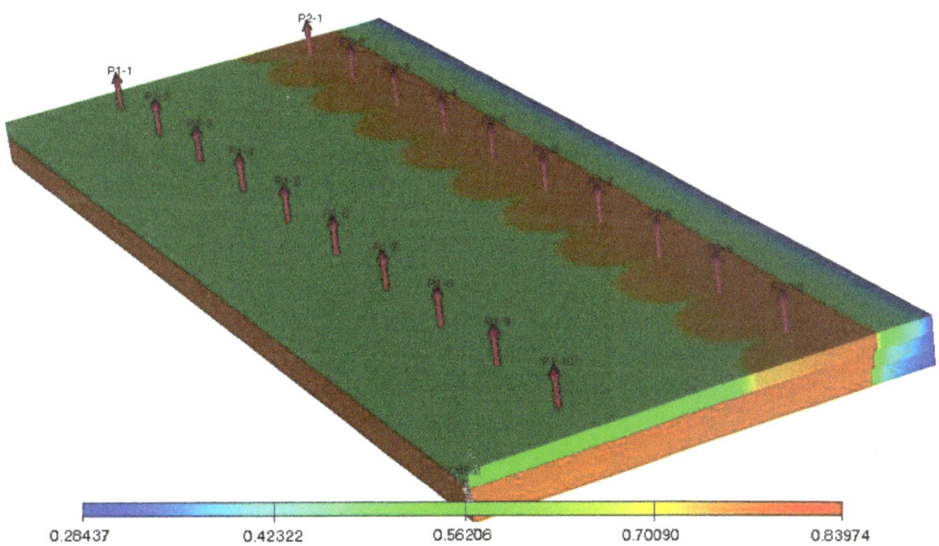

based on its capacity, can also impose oil migration. Figure 18 exhibits oil saturation in a shared oil reservoir with an edge waters that has been producing for 40 years. Although an aquifer may facilitate oil migration to the neighboring country, it will ultimately result in more oil production within part of the reservoir which it exists. As displayed in Fig. 19, at locations where the aquifer has entered the oil zone, oil has been swept toward production wells and consequently, even after 40 years of production, pressure and production rates have been maintained at desirable values of and 800 psi pressure loss and 40,000 STBD (Ahmed and McKinney 2005).

Evaluating joint management versus independent management of a shared reservoir

We intend to investigate the effect of a cooperative plan for reservoir development and production. If the results

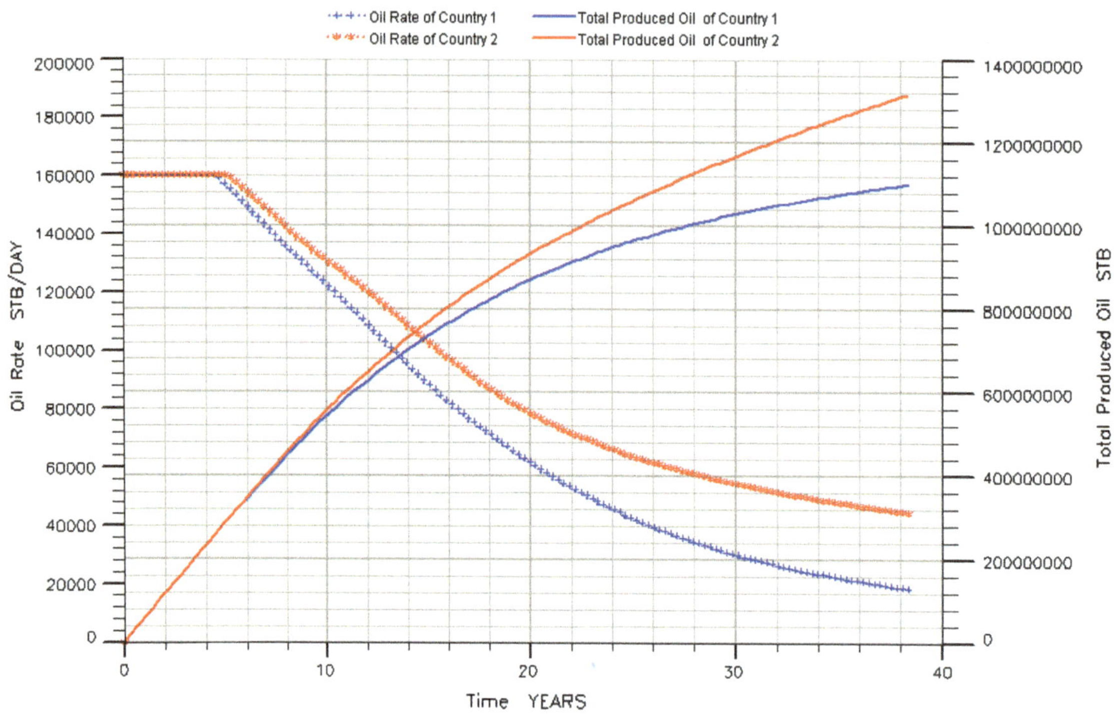

Fig. 19 Comparison of production oil rate and total production from a reservoir shared by a country with an edge aquifer (country 2) and without an aquifer (country 1)

demonstrate that the ultimate recovery of the reservoir will benefit both countries more than they can hope for when competitively exploiting the shared reservoir, then we can be hopeful to see the neighboring countries negotiating collaborative strategies for reservoir development.

Previously, we had assumed 20 flowing wells, 10 across each borderline producing at 160,000 STBD within each territory. Now we define 20 wells with 320,000 barrels of overall daily production configured as demonstrated in Fig. 20 bottom. Simulation results displayed in Fig. 21 demonstrate there isn't much difference between cooperative and competitive production from the shared reservoir. Despite the fact that central wells will allow production from a larger areal extent, but wells in previous configuration already encompassed the region as part of their drainage area. According to Fig. 21, total production and production rates confirm that a collaborative production may result in less production from the reservoir in comparison to independent production. Conclusively, if producers of a shared reservoir optimize production conditions for best ultimate recovery and least harm to the reservoir, there isn't much distinction between independent and cooperative production.

Even so, when countries are exploiting a reservoir independently, their sole objective is to produce more oil and disallow migration of hydrocarbon fluid to neighboring country and consequently not much attention is given to the reservoir or optimized production. The unscientific approaches taken by competing countries will inevitably decrease the ultimate recovery from a shared reservoir and this intensifies the importance of reaching a deal for production.

Conclusions

Initial study on the relationships of fluid flow in porous media, the Darcy and diffusivity equation in particular, suggest that the six influential factor of production from a shared reservoir are wells configuration, time of production, reservoir slope, aquifer performance, rock properties and the fluid properties across reservoir. Out of these 6 influential factors, the first two are determined by exploitation companies and the other four are intrinsic characteristics of the reservoir that with identifying them, we can optimize production by appropriate production schemes.

Simulation results illustrate that cooperative management of the shared reservoir isn't much different to when producing with similar production rates. Independent production may even lead to more produced oil for one country.

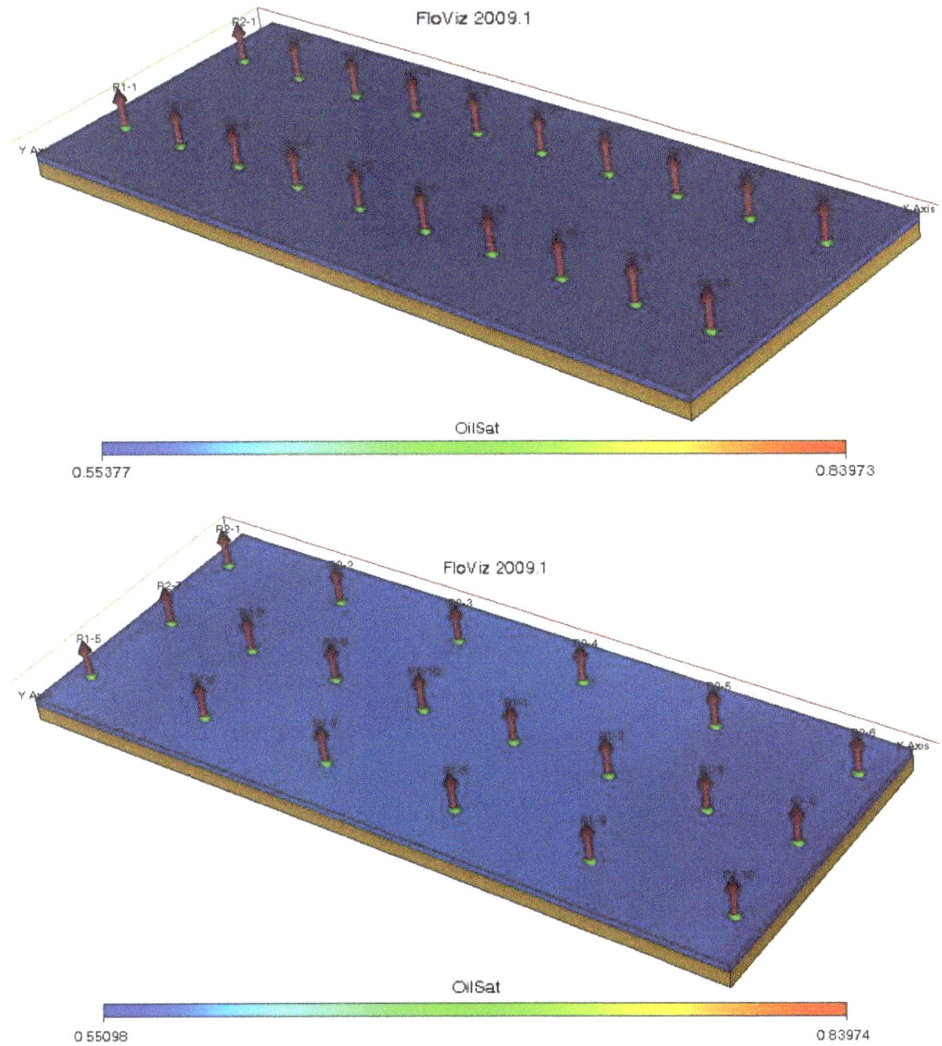

Fig. 20 Comparison oil saturation in a shared reservoir after 40 years of production when the reservoir is independently produced (*top*) and jointly developed (*bottom*)

Investigation on time of production demonstrated that the country that begins production first will produce more oil than its neighboring country with equal number of wells and production rates. The country that begins production at a later time will only be able to save its share from what is left at the time of commencing production and will not be able to make up for the lost time.

We also concluded that the closer the wells are to the borderline, the higher is the chance of cross-border production. Oil tends to migrate towards the producing wells closer to the borderline. It's evident that the effect of time is best observable at the beginning of production but influences of well configuration are noticed at later times.

The effect of the reservoir slope exhibited that oil tends to migrate toward deepest part of the shared reservoir. This effect is usually observed halfway through reservoir production lifetime and only depends on the dip of the reservoir. As the dip increases the gravitational force also increase and migration intensifies.

Furthermore, Aquifer in the form of edge water can induce migration of oil out of the boundary but will hamper pressure decline and maintain flow rates and improve production in the country in which it is active. The benefits of an active edge aquifer are seen after few years of production and endure until reservoir abandonment.

Reservoir heterogeneities such as variations in rock or fluid properties can also influence production and ultimate recovery. The influences of porosity, permeability, viscosity and compressibility can be specified through the constant of the diffusivity equation given by Eq. 6 as $\eta = \frac{0.000264k}{\varphi \mu c_t}$. Considering $t \propto \frac{r^2}{\eta}$ which determines the time and therefore the velocity at which the radius of investigation progresses. Larger values (i.e. higher porosity, lower permeability, lower viscosity or lower compressibility.)

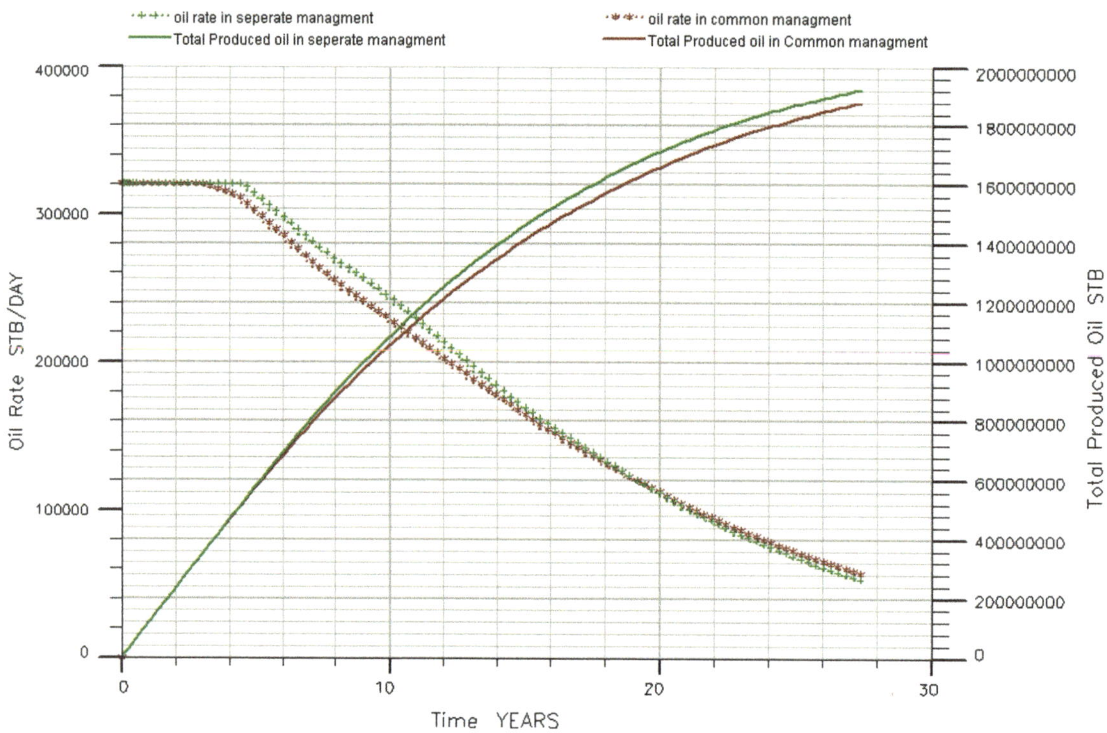

Fig. 21 Comparison of daily production rate and total production in a shared reservoir after 40 years of production by two addressed production scenario

indicate higher production rates and oil tends to migrate toward areas where η is largest.

With all factors discussed, we can say that in a shared reservoir there is high competition between neighboring owners of the reservoir and it seems that the champion of such competition is the country who disregarding the reservoir's natural aspects like aquifer, rock and fluid property, will approach production with optimized engineering schemes. Consequently, independent countries that seek their nation's best interest should preferably negotiate a reservoir development plan with their neighboring country that allows fair exploitation and optimizes production in the long run.

Acknowledgments Special thanks go to the Arvandan Oil and Gas Company for providing required data and implementing commercial Eclipse software of (AOGC) with license agreement.

References

Ahmed T, McKinney PD (2005) Advanced reservoir engineering - unconventional gas reservoirs. Elsevier, Amsterdam, pp 187–290

Al-Hussainy R, Ramey HJ Jr, Crawford PB (1966) The flow of real gases through porous media. J Petrol Technol 18(5): 624–636

Ayala LF, Kouassi JP (2007) The similarity theory applied to the analysis of multiphase flow in gas-condensate reservoirs. Energy Fuels 21(3):1592–1600

Hou J, Luo F, Wang C, Zhang Y, Zhou K, Pan G (2011) Quantitative prediction model for the water–oil relative permeability curve and its application in reservoir numerical simulation. Part 1: modeling. Energy Fuels 25(10):4405–4413

Weems P, Fallon A (2012) Strategies for development of cross-border petroleum reservoirs. King & Spalding: Energy Newsletter, Transactional Transactions

14

A novel, streamline-based injection efficiency enhancement method using multi-objective genetic algorithm

Mohammad Amin Safarzadeh · Mehdi Motealleh · Jamshid Moghadasi

Abstract Water flooding is one of the most important techniques of improved oil recovery. However, two major problems, loss of injected water to the aquifer and unwanted water production, reduce the efficiency of water flooding. An improper allocation of water to injection wells usually is the main reason of these problems. In this paper, both multi-objective genetic algorithm (MOGA) optimization, which is more rapid and accurate than conventional GA, and streamline simulation with the unique advantages of determination of flow path and participation of each injection well in total field oil production, were used in order to appease the mentioned problems. All previous studies have tried to optimize the injection rates based on injection efficiencies that were defined with the application of streamline simulation, while using MOGA optimization and Pareto concept always let us select the best global solutions with regard to the other defined criterions. Final solution in MOGA optimization is a set of correct answers. So, five scenarios, including different restrictions and economic situations were introduced. Best solution was obtained and compared with the common method of water injection optimization exclusively based on improving the efficiencies of injection wells. Results show that MOGA optimization always offers the best solution and all MOGA scenarios have better proficiencies than common optimization methods. Comparison of the proposed methodology in this paper with conventional workflow shows that MOGA, in the best case, can increase total oil production up to 6.5 %, and after considering all limitations, it can increase total oil production up to 4 % and decrease the loss of water to the aquifer to about 26 % in comparison with the common workflows.

Keywords Injection efficiency · Genetic algorithm · Multi-objective optimization · Streamline simulation · Pareto front · Waterflooding

Introduction

Water flooding an oil field has been a widely accepted method for increasing a reservoir's recovery since 1950s. Possible problems associated with waterflood techniques due to unsuitable conditions are destructive effect on fluid transport within the reservoir and early water breakthrough. Therefore, knowing the allocation of flow between well pairs is the starting point of any technique to balance well patterns in waterflood process. Conventionally, allocation numbers have been determined using sophisticated empirical methodologies and then replaced by an immediate and rigorous streamlines solution method (Thiele 2001).

An important component in optimizing field performance is to be able to compare and rank the efficiency of injectors. One of the useful information provided by the streamlines is well allocation factors (WAF). This quantifies the amount of flow in a particular well due to other wells in the system. Since streamlines start at a source and end in a sink, it is possible to determine which injectors (or part of an aquifer) are supporting a particular producer and

M. A. Safarzadeh
Research Institute of Petroleum Industry, Tehran, Iran

M. Motealleh
Computer Aided Process Engineering Group, Tehran, Iran

J. Moghadasi
Petroleum University of Technology, Ahwaz, Iran

vice versa (Thiele 2001). Instead of moving fluids from cell-to-cell, streamline simulation breaks up the reservoir into one-dimensional systems, or tubes. The transport equations are then solved along the streamlines using the concept of time-of-flight. Decoupling the transport problem from the underlying 3D geological model results in a significant computational efficiency while minimizing numerical diffusion and grid orientation effects (Thiele et al. 1997; Samier et al. 2001).

Waterflooding optimization

Thiele and Batycky (2006) proposed an approach to predict well rate targets of injectors and producers to improve waterflood management (Appendix). Their approach is based on streamline simulation and the injection efficiency for injector–producer pairs. Assuming the fixed well status in mature fields with a constraint of total available water, Thiele and Batycky demonstrated improved waterfloods by reallocating injection water from low-efficiency to high-efficiency injectors. The methodology was also implemented in naturally fractured reservoirs (2007) by Iino and Arihara (2007). Alhuthali et al. (2007) proposed a practical approach, based on equalizing the arrival times of the waterflood front at all producers for computing optimal injection and production rates. Almost all of the previous studies were based on optimizing the current injection efficiencies without any knowledge of initial rates while in this study, the injection rates are allocated in the field initially in the best way by the means of multi-objective genetic algorithm.

Multi-objective genetic algorithm optimization

Multi-objective formulations are realistic models for many complex engineering optimization problems. In many real-world problems, objectives under consideration conflict with each other, and optimizing a particular solution with respect to a single objective can result in unacceptable results with respect to the other objectives (Konak et al. 2006). Genetic algorithms are global optimization techniques, which means they converge to the global solution rather than to a local solution. The specific mechanics of the algorithms involve the language of microbiology and, in developing new potential solutions, mimic genetic operations (Marler et al. 2004). A general multi-objective optimization problem can be described as a vector function f that maps a tuple of m parameters (decision variables) to a tuple of n objectives. Formally:

$$\text{Min/Max } y = f(x) = (f_1(x), \ldots, f_n(x))$$
$$\text{subject to } x = (x_1, x_2, \ldots, x_m) \in X, y = (y_1, y_2, \ldots, y_m) \in Y,$$

where x is called the decision vector, X is the parameter space, y is the objective vector, and Y is the objective space. After the first pioneering work on multi-objective evolutionary optimization in the 1980s (Schaffner 1984, 1985), several different algorithms have been proposed and successfully applied to various problems. Zitzler puts forward the Strength Pareto EA (SPEA) which had been applied it to the multi-objective problem successfully (Xie et al. 2005).

Classical multi-objective optimization methods transform multi-objective functions into a single objective function through the evaluation function which has some disadvantages such as essential need for providing weights, non-uniform distributed optimal solution due to local optimal searching algorithm and to repeatedly search algorithm for alternative solution (Xie et al. 2005).

The other general approach in multi-objective optimization is to determine the entire Pareto-optimal solution sets which are often preferred to a single solution, because they can be practical when considering real-life problems (Konak et al. 2006). They give the engineers the option to assess the trade-offs between different designs (Xie et al. 2005; Sbalzariniy et al. 2000). Pareto-optimal sets can be of varied sizes, but the size of the Pareto set usually increases with the increase in the number of objectives (Iino and Arihara 2007).

As GA requires little knowledge about the problem being solved, and they are easy to implement, robust, and inherently parallel, they can be effective regardless of the nature of the objective functions and constraints (Marler et al. 2004; Sbalzariniy et al. 2000). Being a population-based approach, multi-objective GA is well suited to find Pareto-optimal front. The ability of GA to simultaneously search different regions of a solution space makes it possible to find a diverse set of solutions for difficult problems with non-convex, discontinuous, and multi-modal solutions spaces. The crossover operator of GA may exploit structures of good solutions with respect to different objectives to create new non-dominated solutions in unexplored parts of the Pareto front. In addition, most multi-objective GA does not require the user to prioritize, scale, or weigh objectives. Therefore, GA has been the most popular heuristic approach to multi-objective design and optimization problems (Konak et al. 2006).

Methodology

Multi-objective genetic algorithm optimization of injection rate determination procedure was preceded according to the following workflow:

Table 1 Multi-objective GA parameter values used in waterflooding optimization

Crossover fraction	Migration fraction	Pareto front fraction	Maximum generation	Function tolerance
0.8	0.2	0.34	100	10^{-6}

1. Hundred sets of well water injection rates are constructed by GA (genetic algorithm) as an initial random population. For the considered population, injection efficiencies of all wells in each set are calculated using the following equation by means of the streamline simulation as the initial population cost in the GA calculations. Streamline simulation is used to calculate offset oil production rate of each injection well in all sets.

$$IE = \frac{\text{offset oil production rate}}{\text{water injection rate}}$$

Therefore, in this case injection efficiencies would be defined as follows:

$$IE_1 = \frac{\sum_{i=1}^{i=5} \text{offset oil production rate of } P_i \text{ due to } I_1}{\text{water injection rate of } I_1}$$

$$IE_2 = \frac{\sum_{i=1}^{i=5} \text{offset oil production rate of } P_i \text{ due to } I_2}{\text{water injection rate of } I_2}$$

2. Population breeding process is implemented by defining the objective functions as simultaneous maximization of the population cost. This maximization process begins by non-dominated sort of chromosomes and then tournament selection of them. Multi-objective GA parameter are shown in Table 1.
3. In all scenarios, the final population is obtained around 50 generations when the function tolerance limit was reached. The best sets of answers for the objective functions (simultaneous maximization of injection efficiencies) are chosen by the means of Pareto solution set. Members of this improved generation are the most suitable sets of well water injection rates.
4. Although all Pareto-optimal solutions are the correct answers of objective functions, decision-making of the final optimum set must be implemented directly based on the desired criteria. Therefore, various scenarios are presented in this study. These scenarios are generated with consideration of different economic conditions, water production limitations and fluid loss to the aquifer. Any of these scenarios can be used for different approaches of field management and preservation.

Scenario A Minimum distance to average efficiency. The best answer can be the one with the minimum difference between wells injection efficiencies and average injection efficiency. In two-objective function Pareto algorithm solution sets, the best answer could be the nearest answer to the line $Eff_{well1} = Eff_{well2}$.
In this case, the selected injection efficiencies are the most balanced efficiencies.
Scenario B Highest oil production. The best answer can be the one with the highest oil production rate set in Pareto solutions.
Scenario C Lowest water production scenario. The best answer can be the one with the lowest water-cut set in Pareto solutions.
Scenario D Lowest loss water. The best answer can be the one with the lowest edge water loss to the aquifer set in Pareto solutions.
Scenario E Highest defined optimizing parameter scenario. The best answer can be the one with the maximum defined optimizing parameter set in Pareto solutions:

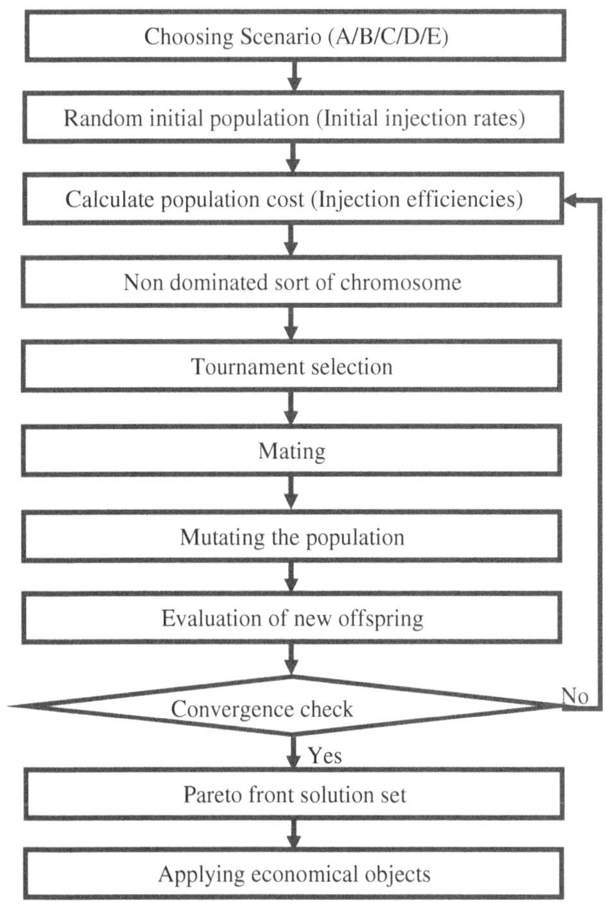

Fig. 1 Workflow diagram of a GA multi-objective optimization process

Defined optimizing parameter

$$= \left(\frac{\text{Oil production rate}}{\text{Water cut} \times \text{Edge fluid loss}} \right)$$

5. The final reservoir condition at the end of the first time step will be generated by streamline simulation of the best answer determined from one of the previous section scenarios.
6. To obtain the most suitable allocation of well water injection rates, considering constant field water injection rate for the following time steps, sections 1–5 are repeated. Workflow diagram of a GA multi-objective optimization process is shown in Fig. 1.

Synthetic model description

Synthetic model, with two injectors and five producers was used to implement different multi-objective GA scenarios and compared with conventional optimization workflow. The average reservoir rock and fluid properties are shown in Table 2. In order to consider well location heterogeneity, wells were located with specified different values of well spacing between 1,000 ft in I2-P3 and 5,000 ft in I1-P1.

Figure 2 shows well locations, oil saturation distribution and producers/injectors relationship through streamlines at the end of the first time step, after 6 months.

Reservoir properties distribution of porosity and permeability are implemented in such way that a non-homogeneous synthetic reservoir is created. To examine the effect of aquifer in multi-objective GA optimization of injection efficiencies a strong aquifer is defined and connected to the top and right sides of reservoir. Producers liquid rate and total injectors water rate target are listed in the Table 3. Initial water injection rate of 500 Stb/day is executed for each injector in the conventional streamline simulation optimization method for 6 years.

Table 2 Reservoir rock and fluid properties

Properties	Value
Dimension	$20 \times 20 \times 4$
Cell size (Dx × Dy × Dz) (ft)	$600 \times 600 \times 120$
Initial water saturation (%)	17
Initial reservoir pressure (Psig)	4,000
Average porosity (%)	14.5
Average permeability (md)	25
Water density (Lb/ft^3)	62.50
Oil density (Lb/ft^3)	50.0
Water viscosity (Cp)	1.0
Oil viscosity (Cp)	0.82
Water compressibility (Psi^{-1})	4.00e-006
Rock compressibility (Psi^{-1})	5.68E-06

Results and discussion

In order to optimize the injection rate portions, genetic algorithm coupled with streamline simulation has been implemented in each timestep. The Pareto front solution set is produced in evaluation of the best answer.

Figure 3 shows the Pareto front solution set for scenario E in the second timestep of the represented synthetic model. Non-optimized injection rates are shown in red and the optimized ones are in blue. Ultimately, a point which has relatively higher oil production with lower water production and lower water loss was chosen from the Pareto front solution set as the best answer.

Figure 4 represents the comparison of field cumulative oil and water production and water loss to the aquifer of different scenarios in the second time step. Scenario E

Fig. 2 Schematic of synthetic model

Table 3 Production and injection control schedule

Well name	Liquid rate (Stb/day)
P_1	75
P_2	100
P_3	100
P_4	125
P_5	90
$I_1 + I_2$	1,000

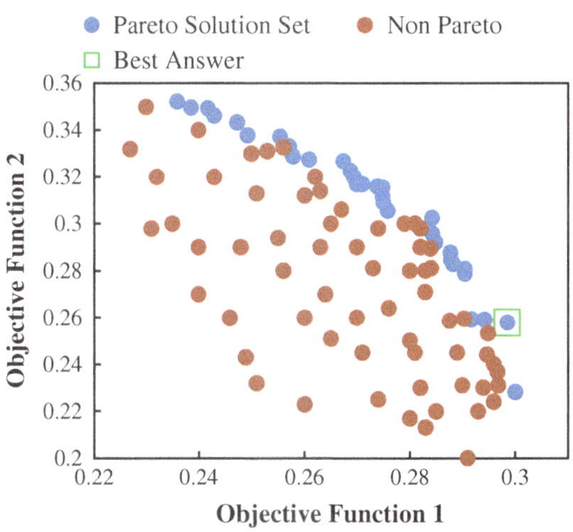

Fig. 3 Pareto-optimal set solution and making the best decision in scenario E

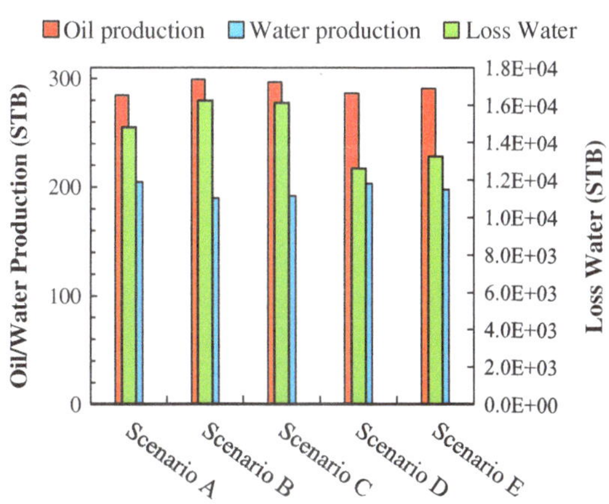

Fig. 4 Comparison of field oil, water production and water loss of different scenarios in the second timestep

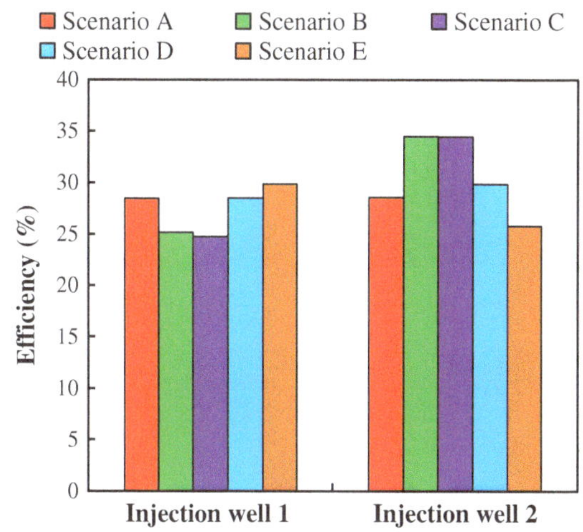

Fig. 5 Comparison of injection efficiencies of different MOGA scenarios in the second timestep

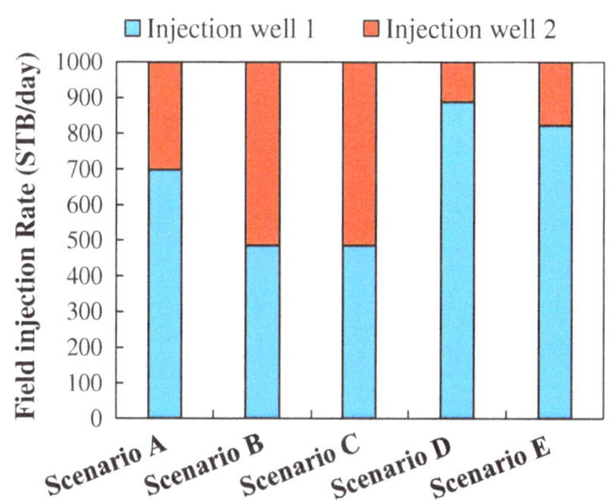

Fig. 6 Allocation of injected water in MOGA scenarios in the second timestep

causes a decrease in water loss of about 20 %, while it has not an undesirable effect on oil and water production. The highest oil production and the lowest water production were obtained in the scenario B which maximizes the total oil production. It may lead to reach the maximum economic benefit from the field after comprehensive assessment.

Injection efficiencies of the injectors for the second timestep are illustrated in Fig. 5. Although average injection efficiency of 30 % in scenario B is more than other scenarios, optimization implementation regardless of production limitation and aquifer performance (similar to the conventional methods) is not recommended. Therefore, it is not unexpected that scenario B has a higher water loss than other scenarios. Furthermore, it reveals that there is no considerable difference in field average pressure in different scenarios.

Optimized injection rates of different MOGA scenarios are shown in Fig. 6 for the second timestep. The maximum oil production can be obtained by selecting the injection rate of 485 and 515 Stb/day for the well 1 and well 2, respectively, while to reduce the amount of losing water to the aquifer and minimizing the operational cost simultaneously, the injection rate of 888 and 112 Stb/day should be chosen for the following wells.

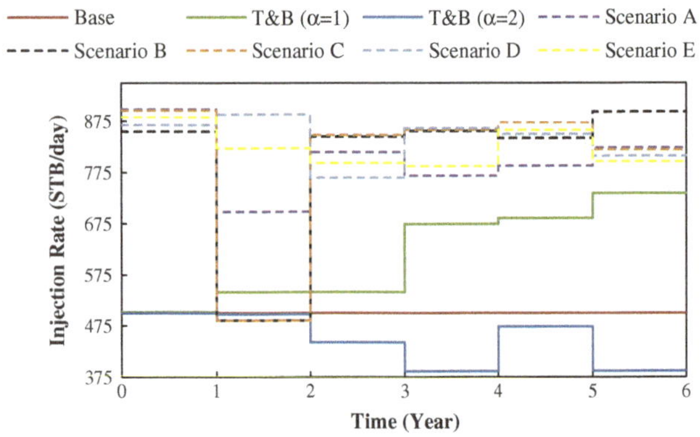

Fig. 7 optimized injection rate in well 1 during 6-year simulation in MOGA scenarios and conventional methods

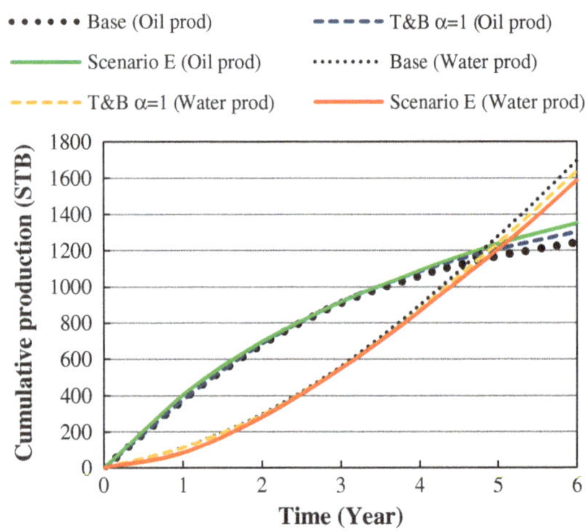

Fig. 8 Cumulative production of oil and water for scenario E and conventional methods

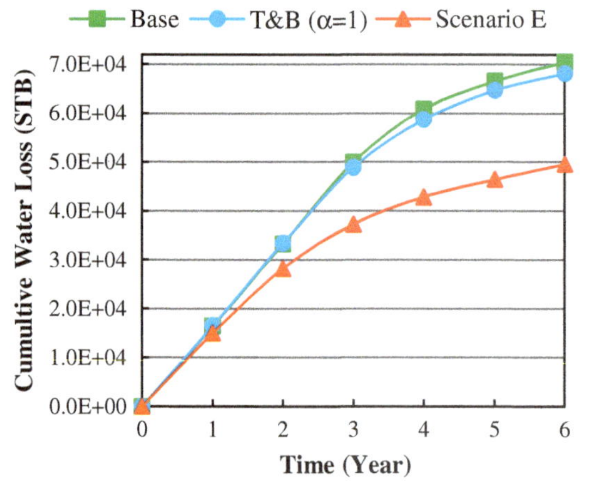

Fig. 9 Reduction of water loss for MOGA scenario E compared with conventional methods

Six-year optimization

A comparison with previous approach has been established to evaluate the advantages of the presented method in this paper. Therefore, in addition to the baseline constant injection rate, two simulation runs with $\alpha = 1$ and $\alpha = 2$ have been executed by Thiele and Batycky approach. In their approach, the best injection rate of the next step was chosen with respect to the field injection efficiency of the current step. Here to obtain the best answer in each timestep, the automated injection rate optimization runs were repeated 30 times. Similar to the other conventional optimization methods, one of the disadvantages of this methodology is that it may converge on the local solution instead of global ones. The other weakness is the impression of the previous timestep in the current timestep solution. On the other hand, using the best answer as a global solution in each timestep of optimization is the unique feature of all MOGA scenarios. Figure 7 shows the optimized injection rate in well 1 in different methods during the 6-year simulation. While Thiele and Batycky approach with $\alpha = 2$ tends to reduce the injection rate of well 1, using this approach with $\alpha = 1$increases the amount of injection rate. In general, all the MOGA scenarios have much more better results than other conventional methods.

From all of MOGA scenarios, scenario E was selected to compare with conventional approaches because of applying most operational limitations in its optimization process. Figure 8 shows the cumulative production of oil and water for 6-year simulation. In MOGA scenario E optimization, the highest oil production with the lowest water production has been observed in comparison with other methods.

In the Thiele and Batycky approach there is no considerable control on decreasing water loss to the aquifer, even though the observed reduction in Fig. 9 by this approach is negligible. However, in MOGA scenario E besides increasing in oil production, significant decline in water loss to the aquifer is observed in Fig. 9.

Evaluation of the injection efficiencies and injection rates in different wells and scenarios in is shown in Figs. 10, 11. Although injection efficiencies are almost the same in different scenarios and different wells, by allocating the injection rates in this paper, higher field performance is achieved.

Cumulative oil and water production and loss of water for 6-year production of all optimization methods are shown in Fig. 12.

In the best case of Thiele and Batycky approach with $\alpha = 1$, ultimate oil recovery improves up to 4.7 % in comparison with baseline, while using various MOGAs at the worst condition cause to increase in ultimate oil production about 7.3 % and in the best case, it increases it about 11 %. Besides, using any MOGAs decrease the total water production. Thiele and Batycky approach with$\alpha = 2$

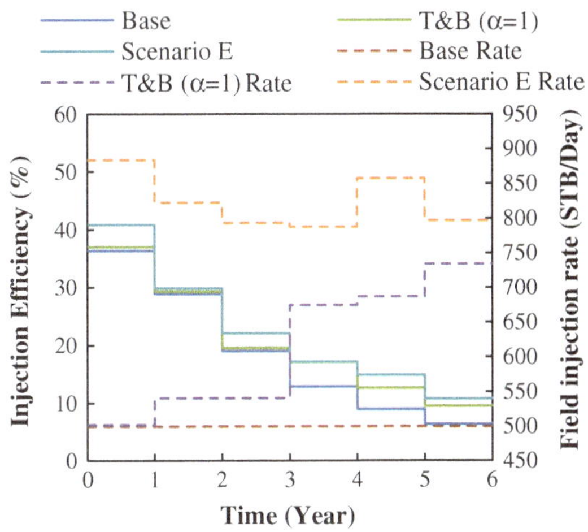

Fig. 10 Optimized injection efficiency and injection rate for MOGA scenarios and conventional method in well 1

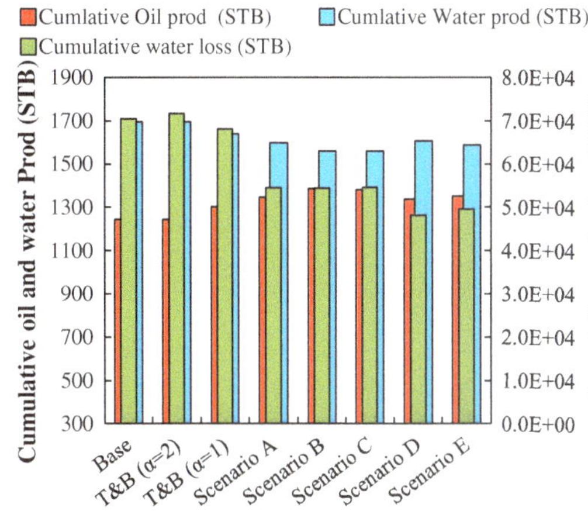

Fig. 12 Total oil and water production in different scenarios

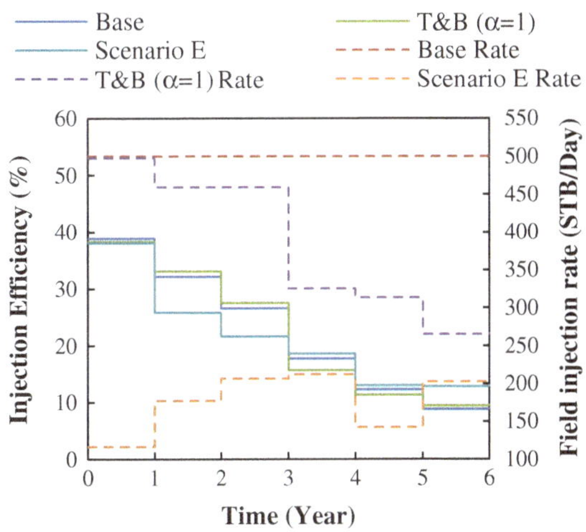

Fig. 11 Optimized injection efficiency and injection rate for MOGA scenarios and conventional method in well 2

results in rise in water loss to the aquifer and with $\alpha = 1$, it can only reduce water loss about 3 %. Multi-objective genetic algorithm scenarios can decrease the amount of water loss about 23–32 %.

Conclusion

Implementing the multi-objective genetic algorithm always offers the best solution duo to its fundamentals and characteristic. Comparison of the proposed approach in this paper with conventional workflow shows that all MOGA scenarios have better proficiencies.

- The primary objective of this work has achieved by using multi-objective algorithm genetic to conclude higher injection efficiency instead of conventional methods.
- To improve the weakness of the other methods, secondary objectives have been defined as various scenarios which cause higher total oil production, lower water production or less water loss to the aquifer.
- Pareto front shows all the answer sets in each time step that implementing different scenarios help to select the best one with different consideration aspects.
- Streamline simulation has been used to produce the allocated offset oil production in injectors and has been improved in each time step by the means of multi-objective genetic algorithm.

Appendix

Thiele and Batycky workflow injection efficiency optimization:

1. Determining injection efficiencies at the current time
2. Re-allocation of water injection with respect to the to the average field efficiency,

$$w_i = \min\left[w_{\max}, w_{\max}\left(\frac{e_i - \bar{e}}{e_{\max} - \bar{e}}\right)^{\alpha}\right], \text{ if } e_i > \bar{e}$$

$$w_i = \max\left[w_{\min}, w_{\min}\left(\frac{\bar{e} - e_i}{\bar{e} - e_{\min}}\right)^{\alpha}\right], \text{ if } e_i < \bar{e}$$

e_i: Injection efficiency
$\bar{e}$: Average field injection efficiency
$w_{\max}$: Maximum weight at $e_{\max}$
$w_{\min}$: Minimum weight at $e_{\min}$

e_{max}: Upper injection efficiency limit
e_{min}: Lower injection efficiency limit

3. Determining new injection rates
   $q_i^{new} = w_i \times q_i^{old}$
4. Normalizing new injection rates and determining the production rates using streamline simulation (Iino and Arihara 2007; Thiele et al. 2003).

References

Alhuthali A, Oyerinde A, Datta-Gupta A (2007) Optimal waterflood management using rate control, Society of Petroleum Engineers, SPE-102478-PA

Konak A, Coit DW, Smith AE (2006) Multi-objective optimization using genetic algorithms: a tutorial. Reliab Eng Syst Saf 91:992–1007.

Iino A, Arihara N (2007) Use of streamline simulation for waterflood management in naturally fractured reservoirs, SPE 108685. In: International oil conference and exhibition. Mexico

Marler RT, Arora JS (2004) Survey of multi-objective optimization methods for engineering. Struct Multidisc Optim 26:369–395.

Samier P, Quettier L, Thiele M (2001) Applications of streamline simulations to reservoir studies. In: SPE reservoir simulation symposium. Texas, USA

Sbalzariniy IF, Sibylle M, Koumoutsakosyz P (2000) Multiobjective optimization using evolutionary algorithms, Center for turbulence research

Thiele MR (2001) Streamline Simulation. In: 6th international forum on reservoir simulation, Austria

Thiele MR, Batycky RP, Blunt MJ (1997) Streamline-based 3d field-scale compositional reservoir simulator, SPE 38889. In: SPE annual technical conference and exhibition. Texas, USA

Thiele MR, Batycky RP, Streamsim Technologies (2003) Water injection optimization using a streamline-based workflow, SPE 84080. In: SPE annual technical conference and exhibition. Colorado, USA

Xie Q, Li S, Yang G (2005) Studies on fast pareto genetic algorithm based on fast fitness identification and external population updating scheme, Chinese academy of science

15

Impact of well configuration on performance of steam-based gravity drainage processes in naturally fractured reservoirs

Hamed Akhondzadeh · Ali Fattahi

Abstract In spite of the immense resources of heavy oil and bitumen laid in naturally fractured reservoirs (NFR), the study of steam-assisted gravity drainage (SAGD) in this type of reservoirs is very limited and there is no study examining the impact of well configuration on the process performance. This numerical study investigates various steam-based gravity drainage process well patterns in NFR including conventional SAGD, off-set SAGD, staggered SAGD and vertical-injection SAGD. Furthermore, to obtain the best well location, sensitivity analysis of well spacing was carried out for each well pattern and the impacts of oil viscosity, preheating and fracture permeability and water saturation were assessed as well. The best well location case for staggered SAGD represented the ultimate oil recovery factor of 59 % among all the cases with the SOR of 4.1, while the best off-set SAGD case showed a low SOR of 3.7 with a moderate recovery factor of 57 %. With the exception of off-set well configuration, all well patterns illustrated their most efficient performance while having 8 m vertical well spacing. In cases of off-set and staggered patterns, horizontal well spacing is a crucial factor for oil production initiation. Original fracture water saturation is the key parameter to establish early communication between the wells and gives rise to fast steam chamber expansion as well. We found that preheating is a necessary part of the process in the off-set pattern due to horizontal well spacing. However, preheating operation does not play a significant role in processes with no horizontal well spacing even in high vertical well intervals and high oil viscosities, making SAGD process in NFR different from that in sandstones. We attributed these behaviors to zero irreducible water saturation in fracture which makes initial water saturation mobile. At the end, the term $\frac{S_{wf} * K_f}{(1+L_{hw}) * \mu_o}$ (where S_{wf}: fracture water saturation, K_f: fracture permeability, L_{hw}: Horizontal well spacing and μ_o: oil viscosity) was introduced as the governing index of the oil production threshold.

Keywords Heavy oil · Steam-based gravity drainage · Naturally fractured reservoir · Enhanced heavy oil recovery · Well configuration

Introduction

Due to the current decline in conventional oil reserves, the commodity prices are increasing globally. As a result, oil industry has turned its attention to production techniques for known heavy oil deposits to solve this problem. Viscous oil trapped in carbonates (over 1.6 trillion bbl) is perceived as a huge potential resource for future energy supply which needs application of new technologies. While steam injection in sandstone reservoirs is known as a successful technology, there is a limited study on this process in fractured reservoirs.

There have been some thermal projects in fractured formations previously, all of which have been in pilot scale. Steam drive in Yates Field in Texas and Garland Field in Wyoming both in US, Cyclic steam stimulation in Cao-32

H. Akhondzadeh
Faculty of Petroleum and Renewable Energy Engineering, Universiti Teknologi Malaysia, No 63, Lane 13, Jahanara St., Shiraz, Iran

A. Fattahi
Faculty of Petroleum and Renewable Energy Engineering, Universiti Teknologi Malaysia, No 10, level 2, Sanei Bon bast, Fifth Ave, Fatemi St., Tehran, Iran

Field, China are some of the pilot scale experiences. Qarn Alam in Oman steam injection plant passed pilot test successfully and it was announced as a first full field scale steam injection in fractured media by shell as a partner company, Alvarado and Manrique (2010). Recently, Laricina Energy planned to apply SAGD pilot project at Saleski, Grosmont, Alberta, Canada, Cimolai et al. (2008).

SAGD mechanism in sandstone

The SAGD process involves two parallel long horizontal wells with one being located at the bottom as the producer and the other located at the top to inject steam continuously. Due to the density difference of oil and steam, steam tends to move upwards and sideways. Steam condenses at steam chamber interface and oil drains by gravity to the production well, Butler (1991). The field experience of SAGD process in UTF project in Alberta, Canada was successful (Edmunds et al. 1994 and Rouke et al. 1994) and there are many other SAGD processes being conducted in commercial stage in Alberta oil sands, Butler (2001).

Previous experiences and simulations have shown that well placement affects the recovery factor and breakthrough time. Some reservoir and operating parameters of three major oil-sand projects in Alberta, Canada were studied by Shin and Polikar (2005) numerically, one of which was the injector to producer spacing. The results showed that an increase in the injector to producer spacing leads to an increment in the recovery factor, however, it prolongs the breakthrough time. For example, in 5 m well spacing, the recovery factor was 42 % and increased to 49 % for 15 meter well spacing.

Chen and Fong (1997) simulated a Saskatchewan type reservoir with different well pattern including conventional SAGD, 2-offset SAGD and 3-staggered SAGD. They suggested that an incremental amount of 10–15 % oil can be recovered by offsetting the producer from injector. An additional 5–10 % oil recovery over the off-set recovery is possible by staggering the well in SAGD process, however, at the expense of deteriorating SOR.

Vertical wells can also be used for steam injection above horizontal producers. There are several advantages in the use of vertical injection well. Firstly, this sort of well is cheaper and simpler to drill. Furthermore, exploration wells can be used as this kind of injectors since they have been drilled vertically at the beginning of the development plan. Also, it is possible to change the point of steam injection vertically as the project matures. Gao et al. (2002) studied various well patterns for a tar sand reservoir in china, numerically. The patterns included steam injection as cyclic steam stimulation and steam flooding in vertical wells as well as horizontal wells and finally the application of SAGD through vertical injection wells. They concluded that the latest one has the best oil recovery performance with oil recovery factor of 46.5 % and an OSR of 0.27. Their sensitivity analysis revealed that off-setting the vertical injection well and applying 20 m vertical well spacing between the injector and producer in this particular well configuration increases oil recovery performance.

Sawhney et al. (1995) carried out an extensive experimental, theoretical and scale up study of vertical injection SAGD. They concluded that horizontal injection SAGD is more efficient than vertical injection SAGD in the case of Cold Lake reservoir.

Numerical study of well configuration impact on oil recovery in gravity drainage processes for UTF project was carried out by Tamer and Gates (2009). Well configurations in that study included normal SAGD, vertical wells SAGD, vertical injection SAGD and their offsetting. In the term of oil recovery performance, off-set SAGD demonstrated the best results followed by normal SAGD with a negligible difference. Vertical wells SAGD was more efficient than vertical injection SAGD, economically.

Thermal processes in fractured reservoirs

Preliminary studies suggest that recovery efficiencies in fractured reservoirs may be increased by as much as 50 % with the application of steam compared to that of low temperature processes. The key mechanisms enhancing oil production at high temperatures are the viscosity reduction, differential thermal expansion between oil and the pore volume and the appearance of some gases within matrix blocks. However, other mechanisms may also contribute to the increased oil production (Mollaei and Maini 2010). These mechanisms are relatively independent of oil gravity, making steam injection into fractured reservoirs equally attractive to light and heavy oil deposits (Reis 1992).

Britton et al. (1982) developed a new in situ steamflood method and assigned it the fractured assisted steamflood technology process (FAST), and tested the method in a South Texas tar sand deposit containing a viscous tar of 2 API. The first phase of the process consisted of horizontally fracturing from the injector toward the production wells and stimulating the fractures with high pressure steam. The horizontal fracture extended from the injection well to the producers, and steam was injected at high rates and pressures to hold the fracture open and to preheat the formation (fracture preheating phase). The results of the pilot showed that this process can be efficient and an average oil recovery of more than 50 % was obtained.

After several steamdrive patterns evaluated in the fractured zones of the Lacq Superieur field, France, a high rate steam injection in a very low permeability zone through a 350 m horizontal well was performed. Sahuquet et al.

(1990) simulated a horizontal injector and vertical producer process in this field because several tests showed that steam injection through a vertical well was not efficient due to the low permeability. The particularities of the Lacq Superieur operation is that the horizontal well is used as the injector to have a high steam injectivity in a low permeability carbonated facies and that the drain location is optimized to sweep a large area delimited by existing producing wells.

Experimental, numerical and pilot test of steam injection in Qarn Alam Field in Oman resulted in 27 % oil recovery factor with an expected OSR of 0.3. Steam is injected in the crest of the reservoir and oil drains into horizontal producers. Oil recovery takes place in a process called thermally assisted gas oil gravity drainage (Shahin et al. 2006).

Although there are some studies investigating the effect of fracture on SAGD performance, almost in all of them the role of fracture could be considered just as a heterogeneity that acts as a path for conduction of fluids, Bagci (2004) and Fatemi (2009).

According to authors' literature review, there are just a few studies applying SAGD in fractured reservoirs. Sedaee and Rashidi (2006) simulated SAGD in an Iranian carbonate fractured reservoir. They mentioned that SAGD is the best thermal process, however, they did not present the results of the simulation. Das (2007) applied SAGD and CSS in fractured reservoir and compared them, numerically. Oil recovery was 34 % with the SOR of 3 at the year 8. Furthermore, he carried out sensitivity analysis of some reservoir parameters including wettability and capillary pressure, fracture spacing, oil viscosity and well pattern. More steam was injected in the case of wider fracture spacing and it was attributed to the invasion and drainage from matrix and finally, he recommended more study. Matrix permeability did not have significant effect on the process performance. Water-wet rock illustrated 30 % increased oil recovery factor compared to that of the oil-wet case (without capillary pressure) in 8 years. He discerned that lower well spacings result in better performance for highly viscose oil and in the case of low viscose oil, staggered well pattern or higher well spacing has better performance. One of the abnormal features is constant steam injection rate of 1,200 bbl/day subsequent to around 5 months, making it unusual in SAGD process since SAGD steam injection rate comes in bell-curved shape. Bagci et al. (2007) conducted a numerical study on SAGD involving steam alternating solvent (SAS) process in fractured reservoirs. They found that oil recovery performance of this process is higher than that of pure SAGD. This process was optimized at the injection schedule of 1-year steam and a half-year solvent injection, alternatively. Moreover, the examined fracture orientation did not lead to any differences in oil production performance. However, the oil production performance was benefited by higher operating pressures and permeabilities.

Through numerical study, Murty et al. (2009) assessed various thermal processes to find the most efficient for shallow limestone fractured reservoir (rubble zone) in Bahrain. It was concluded that SAGD is not suitable due to low matrix permeability and low fracture porosity. Edmunds et al. (2009) analyzed core samples and past CSS pilots in Grosmont, Alberta and concluded the possibility of applying SAGD because of high bulk permeability, being in the order of higher than 10D. The reservoir rock texture includes centimeter-scaled karst and fractures, extensively. Moreover, single porosity model was proposed as approach to steam injection simulation since it was successful in history matching of CSS pilot. Yuan et al. (2010) created a 3D model for the Buffalo Creek Pilot Phase II of grosmont carbonate heterogeneous fractured reservoir and history matched the CSS pilot successfully. Then SAGD process performance was forecasted for layer C of this field. Their simulation showed that the lateral expansion of steam chamber with excellent performance reached its peak production rate of 264 m^3/day at 3 years and subsequently stabilized at around 87 m^3/day. A dual permeability/dual porosity approach was used in this study. Cimolai et al. (2010) gave solution for Grosmont type layered carbonate. SAGD simulation was started in layer D underling by layer C. Subsequently, a producer in the later layer at bottom, separated by marlstone, started to produce while this zone had been heated by conduction of overburden, layer D, by expanded steam chamber.

Solanki et al. (2010) discussed the first solvent added SAGD process called solvent cyclic-SAGD that has been planned to apply in Grosmont, Alberta. Subsequent to preheating phase, steam chamber creation and reaching overburden, a heavy solvent will be co-injected followed by the co-injection of a lighter hydrocarbon such as propane. When the steam chamber starts the blow down phase, a non-condensable gas such as methane will be injected to maintain the steam chamber pressure.

Laricina Company (Laricina Company Operation Report 2010) started to apply SAGD (1,800 b/day capacity) at Saleski, December 2010 as the world's first SAGD carbonate project and oil production was initiated in the spring of 2011. The commercial phase will be initiated in late 2013 with 12,500 b/day capacity.

Fractured reservoirs are made up of two systems; matrix and fracture. The flow toward the wellbore is considered to take place in the fracture network, while the matrix continuously feeds the system of fracture. Moreover, in some conditions vugs are included in the reservoir formation rock. These differences change the oil recovery mechanism and steam injection process rather than sandstone. Accordingly, although thermal recovery in sandstone is a

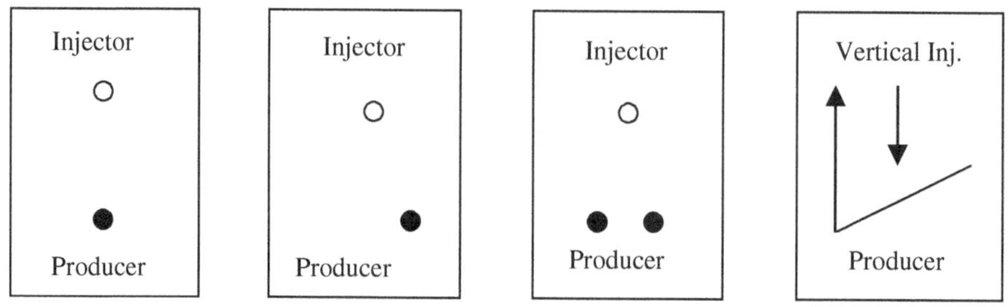

Fig. 1 Well patterns from *left* to *right*: conventional SAGD, offset SAGD, staggered SAGD and vertical-injection SAGD

mature process, it is in very early stages of application in fractured reservoirs and need extensive study. Studies which focused on SAGD in fractured reservoirs are quite rare and there is only few published numerical study. Furthermore, there is not any study investigating the impact of well configuration on SAGD in fractured reservoirs. In this paper, the impact of various well configurations on steam-based gravity drainage process is studied. The cases of investigation are conventional SAGD, off-set SAGD, staggered SAGD and vertical injection SAGD, which are presented in Fig. 1 and the sensitivity analysis of well spacing was carried out for each pattern. It should be mentioned that four different vertical well spacing were defined for each well pattern and these vertical intervals were the same in all well patterns. In addition, higher oil viscosity case was simulated and the conditions the preheating needed were analyzed. Higher fracture water saturation and permeability were also examined to assess the expected beneficial effects of these parameters.

Simulation model

The CMG STARS was used to simulate the mentioned processes. The model dimensions in all processes were the same with $106 \times 90 \times 33$ m^3. The length of horizontal wells was considered 90 m to reduce the simulation run time. In the case of vertical injection SAGD, the grid system was $55 \times 15 \times 14$ and for the other processes was $55 \times 1 \times 14$. The reservoir and fluid properties brought in Table 1 approximately represent the typical characteristics of fractured reservoir. Dual porosity with vertical subdivision option was used to capture gravity drainage precisely. Matrix block size of 1 m was used for all models with 5 divisions vertically. Oil-wet with gas/oil capillary pressure was defined as the reservoir rock and Cold Lake type dead oil was assumed. All processes were under the operating pressure of 2,550 kPa. No preheating was applied at the start of processes.

Table 1 Reservoir and fluid properties of simulation model

Reservoir pressure (kPa)	2,500
Reservoir thickness (m)	30
Matrix permeability (mD)	50
Fracture permeability (mD)	2,500
Matrix porosity	0.25
Fracture porosity	0.02
Matrix oil saturation	0.75
Fracture oil saturation	0.7
Oil viscosity at 75 F (cp)	40,000
Rock compressibility	9.6×10^{-6}/kpa
Formation heat capacity	2,350 kj/m^3 k
Rock thermal conductivity	6.6×10^5 J/m day C
Water thermal conductivity	1.15×10^4 J/m day C
Oil thermal conductivity	5.35×10^4 J/m day C
Gas thermal conductivity	1.4×10^2 J/m day C

Results and discussion

Oil recovery mechanism

Before study of process performance, heating and oil recovery of all processes are discussed to some extent. The following mechanisms mentioned here is the result of study presented in a previous author's paper (Fattahi et al. 2012) which is thoroughly discussed in that paper. In oil-wet FR, steam enters fractures and then heats the matrix block, however, oil cannot move out from matrix until fractures are filled with steam, therefore oil drains from the matrix to fracture due to density difference between oil in matrix and steam in the fracture. The extracted oil from matrix to fracture is drained to producer via fracture by gravity. However, at the bottom part of steam chamber, fracture oil saturation is high, and consequently it prevents expulsion of oil from matrix, although it is inside the steam chamber. Gas/oil capillary pressure prevents oil extraction from matrix before the oil saturation reaches its residual amount.

In this process, a large amount of water was produced at the start of process while oil production is not occurred yet. Produced water reached its peak in few months of operation, while oil production peak occurred at 5 years. Similarly, a large amount of steam is injected corresponding to produced water. Due to the presence of mobile water in fracture, water is produced from the start of process largely and makes empty space for large steam injection, while in sand stone, the heated oil should be produced to make empty space for steam injection, making the process slow and providing enough time to sideways expansion. This issue also creates larger steam chamber in FR compared to sandstone at early time.

Conventional SAGD pattern

To investigate the impact of well spacing, four different intervals of 2, 7, 12 and 17 m were considered for conventional SAGD pattern. Figure 2 illustrates the cumulative oil production and cumulative steam oil ratio (CSOR) in the four cases of conventional SAGD. Each curve's name for both parameters is ended with the distinct well spacing of that case in meter.

As it is seen in the figure, the ultimate cumulative oil production has been almost the same at 38 Mm^3 with negligible variations for all cases and given the original oil of 66 Mm^3, all cases resulted in a recovery factor of 57 %. It should be noted that although the cumulative produced oil at the end of the period is almost the same for all cases, the early oil rate versus well spacing shows a bell-shaped curve, hitting its peak at 7 m case. This happens as the result of the communication establishment interval which lasts longer for higher well spacings, hence postponing desirable oil production rates to near 4 months after the initiation of the operation for the 17 m case. On the other hand, a high well spacing results in higher amounts of oil being swept toward the production well initially. The 7 m case early oil production superiority than others therefore, is owing to the fact that applying this spacing for the wells would locate them far but yet near enough to make a communication shortly. Accordingly, being benefited from both aspects, a large amount of oil is swept within a short course of time in the 7 m case.

Unlike the cumulative produced oil, cumulative SOR differs somewhat for the conventional SAGD cases. After a high peak in all the CSOR curves at very early time showing the duration of the wells communication establishment, all falls sharply to their nadir owing to the early expansion of the steam chamber in the reservoir sweeping large amount of oil. There are two salient factors governing the shape of the dip in the SOR curve. On the one hand, the higher the well spacing, the bigger the steam chamber and in turn, resulting in lower SOR quantities, placing the case 12 m at the lowest SOR value among the three first cases. On the other hand, too far distant wells would result in early heat loss through overburden, hence increasing the value of the dip point in SOR curve and giving the case 17 m a higher SOR at its turning point than the 7 and 12 m cases. Subsequently, all curves are followed by upward trends toward the end of the operation. In spite of the higher early-time CSOR observed in this case, the 2 m instance resulted in the lowest ultimate CSOR of 3.8 and higher well spacings revealed slightly less favorable results of this value. However, no large distance was seen between the lowest and highest amounts of ultimate cumulative oil recovery and CSOR, the 7 m case yet takes the first place of performance among the 4 cases according to the two evaluated parameters along with the early oil recovery rate profile.

To be noted in assessing the results is that such moderately viscous oil, although not in high rates, has moved toward the production well without applying pre-heat in the SAGD process with almost the same operating pressure in wells, while in sandstone reservoir a preheat period is often necessary to initiate the process. This result was attributed

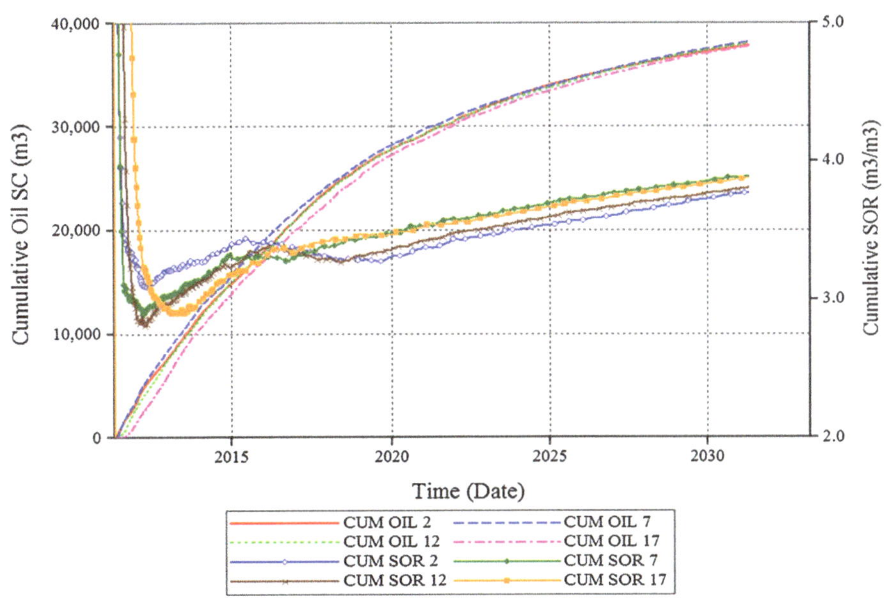

Fig. 2 Cumulative oil production and cumulative SOR for conventional SAGD cases with different well spacing

to the cross-like fracture relative permeability curve of this sort of reservoir. Accordingly, a sensitivity analyses was performed to investigate the mentioned claim.

Figure 3 demonstrates the performance of two conventional SAGD cases having the same operating and reservoir conditions. Considering the "conventional SAGD 7 m" as the base case, a new case was defined in which the relative permeability curve for fracture network was the same as that of matrix and was captioned "conventional SAGD 7, same Kr" in the figure. Similar to the well spacing study, the main performance comparison elements are cumulative oil production and SOR in this sensitivity analysis. The adverse effect of having non-zero endpoints in the saturation axis of relative permeability curve than the cross-shaped one for fracture was much severe that makes it difficult to draw an analogy between the poor performance of this case and the "conventional SAGD 7 m" case. Whereas the "conventional SAGD 7 m" case has started to produce oil almost from the beginning of the process, a 20-month delay is seen in the cumulative oil production curve of the "conventional SAGD 7, same Kr" case. This case then started to produce oil with unfavorable rates toward the end of the period terminating with just 14 Mm^3 cumulative oil which corresponds to as low as 21 % recovery factor. Likewise, very undesirable results for CSOR were observed in this instance. While the CSOR was less than 3.9 during the operation and ended with this quantity in the "conventional SAGD 7 m" case, the other case never fell beyond 4 throughout the operation time and ended with 6.2.

Accordingly, as it was expected, cross-shaped fracture relative permeability curve in fractured reservoirs was concluded to result in the free movement of the initial water inside the fracture and this is the principal reason for the early initiation of the oil production in such media. Furthermore, this feature has a significant effect on the steam chamber expansion throughout the operation period as well.

Off-set SAGD pattern

Figure 4 depicts the cumulative oil production and CSOR for four different cases of off-set pattern with all having equal vertical well spacings with those in the conventional SAGD study. Horizontal spacings were considered the same as the vertical interval for each distinct case. Similar to conventional SAGD cases, each parameter's label is ended with the specific distance between the production and the injection wells for that particular case.

As illustrated in Fig. 4, all cases except "off-set SAGD 3 m" case have initiated to produce oil after a remarkable delay. While there was no noticeable difference between the cumulative oil production and CSOR values for conventional SAGD cases, there are substantial alterations seen in these two factors for off-set pattern instances. The case having 3 m spacing between the wells started to produce oil from very early time and ended with 38 Mm^3 cumulative produced oil and the CSOR of 3.7. This instance was closely followed by the second small spacing case of 9 m which ended with the same CSOR and slightly lower cumulative oil production. The latter case, however, delayed to produce for over a year after the beginning of the process.

The two other cases revealed less favorable results particularly in cumulative oil production. While "off-set SAGD 16 m" case began to produce oil after 3 years of operation, this instance ended with somewhat lower cumulative oil production than the two previous off-set

Fig. 3 Cumulative oil production and cumulative SOR for relative permeability sensitivity analysis

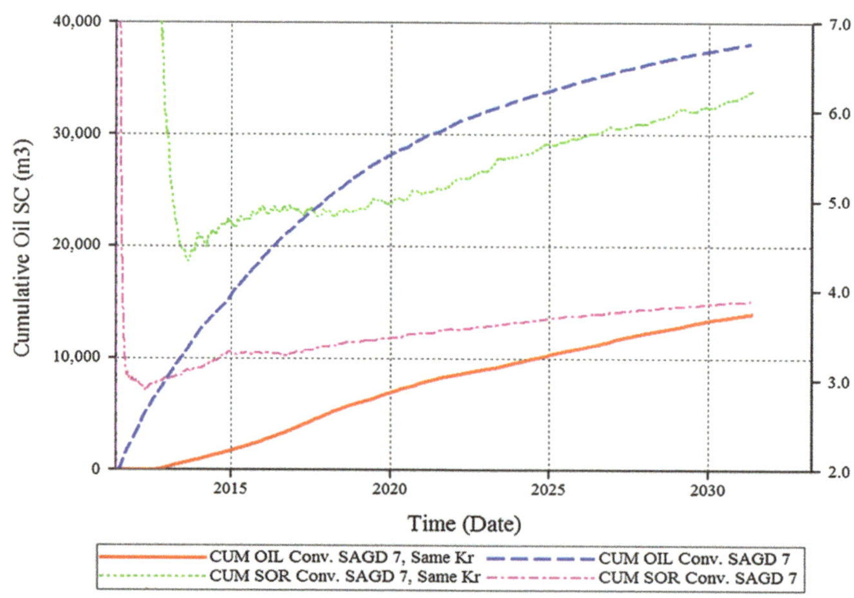

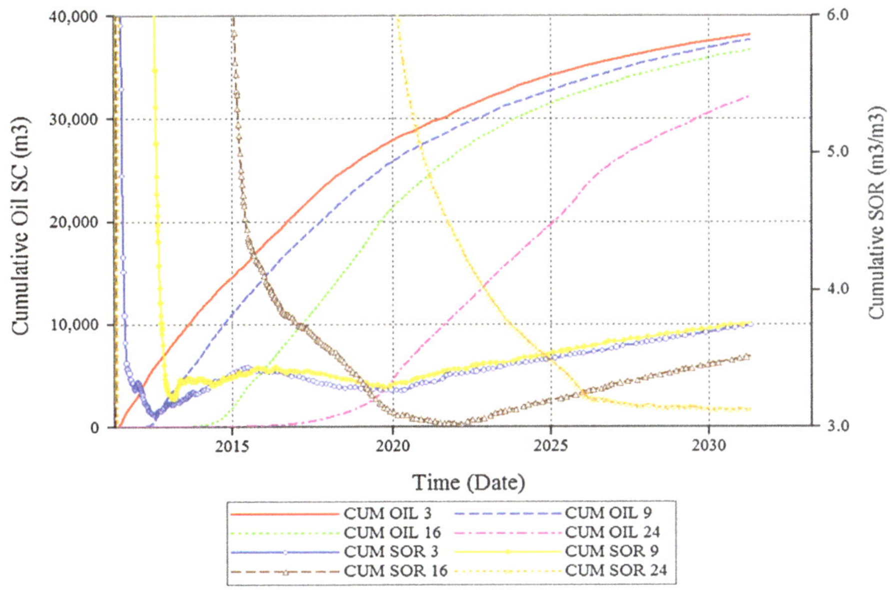

Fig. 4 Cumulative oil production and cumulative SOR for off-set SAGD cases with different well spacing

cases with 36.7 Mm^3. The last off-set case having 24 m distance between the wells demonstrated the worst results. The oil production curve remained at negligible amounts for less than 6 years for this case and the ultimate oil production was far less than the three previous instances with 32 Mm^3. The CSOR at the end of the defined period was 3.5 and 3.1 for the two last cases, respectively. As is obviously seen in the Fig. 4, the cumulative oil production curve in the three first off-set cases has approached maturity in the given period, however, in the last case this curve has a long way to reach that level and that is the reason of CSOR curve getting terminated with downward trend, on the contrary to the three other cases. The recovery factor of the first off-set example, having 3 m distance between the wells, was the same as that of conventional SAGD cases being 57 %. The three other off-set cases resulted in recovery factors of 56, 55 and 48 %, respectively.

The delay in the oil production in off-set SAGD cases lasts significantly longer than that of conventional SAGD. This long-lasting delay is the result of the horizontal spacing of the wells. The more the horizontal spacing between the wells, the more the inclination from the vertical orientation between them (if the vertical spacing is constant) and in turn, the later the communication between the wells is established. It is obvious that the vertical spacing between the wells also has a direct relationship with the communication establishment period. However, the later parameter's effect is much less intense compared to that of horizontal distance.

Staggered SAGD pattern

The cumulative oil production profile for the four defined staggered cases is brought in Fig. 5. The vertical spacing in these cases is exactly the same as the previous well patterns, while the horizontal spacings differ from those in off-set configuration. Similar to previous sections, the injector-to-producer well spacings are brought in meter at the end of each case's title. The oil production quantities in each curve represent the summation of cumulative produced oil through both producers.

One of the most conspicuous features in the cumulative oil production figure of different staggered SAGD cases is the early time separation of 8 m case curve from the others which reveals higher early oil rates in this case. Furthermore, this well configuration in SAGD process has led to sweeping the most oil at the end of the operation among the four staggered cases as well as among all the previously discussed cases with 39.3 Mm^3, equivalent to a recovery factor of 59 %. A corresponding behavior is seen in the figure of CSOR, Fig. 6, for this case. The most efficient early time CSOR among the staggered cases was observed in this example, however, this case took a middle place regarding the CSOR at the terminal time. Of importance in this case is taking the highest place of cumulative oil production throughout the process among all the cases making it the most favorable staggered well spacing.

The least efficient staggered SAGD performance was witnessed while applying 2 m distance between the injection well and each of the producers in staggered well configuration pattern. Whereas this case started to produce oil from almost the beginning of the operation, it ended with 38.2 Mm^3 cumulative produced oil being the least among all the staggered cases. To render this case as the least efficient one in this particular well pattern, the CSOR is considerably higher than the others almost throughout the operation period ending at 4.3.

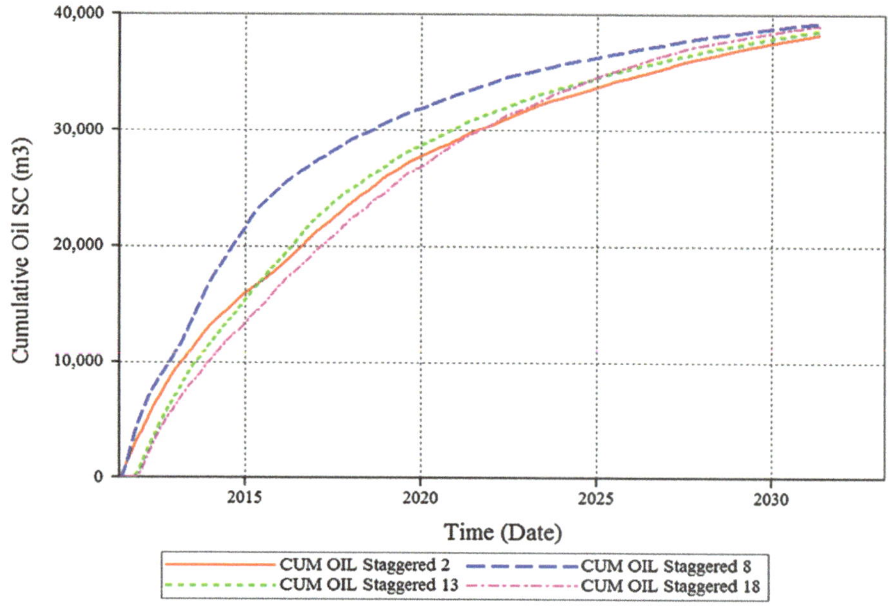

Fig. 5 Cumulative oil production for staggered SAGD cases with different well spacing

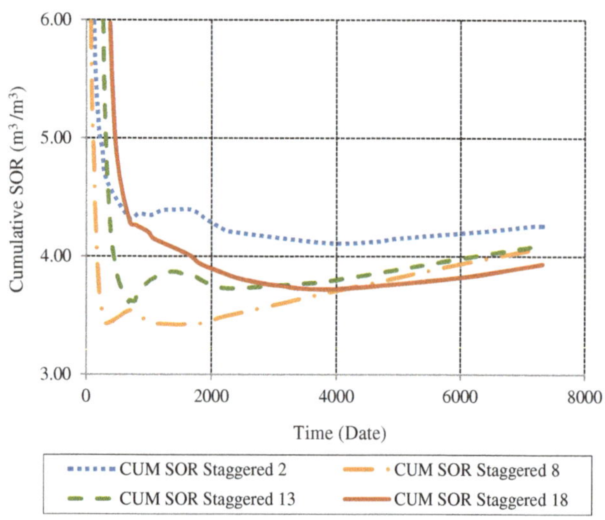

Fig. 6 Cumulative SOR for staggered SAGD cases with different well spacing

The case applying 18 m distance between the injection and the production wells in staggered pattern revealed one of the most desirable results. While the ultimate cumulative oil of this case follows the highest of this value in staggered pattern seen in the 8 m case, the final CSOR for this case has been lower with 3.9, making this well configuration one of the most promising case among the studied staggered SAGD instances. A delay in oil production at the beginning of the process in this case should be noticed. The 13 m case has illustrated middle results both in its cumulative oil production and CSOR.

Vertical-injection SAGD pattern

As the last well configuration, the effect of vertical well spacing in vertical-injection SAGD was studied in three distances of 5, 10 and 15 m and then compared to the other patterns. Results of the simulation are presented in Fig. 7 in the terms of cumulative oil recovery and CSOR. In the early time of process within 5 years, 15 m well spacing case showed the highest oil recovery among all the three well spacings, however, this well interval case revealed less favorable oil production later, reaching the lowest ultimate oil recovery factor of 48 %. The case of 10 m well spacing extracted the most oil at the end of the process, resulting in a terminating oil recovery factor of 50 %. This case also represented the best final SOR of 3.5. As a simple rule, it can be stated that a medium well spacing in vertical-injection SAGD leads to the best performance.

Viscosity sensitivity analysis

A viscosity sensitivity analysis was carried out on the off-set and conventional well patterns to observe the impact of higher viscosities on the process. In the off-set well pattern, the second case was chosen for this sensitivity analysis. The reason for choosing "off-set SAGD 9 m" case was that the adverse effect of higher viscosity of heavy oil was to be investigated, therefore, the first case, having just 3 m spacing between the wells was perceived to be an incompetent instance to highlight that negative effect. On the other hand, the injection and production wells in the next two cases of off-set pattern study were conjectured to be located too far to be able to sweep such heavy oil having a

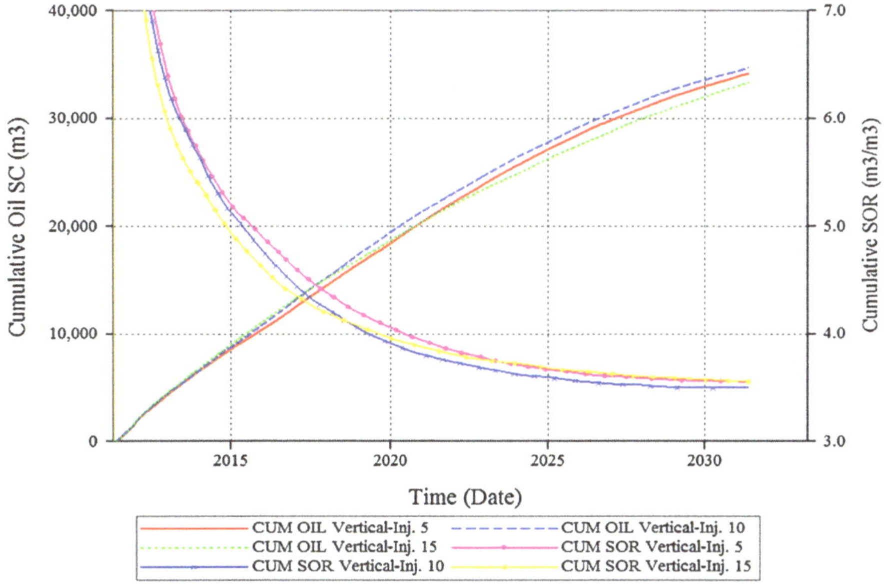

Fig. 7 Cumulative oil production and cumulative SOR for vertical-injection SAGD cases with different well spacing

viscosity in order of 300,000 cp at the original reservoir temperature.

Accordingly, five cases were defined for the viscosity analysis study including "off-set SAGD 9 m" example and a higher-viscosity alternative of this case as the base case for low and high viscosity to compare the results and evaluate the negative effects. The third case of investigation in this sensitivity analysis was the higher viscosity base case with a preheat period of 90 days in both wells to observe how this operation benefits the process.

Previously, it was suggested that the production of heavy oil in such fractured reservoirs is feasible owing to the relative permeability curve and relatively higher saturation of water in fracture in this kind of reservoir. While the former was discussed before, the effects of the water saturation in the fracture section of fractured reservoirs will be discussed in this sensitivity analysis by defining a new alternative from the base case of higher viscosity, having 50 % water percentage in fracture section as initial water saturation rather than the 30 % water saturation in all other cases. The last case of this sensitivity analysis was a higher fracture permeability as an alternative of the high viscosity base case to examine the expected positive effect of a higher permeability.

Figures 8 and 9 illustrate the cumulative oil and cumulative steam oil ratio for the four mentioned cases defined for viscosity sensitivity analysis, respectively. As can be seen in Fig. 8, while the delay before producing oil has been about a year for the "off-set SAGD 9 m" case, the base case of higher viscosity possessing almost tenfold viscosity has not been able to produce the heavy oil for over 3 years. It should be noted that the entire reservoir parameters including the water saturation and the relative permeability curve of the reservoir have not been changed, therefore, the unfavorable result of the production postponement for more 2 years over "off-set SAGD 9 m" case is attributed only to the high viscosity of oil. Whereas final cumulative oil production has been 37.6 Mm^3 for the former case leading in a recovery factor of 56 %, the base case of higher viscosity has revealed substantially less promising result of 25.5 Mm^3 ultimate oil production, corresponding to a recovery factor of 38 %.

As an approach to enhance the performance of the off-set SAGD process, a 90-day preheating period was introduced to the higher viscosity case. Desirably, the communication establishment between the wells lasted for half a year to achieve in this case, far sooner than the high viscosity base case and even noticeably faster than "off-set SAGD 9 m" case. The preheated case, however, resulted in a terminal cumulative oil production of 31 Mm^3 equivalent to 47 % recovery factor, being significantly higher than "off-set 9 m, High Visc" case, emphasizing the importance of preheating in early oil production as well as higher ultimate oil recovery factor. It should be noted that ultimate oil recovery factor of "off-set 9 m, High Visc-Preheat" is yet considerably lower than that of "off-set SAGD 9 m" case highlighting the negative effect of higher oil viscosities in oil recovery performance in steam based gravity drainage processes.

As the next case of investigation in this sensitivity analysis, the effect of higher water saturation was examined by defining an alternative of the high viscosity base case with just varied water saturation in the fracture equal to 50 % from the 30 % in the base case. As expected, much better results of cumulative oil production and CSOR were obtained in this case in comparison with the high viscosity

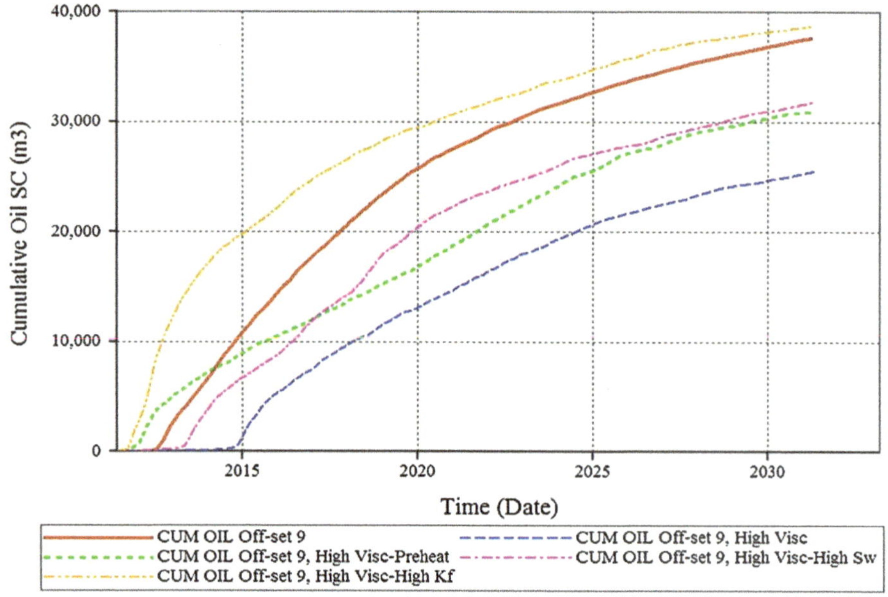

Fig. 8 Cumulative oil production for viscosity sensitivity analysis

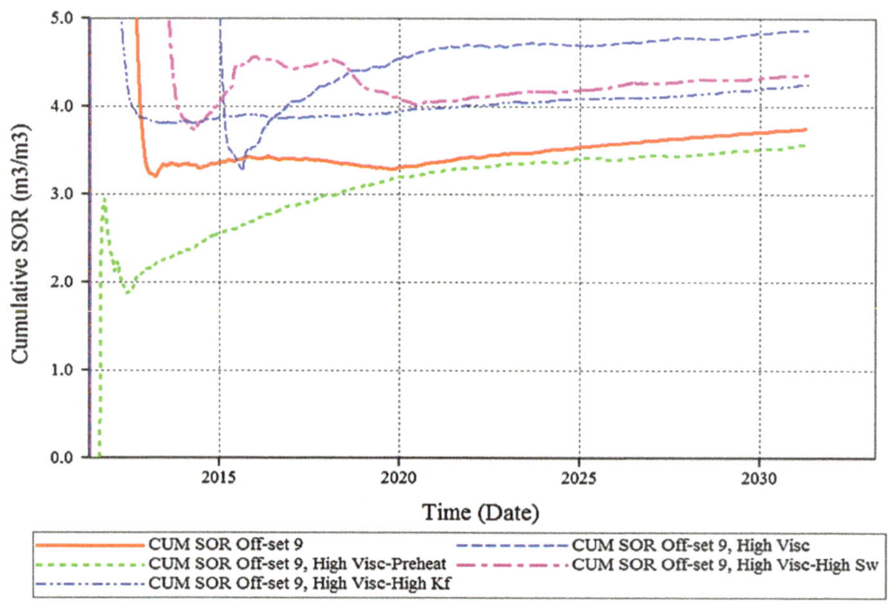

Fig. 9 Cumulative SOR for viscosity sensitivity analysis

base case. Although this instance commenced to produce oil significantly later than the preheated case, the delay was considerably shorter than the high viscosity base case, "off-set 9 m, High Visc". It is notable that the terminal cumulative oil production of this case has been 6.3 Mm^3 higher than that of "off-set 9 m, High Visc" case and in opposite to the result of communication establishment period, the cumulative oil production in this case has slightly surpassed the preheated case and ended with almost 1 percent more recovery factor over that case with 48 %.

To investigate the impact of fracture permeability on process initiation, the fracture permeability of "Off-set 9, High-Visc" which has been the high viscosity base case was increased fourfold to 10D. Figure 8 illustrates the great advantageous impact of the higher fracture permeability on the threshold of oil production. While in the case of "Off-set 9, High-Visc" oil production started after 4 years, this delay decreased to only 3 months in the case of higher fracture permeability. Moreover, final oil recovery reached 58 % at the end of 20 years compared to 38 % for the lower fracture permeability case.

The cumulative steam oil ratio of the five discussed cases of viscosity sensitivity analysis is seen in Fig. 9. The worst result of CSOR is seen in the curve of "off-set 9 m, High Visc". As it is obviously noticed, CSOR in this case has fallen below 5 after almost 4 years of operation by around 2015, and in accordance with the cumulative oil production curve, suggesting that no significant oil production has obtained prior to that time. Moreover, this curve has resulted in the least favorable ultimate CSOR of 4.9, while this value for "off-set SAGD 9 m" case has been 3.8 indicating the negative effect of higher oil viscosity. The higher initial water saturation case illustrated more desirable results than the high viscosity case but yet not enough promising by resulting in an ultimate CSOR of 4.4. The preheated case demonstrated the most conspicuous behavior compared to other cases. This case never exceeded the CSOR of 4 throughout its operation period. The very promising behavior of this case was certainly attributed to the beneficial effects of the preheating period. The preheated case ending at CSOR 3.6, revealed the most desirable result for this value over all the other cases of this sensitivity analysis including "off-set SAGD 9 m" case. For the higher fracture permeability, the ultimate SOR reduced effectively to 4.2 from 4.9 in the base case of high viscosity.

A viscosity analysis on conventional SAGD pattern was examined for various well spacings of higher viscosity to compare the results with the previously discussed low viscosity cases of this configuration. The results' trend based on different well spacings was the same with the low viscosity conventional SAGD. However, it is notable that even in the worst case of conventional SAGD high viscosity, the delay in the oil production initiation was just in order of months while it was in order of years in the higher spacing cases of Off-set pattern. The results of conventional SAGD high viscosity were not brought since no significant alteration was seen between these cases and the conventional SAGD low viscosity cases.

Discussion

All of the simulations results of CSOR, cumulative oil production and ultimate recovery factor are brought in Table 2 and the best case of each well pattern were compared. Staggered SAGD resulted in the highest oil recovery factor of 59 % among all of the patterns; however, it leads to an undesirable SOR of 4.1. It should be noticed that in this pattern three wells are involved for each distinct zone rather than the two in the other well configurations, hence suffering the economy of the process. The Off-set case performance is better than conventional SAGD in the both terms of recovery factor and SOR being 57 % and 3.7, respectively. Having SOR of 3.5, Vertical–Injection SAGD showed the best thermal efficiency, while on the contrary its recovery factor is the lowest showing the same trend with that in sandstone. Regarding the economic analysis in the Vertical–Injection SAGD, one should take it into consideration that firstly, the capital expenditures (CAPEX) for drilling each vertical well is far lower than a horizontal one and secondly, the previous exploration wells can be utilized as the vertical injector, hence benefitting the process by lowering the CAPEX.

According to the investigated parameters and previous discussions on sensitivity analysis of fracture water saturation, fracture permeability and oil viscosity in all of the well patterns, we introduced a governing term for the oil production initiation. While the term is $\frac{S_{wf} * K_f}{(1+L_{hw}) * \mu_o}$ (where S_{wf}: fracture water saturation, K_f: fracture permeability, L_{hw}: horizontal well spacing and μ_o: oil viscosity), higher values of this term result in faster well communication establishment and in turn, the oil production commencement. It is obviously observed that higher values of fracture water saturation and fracture permeability lead to sooner oil production threshold. Conversely, the increase in the amounts of horizontal well spacing and oil viscosity would delay the oil production initiation. The 1 in the parenthesis is to prevent the fraction to be infinitive since in the conventional SAGD L_{wh} is zero.

Table 2 Cumulative SOR, cumulative oil production and recovery factor for all the simulation cases

	SOR	Cumulative oil, 1,000 m^3 (Mm^3)	Recovery factor (%)
Conventional SAGD 2	3.8	37.8	57
Conventional SAGD 7	3.9	38.1	57
Conventional SAGD 12	3.8	38.0	57
Conventional SAGD 17	3.9	37.7	57
Conv. SAGD 7, same Kr	6.2	14.0	21
Off-set SAGD 3	3.7	38.1	57
Off-set SAGD 9	3.7	37.6	56
Off-set SAGD 16	3.5	36.7	55
Off-set SAGD 24	3.1	32.2	48
Off-set 9 High Visc	4.9	25.5	38
Off-set 9 High Visc-Preheat	3.6	31.0	47
Off-set 9 High Visc-High S_w	4.4	31.8	48
Off-set 9 High Visc-High K_f	4.2	38.7	58
Staggered SAGD 2	4.3	38.2	57
Staggered SAGD 8	4.1	39.3	59
Staggered SAGD 13	4.1	38.6	58
Staggered SAGD 18	3.9	39.0	59
Vertical-Inj SAGD 5	3.5	34.2	51
Vertical-Inj SAGD 10	3.5	34.7	52
Vertical-Inj SAGD 15	3.6	33.4	50

Conclusion

The following remarks are concluded from this study:

- Staggered SAGD resulted in the highest oil recovery factor as well as CSOR compared to other well configurations. The Off-set pattern performance showed better results than conventional SAGD in the both terms of ultimate recovery factor and CSOR. Although the best thermal efficiency was seen in the vertical-injection SAGD pattern, undesirable cumulative oil production of this pattern deteriorated this promising result. Overall, selecting the most efficient and promising pattern was left to economic analysis.
- Generally, the second vertical well spacing cases in all patterns (7–10 m) are the most favorable intervals between the wells with the exception of Off-set configuration.
- In conventional SAGD process in NFR, preheating is not needed even in the cases of large well spacing and high oil viscosity, making it different from SAGD in sandstone.
- In the cases of off-set SAGD, horizontal well spacing delays the well connection establishment and early oil production, making preheating an essential part for this pattern.
- On the contrary to sandstone which connate water does not benefit the process due to its immobility, initial fracture water saturation is the key point in soon well connection establishment and fast steam chamber expansion rate in NFR. This conspicuous role of initial fracture water saturation was attributed to its high mobility due to zero irreducible water saturation in fracture.
- Finally, a governing term for the oil production initiation was introduced as $\frac{S_{wf}*K_f}{(1+L_{hw})*\mu_o}$ where higher values of this term result in faster well communication establishment and in turn, the oil production commencement.

References

Alvarado V, Manrique E (2010) Enhances oil recovery: an update review. Energies 3:1529–1575.

Bagci AS (2004) The effect of fractures on the steam-assisted gravity drainage process. Energy Fuels 2004(18):1656–1664

Bagci AS, Samuel OM, Macky E (2007) Recovery performance of steam-alternating-solvent (SAS) process in fractured reservoirs' The paper 2007-040 was presented at the Petroleum Society's 8th Canadian international petroleum conference (58th annual technical meeting), Calgary, Alberta, Canada, 12–14 June 2007

Britton MW, Martin WL, Leibrecht RJ, Harmon RA (1982) The Street Ranch Pilot Test of fracture-assisted steamflood technology. Paper SPE 10707 presented at the 1982 SPE California regional meeting, San Francisco, CA, 24–26 March 1982

Butler RM (1991) Thermal recovery of oil and bitumen. Prentice Hall, New Jersey

Butler RM (2001) Some recent developments in SAGD. J Can Pet Technol 40(1):18–22 (distinguished authors series)

Chen MYS, Fong J (1997) Effects of well placement and critical operation conditions on the performance of dual well SAGD well pair in heavy oil reservoir. Paper SPE 39082 presented at the Fifth Latin American and Caribbean petroleum engineering conference and exhibition held in Rio de Janerio, Brazil, 30 Aug-3 Sept 1997

Cimolai, Edmunds, Barrett, Solanki SC, Kantzas, Duan (2008) Bitumen mobilization in an oil-wet, fractured grosmont carbonate. Available at http://www.laricinaenergy.com/investor/tech/GPS%20Bitumen%20Mobilization.pdf

Cimolai MP, Solanki SC, Esmunda NR (2010) Passive reservoir heating for improving bitumen recovery. J Can Pet Technol 49(11):30–42

Das D (2007) Application of thermal recovery processes in heavy oil carbonate reservoirs. Paper SPE 105392 presented at the 15th SPE Middle East oil and gas show and conference, Bahrian, 11–14 March 2007

Edmunds NR et al (1994) Review of phase a steam-assisted gravity-drainage test. SPERE May 1994 p 119

Edmunds N, Barrett K, Solanki S et al (2009) Prospects for commercial bitumen recovery from the grosmont carbonate, Alberta. J Can Pet Technol 48(9):26–32

Fatemi SM (2009) Simulation study of steam assisted gravity drainage (SAGD) in fractured systems. Oil Gas Sci Technol Rev IFP 64(4):477–487

Fattahi A, Akhondzade H, Nawi M, Hussin M, Namazi S (2012) Impact of reservoir heterogeneity on steam assisted gravity drainage in heavy oil fractured reservoirs. Energy Explor Exploit 30(4):553–566

Gao Y, Liu S, Zhang Y (2002) Implementing steam assisted gravity drainage through combination of vertical and horizontal wells in a super-heavy crude reservoir with top-water. Paper SPE 77798 was presented at the SPE Asia Pacific oil and gas conference and exhibition held in Melbourne, Australia, 8–10 Oct 2002

Laricina Company Operation Report (2010) Available at http://www.laricinaenergy.com/operations/saleski.html

Mollaei A, Maini B (2010) Steam flooding of naturally fractured reservoirs: basic concepts and recovery mechanism. J Can Petrol Technol 49(1):65–70

Murty C, Kunar RK, Bapco K et al (2009) Thermal EOR pilot plan for a shallow limestone reservoir (rubble zone) in Bahrain field, SPE 120569, Middle East oil and gas show and conference, Bahrain, p 15–18

Reis JC (1992) An analysis of oil expulsion mechanism from matrix blocks during steam injection in naturally fractured reservoirs. Situ 16(1):43–73

Rouke JC et al (1994) UTF project status and commercial potential. Petroleum Society of CIM and AOSTRA paper 94-40 presented at the 1994 annual technical meeting of the Petroleum Society of CIM, Calgary, 12–15 June 1994

Sahuquet BC, Spreux AM, Cwe B, Guittard MP (1990) Steam injection in a low-permeability reservoir through a horizontal well in Lacq Superieur field. Paper SPE 20526 presented at the 51 h annual technical conference and exhibition of the Society of Petroleum Engineer held in New Orieane, LA, 23–26 Sept 1990

Sawhney GS, Liebe H, Butler RM (1995) Vertical injection wells for SAGD: a practical option or not? J Can Pet Technol 34(1):9. https://www.onepetro.org/journal-paper/PETSOC-95-01-06

Sedaee B, Rashidi F (2006) Application of the SAGD to an Iranian carbonate heavy oil reservoir. In: Proceedings of SPE Western Regional/AAPG Pacific Section/GSA Cordilleran Section Joint Meeting, 8–10 May, Anchorage, Alaska, USA

Shahin GT, Moosa R, Kharusi B, Chilek G (2006) The physics of steam injection in fractured carbonate reservoirs: engineering development options that minimize risk. Paper SPE 102186 was presented at the 2006 SPE annual technical conference and exhibition held in San Antonio, Texas, USA, 24–27 Sept 2006

Shin H, Polikar M (2005) Optimizing the SAGD process in three major Canadian oil-sands areas. Paper SPE presented at the 2005 SPE annual technical conference and exhibition held in Dallas, Texas, USA 9–12 Oct 2005

Solanki SC, Barrett K, Riva D et al (2010) The First SC-SAGD pilot in the Grosmont. Paper WHOCV 11-615, World heavy oil conference, March 2010

Tamer M, Gates ID (2009) Impact of well configuration on performance of steam-based gravity drainage process. Paper 2009-049 was presented at Canadian international petroleum conference (CIPC), Calgary, Alberta, Canada, 16–18 June 2009

Yuan J-Y, Jiang Q, Russel-Houston J et al (2010). Evolving recovery technologies directed towards commercial development of the grosmont carbonate reservoirs, CSUG/SPE 137941. Canadian unconventional resources and international petroleum conference, Calgary, Alberta, Canada, 19–21 Oct 2010

Effect of drillpipe rotation on cuttings transport using computational fluid dynamics (CFD) in complex structure wells

Xiaofeng Sun · Kelin Wang · Tie Yan · Shuai Shao · Jianjun Jiao

Abstract Effective cuttings transport is one of the most major concerns in designing hydraulic parameters during drilling complex structure wells. It is important and essential to study critical parameters how to influence cuttings bed development, especially the drillpipe rotation effects on cuttings transport behavior. In this study, cuttings transport simulations were carried out for inclinations from 45° to 90°, for pipe rotation speeds from 80 to 240 rpm, and for flow rates from 30 to 50 L/s. Annular cuttings volume, pressure drop, and cuttings distribution are recorded simultaneously within the simulation proceeding. The CFD simulations show that pipe rotation makes cuttings presented in asymmetric distribution along the wellbore and significantly improves the drag effects on drilling fluid in the tangential direction. At a low or medium flow rate, pipe rotation has a significant impact on annular cuttings volume and pressure drop. However, there is no additional contribution of pipe rotation after reaching critical speed at high flow rates. Buckingham-π theorem combined with least square method was applied to establish the empirical correlations for estimating cuttings concentration and annular pressure drop.

Keywords CFD · Hole cleaning · Cuttings transport · Drillpipe rotation · Complex structure well

X. Sun · K. Wang · T. Yan · S. Shao
College of Petroleum Engineering, Northeast Petroleum University, Daqing 163318, China

J. Jiao
Tianjin Branch of CNPC Offshore Engineering Co. Ltd, Tianjin 300450, China

Introduction

Complex structure well includes horizontal well, extended reach well, and multilateral well, which has been widely used for offshore oil exploration and exploitation and improving the recovery of oil field. However, hole cleaning is often difficult to be handled during drilling inclined and horizontal sections of complex structure wells, and cuttings are easily accumulated to develop cuttings bed, especially in the inclinations around 40°–60°, which may lead to various problems such as high drag and torque, pipe stuck, weight stacking, higher hydraulic requirements and excessive ECD, and cementing issues. Therefore, it is important and essential to study the key factors of cuttings bed development in annulus, especially if drillpipe is rotating. Numerous researches related to the effect of pipe rotation on hole cleaning have been carried out by experimental investigations and CFD simulations in the past several years.

In experimental investigations, Peden et al. (1990) observed that pipe rotation has a significant effect on the minimum transport velocity in medium or highly viscous fluids and contributes to cuttings transport in small annulus or positive eccentricity. Sifferman et al. (1992) pointed out that drillpipe rotation reduces annular cuttings buildup under certain conditions, and the effect is greatest at inclination near horizontal, for small cuttings and low rate of penetrations. Bassal (1996) indicated pipe rotation enhances hole cleaning more when the used mud has a higher viscosity with smaller cuttings sizes or for hole angle at 65°. Sanchez et al. (1999) analyzed the effects of pipe rotation–flow rate, pipe rotation–mud rheology, pipe rotation–inclination, and pipe rotation–cuttings size. The results show that in lower inclinations, no critical range of rotary speeds is determined but higher meant better, and orbital motion

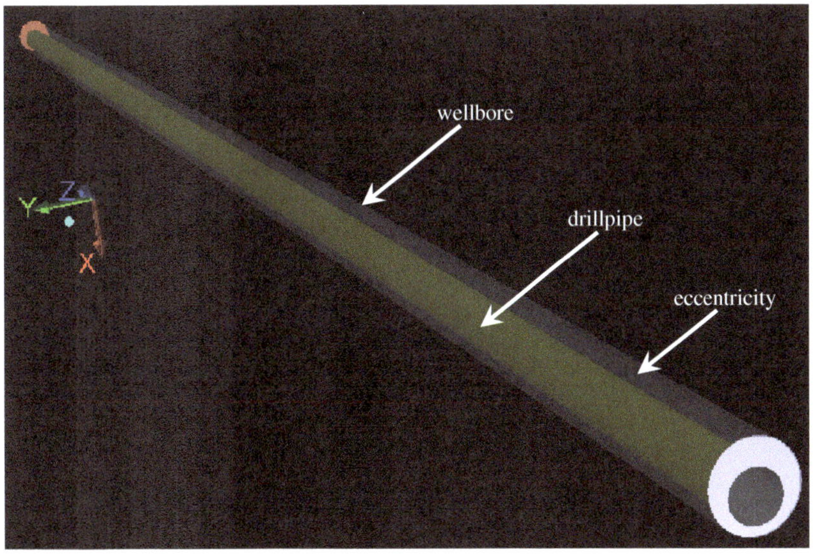

Fig. 1 Physical model of inclined section

can significantly improve cuttings transport. Duan et al. (2008) performed the experiments about different cuttings sizes under pipe rotation. It was observed that small cuttings transport is mainly dominated by pipe rotation and fluid rheology, and improvement by pipe rotation in the transport efficiency of small cuttings is up to twice as large as the improvement in large cuttings transport. Ozbayoglu et al. (2008) and Sorgun (2010) indicated pipe rotation has a significant improvement on cuttings transport, especially when the pipe is making an orbital motion. But drillpipe rotation has no additional contribution to hole cleaning after a certain rotation speed. Also, the empirical correlations for 60°–90° were developed for predicting the stationary cuttings bed height and frictional pressure drop.

In CFD simulations, Bilgesu et al. (2002, 2007) were one of the first researchers to analyze cuttings transport parameters using CFD. The simulations show drillpipe rotation can improve the cuttings transport, and the effect is more pronounced for smaller particle size. Mishra (2007) found that hole cleaning efficiency is improved more significantly for small cuttings than for large cuttings as pipe rotation is increased, and pipe rotation at a fixed velocity can decrease pressure loss. Wang et al. (2009) indicated the solid flow becomes favorable, and the solid volume is reduced with the increase of rotary speeds. The correlation was also developed to estimate cuttings bed height. Li et al. (2010) found that pipe rotation between 80 and 120 rpm has a significant effect on hole cleaning.

Although some achievements have been made, but the pipe rotation effects on cuttings transport are difficulty be accurately depicted due to the complexity of cuttings–fluid coupling. The purpose of this paper mainly further discussed the cuttings transport behavior with pipe rotation using CFD.

Physical model of inclined section

Complex structure well usually contains the two or three buildup sections. The 215.9 mm wellbore diameter and 127 mm outer diameter of drillpipe are adopted to simulate the actual annulus. A 12-m-long interval is simulated without considering borehole enlargement and shrinkage. The eccentricity is 0.5, and eccentric distance is 44.45 mm. The gravity effect, horizontal and vertical vibrations are also considered, and the diagram of the physical model is shown in Fig. 1.

Gridding

To simulate the effect of drillpipe rotation on transient cuttings transport in annulus, moving mesh is applied for the flow domain in annulus. As shown in Fig. 2, all the annulus can be divided into both outer and inner flow domain. The flow domain near drillpipe is rotating with a certain speed to simulate the pipe rotation effect, and the outer flow domain is static. In addition, inner flow domain transfers pressure, velocity, cuttings concentration, and other parameters through interface.

Mathematical model of CFD simulations

To observe the effect of different rotation speeds on the drilling fluid–cuttings two-phase flow behavior, the Euler multiphase flow in rotary coordinate system is chosen as the simulated model. Standard function is used to calculate the momentum equation for low Reynolds number zone near drillpipe surface. The momentum equation for fully turbulent zone is determined by Realizable κ-ε model (Shih

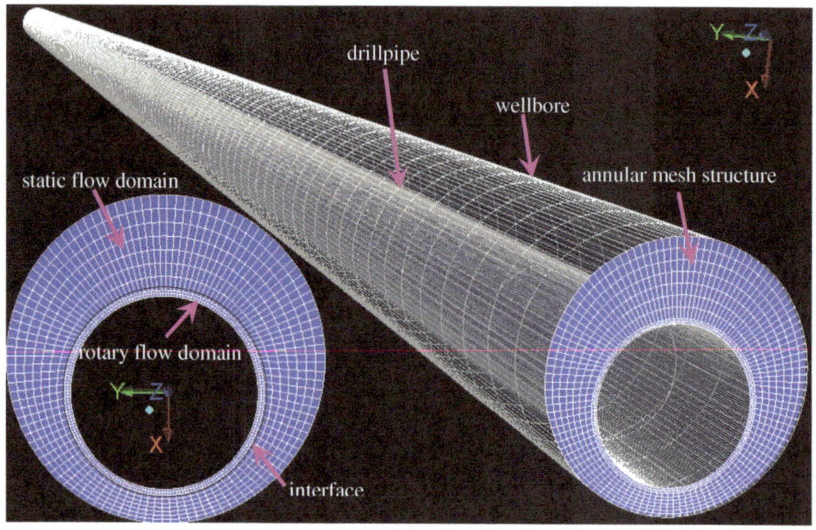

Fig. 2 Side view and profile of annular grid structure

et al. 1995). Also, first order upwind scheme of finite volume method is applied to discretize the governing equations, and the value that settling cuttings volume in annulus tends to be stable is considered as convergence criterion. These governing equations can be listed as follows:

Mass equation of liquid phase:

$$\frac{\partial}{\partial t}(\alpha_l \rho_l) + \nabla \cdot (\alpha_l \rho_l \vec{v}_l) = \Delta m_{sl} \tag{1}$$

Mass equation of solid phase:

$$\frac{\partial}{\partial t}(\alpha_s \rho_s) + \nabla \cdot (\alpha_s \rho_s \vec{v}_s) = \Delta m_{ls} \tag{2}$$

Momentum equation of liquid phase:

$$\begin{aligned} &\frac{\partial}{\partial t}(\alpha_l \rho_l \vec{v}_l) + \nabla \cdot (\alpha_l \rho_l \vec{v}_l \vec{v}_l) + \rho_l (2\vec{\omega} \times \vec{v}_l + \vec{\omega} \times \vec{\omega} \times \vec{r}) \\ &= -\alpha_l \nabla p + \mu \nabla^2 \vec{v}_l + \alpha_l \rho_l \vec{g} \\ &\quad + K_{sl}(\vec{v}_s - \vec{v}_l) + \Delta m_{sl} \vec{v}_{sl} - \Delta m_{ls} \vec{v}_{ls} \end{aligned} \tag{3}$$

Momentum equation of solid phase:

$$\begin{aligned} &\frac{\partial}{\partial t}(\alpha_s \rho_s \vec{v}_s) + \nabla \cdot (\alpha_s \rho_s \vec{v}_s \vec{v}_s) + \rho_s (2\vec{\omega} \times \vec{v}_s + \vec{\omega} \times \vec{\omega} \times \vec{r}) \\ &= -\alpha_s \nabla p + \mu \nabla^2 \vec{v}_s + \alpha_s \rho_s \vec{g} \\ &\quad + K_{sl}(\vec{v}_l - \vec{v}_s) - \Delta m_{sl} \vec{v}_{sl} + \Delta m_{ls} \vec{v}_{ls} \end{aligned} \tag{4}$$

where "*s*" and "*l*" are the subscripts that represent solid- and liquid-phase parameter. Δm_{sl} is that the mass transfers from solid phase to fluid phase. Similarly, the position exchange of subscript indicates the mass transfers from fluid phase to solid phase. Also, $\Delta m_{sl} = -\Delta m_{ls}$; v_{ls} is the velocity between phases. When $\Delta m_{ls} > 0$, $v_{ls} = v_{sl}$. When $\Delta m_{ls} < 0$, $v_{ls} = v_s$ and $v_{ls} = v_{sl}$; α is the phase volume fraction; K_{sl} is the solid–liquid momentum interchange coefficients, and $K_{sl} = \alpha_s \rho_s f/\tau_s$. τ_s is the relaxation time of particles, and $\tau_s = \rho_s d_s^2/(18\ \mu)$. Here, $K_{ls} = K_{sl}$, and *f* which can be obtained from Syamlal–O'Brien model (Syamlal et al. 1989) is the function of drag force.

Boundary conditions and initial conditions

Boundary conditions of the transient simulation apply velocity inlet and pressure outlet, and setting different values of inlet velocity and outlet pressure can be used to simulate the different rotation speeds and well depths, respectively. The density of drilling fluid is 1,200 kg/m^3, and the viscosity is 30 mPa s. Cuttings density is 2,500 kg/m^3, and the diameter is 8 mm. The volume fraction of cuttings is 1 %. The simulations are divided into 12 groups, and the more detailed parameters are shown in Table 1.

Results and discussions

Effects of pipe rotation on velocity profile

The purpose of analyzing the velocity profile is that the effects of pipe rotation on annular pressure drop are actually caused by redistribution in annulus. The redistribution of velocity also has a significant effect on cuttings transport. As shown in Fig. 3, CFD simulations show that tangential velocity is decreases by exponential function along the tangential direction and the ratio of reduction become small with the increase of rotation speed. It is generally considered that the tangential velocity of drilling fluid is

Table 1 The variables required by CFD simulations

Groups	Inclination (°)	Flow rate (L/s)	Pipe rotation speed (rpm)
1–9	45	30	80, 100, 120,140, 160, 180, 200, 220, 240
10–18	45	40	80, 100, 120, 140, 160, 180, 200, 220, 240
19–27	45	50	80, 100, 120, 140, 160, 180, 200, 220, 240
28–36	60	30	80, 100, 120, 140, 160, 180, 200, 220, 240
37–45	60	40	80, 100,120, 140, 160, 180, 200, 220, 240
46–54	60	50	80, 100, 120, 140, 160, 180, 200, 220, 240
55–63	75	30	80, 100, 120, 140, 160, 180, 200, 220, 240
64–72	75	40	80, 100, 120, 140, 160, 180, 200, 220, 240
73–81	75	50	80, 100, 120, 140, 160, 180, 200, 220, 240
82–90	90	30	80, 100, 120, 140, 160, 180, 200, 220, 240
91–99	90	40	80, 100, 120, 140, 160, 180, 200, 220, 240
100–108	90	50	80, 100, 120, 140, 160, 180, 200, 220, 240

generated near drillpipe surface with drillpipe rotation. The cuttings are tangentially dragged by viscosity force of drilling fluid, which can improve hole cleaning. However, due to the reduction of tangential velocity of drilling fluid, the drag force acting on cuttings gradually becomes less and some cuttings can not be effectively transported into main flow field. As a result, these cuttings will settle at the bottom of annulus and eventually cuttings bed is formed. This indicates high rotation speed can provide more energy for suspending the cuttings, makes most cuttings fully suspended, and decrease cuttings bed height.

Distribution of annular cuttings with pipe rotation

The simulations on the distribution of cuttings in annulus for angle at 45° and 60° were conducted, and the following conclusions can be reached by comparing Fig. 4 with 5:

- Cuttings present asymmetrical distribution on either side of drillpipe, and cuttings bed is easy to be formed when low mud flow rate is used.
- As drillpipe rotation speed increases, cuttings concentration in annulus decreases for any pipe rotation speed, but the level of enhancement is more pronounced with the decrease at flow rate.

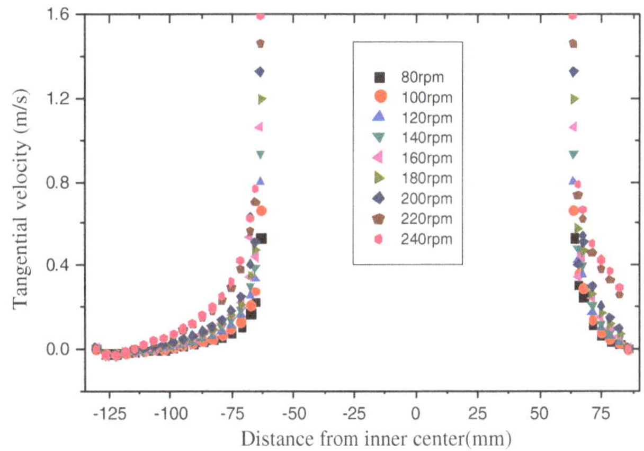

Fig. 3 Tangential velocity profile in an eccentric annulus with pipe rotation

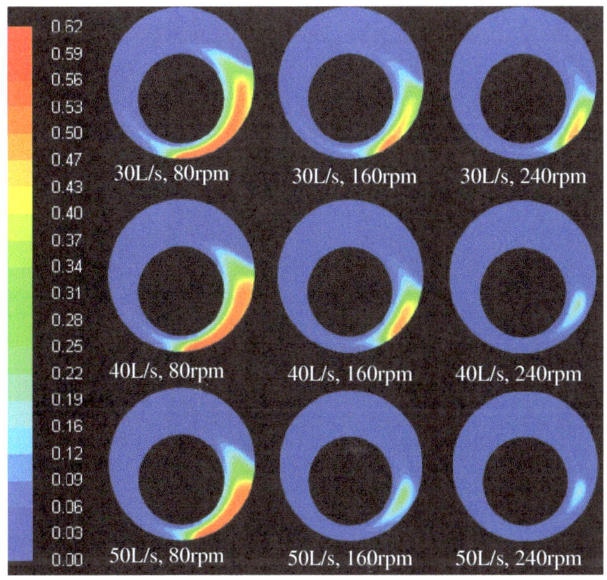

Fig. 4 Annular cuttings concentration contour at the angle of 45°

- Lower cuttings concentration near drillpipe are presented, which indicates that drilling fluid can obtain high tangential velocity under the drillpipe rotation, and cuttings are suspended by high drag force and transported by axial fluid flow.
- In the low flow rate, cuttings are more transported at the angle of 60° than 45°, but inclination shows only a minor effect on cuttings concentration in high flow rate and rotation speed.
- Pure drilling fluid region, dispersed layer, and stationary cuttings bed are emerged in low rotary speed or flow rate. As the flow rate is increased, the intensity of turbulent eddies grows in strength and eventually the particles of the cuttings bed are totally lift into the dispersed layer and there only exists pure drilling fluid layer and dispersed layer. Finally, with further

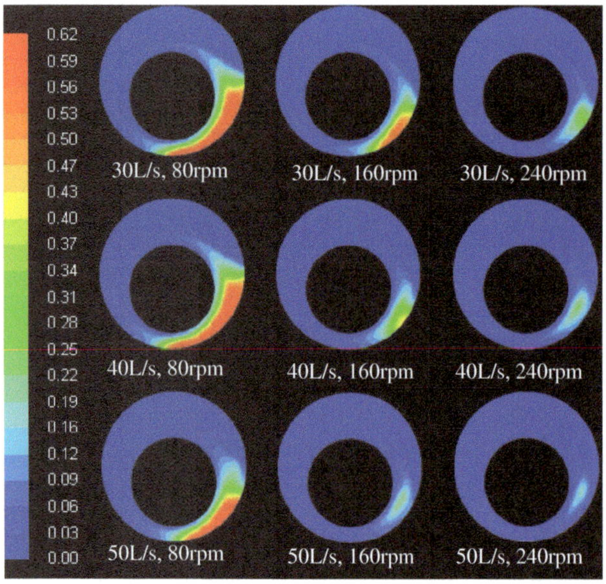

Fig. 5 Annular cuttings concentration contour at the angle of 60°

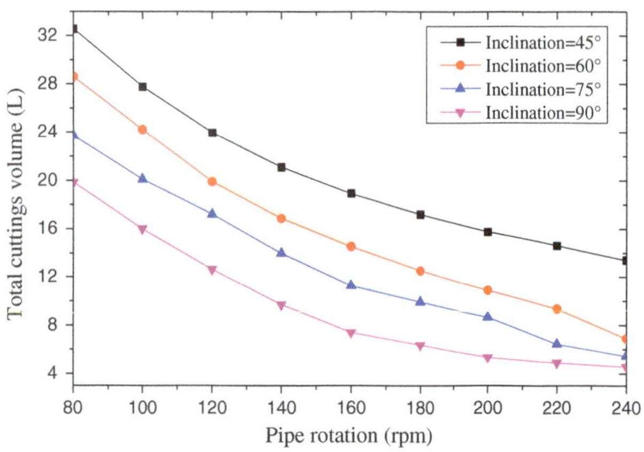

Fig. 6 Effect of pipe rotation on cuttings volume for different inclinations and flow rate 30 L/s

increases in the flow rate, all particles will be transported in a heterogeneous flow.

Effects of pipe rotation on cuttings concentration and pressure drop

As seen from Figs. 6, 7, 8, 9, at fluid rates from 30 to 50 L/s and inclinations horizontal to 60°, pipe rotation can significantly decrease the total volume of annular cuttings. Even though this is the case, still a decrease in the cuttings volume is observed with the increase of rotation speed. However, the additional contribution of drillpipe rotation on decreasing cuttings volume is almost negligible after a certain rotation speed when flow rate reaches 50 L/s or inclination is close to horizontal. Considering the limited rotation speed in real drilling, pipe rotation has a significant

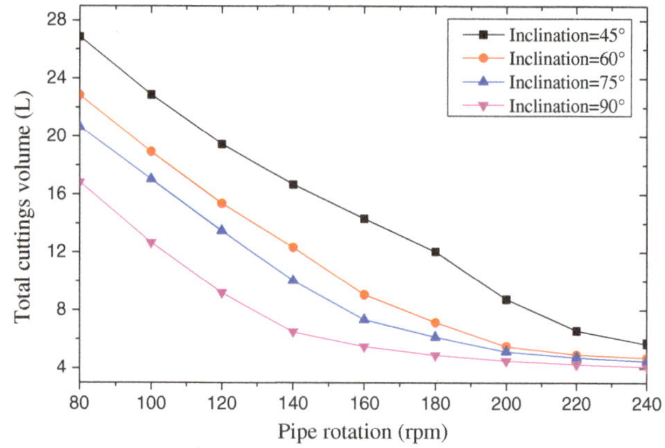

Fig. 7 Effect of pipe rotation on cuttings volume for different inclinations and flow rate 40 L/s

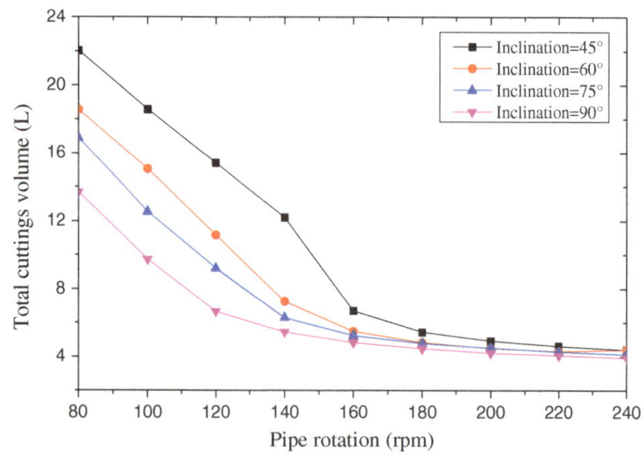

Fig. 8 Effect of pipe rotation on cuttings volume for different inclinations and flow rate 50 L/s

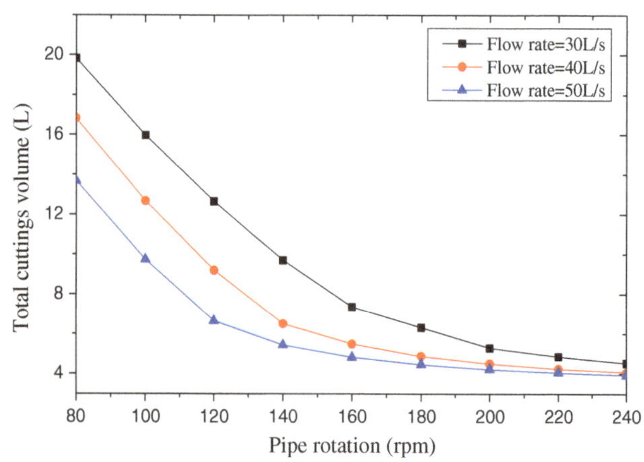

Fig. 9 Effect of pipe rotation on cuttings volume for different flow rates in horizontal wellbores

impact on hole cleaning in low flow rates. In addition, the pipe rotation effect on hole cleaning is also affected by inclination. In particular, in the lower pipe rotation speeds, the cuttings volume in the annulus is more for lower

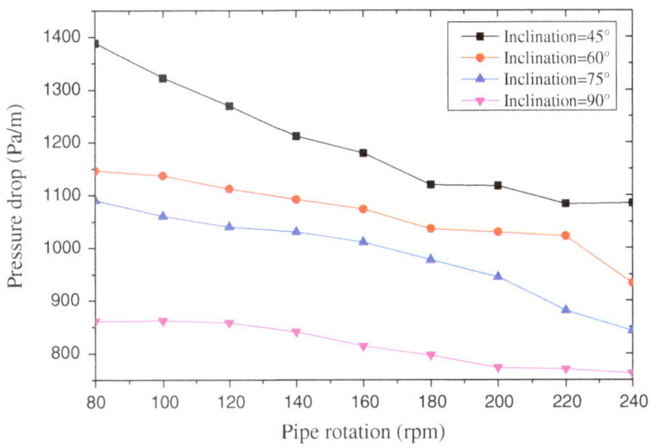

Fig. 10 Effect of pipe rotation on annular pressure drop for different inclinations and flow rate 30 L/s

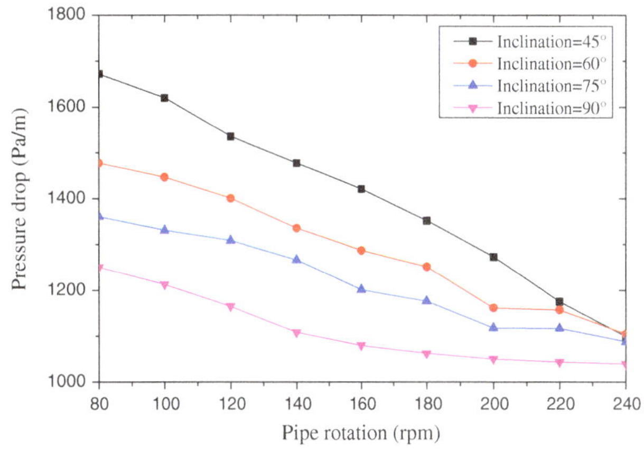

Fig. 11 Effect of pipe rotation on annular pressure drop for different inclinations and flow rate 40 L/s

inclinations than for higher inclinations, which means that cuttings bed is easier to be formed in the low inclination. This conclusion is consistent with Li et al.'s (2001) observation that the toughest section for hole cleaning is the build section rather than the vertical or the horizontal section.

As shown from Figs. 10, 11, 12, the influences of drillpipe rotation for different flow rates and inclinations on annular pressure drop are depicted. For any flow rate and inclination, as the pipe rotation speed is increased, annular pressure drop is decreased. The pressure drop almost reduces by half when pipe rotation increases from 80 to 160 rpm with 50 L/s flow rate at 45°. But when keeping flow rate 50 L/s, the pressure drop almost remains constant after pipe rotation speed reaches 180 rpm. Also, in the low rotation speed, producing pressure drop becomes more at the lower angle than at higher angle.

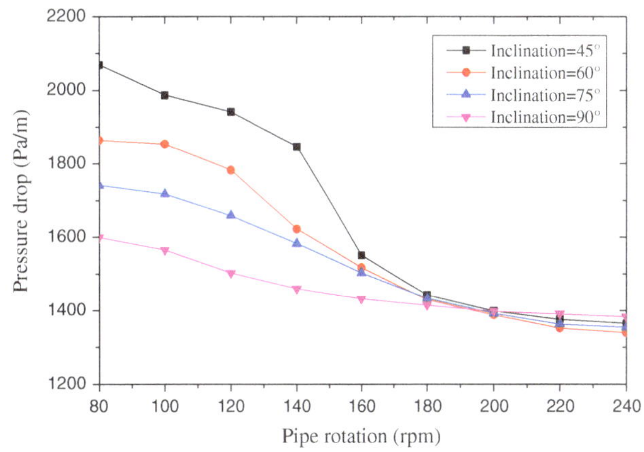

Fig. 12 Effect of pipe rotation on annular pressure drop for different inclinations and flow rate 50 L/s

Empirical correlations

It is considered that the ratio of cuttings volume and wellbore volume is function of inclination, flow rate, density and size of cuttings, viscosity and density of drilling fluid, annular hydraulic diameter, pipe rotation, and gravity. Major parameters affecting cuttings concentration can be defined as:

$$\frac{V_b}{V_W} = f(\theta, v, \omega, \rho_l, \rho_s, \mu, D_h, g, d_c) \quad (5)$$

Based on Buckingham-π theorem, seven dimensionless groups are determined as follows (Ozbayoglu et al. 2008):

$$\pi_1 = \frac{\rho_l v D_h}{\mu} \quad (6)$$

$$\pi_2 = \frac{v^2}{g D_h} \quad (7)$$

$$\pi_3 = C \quad (8)$$

$$\pi_4 = \theta \quad (9)$$

$$\pi_5 = \frac{d_c}{D_h} \quad (10)$$

$$\pi_6 = \frac{\rho_l v d_c}{\mu} \quad (11)$$

$$\pi_7 = \frac{\omega D_h}{v} \quad (12)$$

where V_b represents cuttings volume (m^3), V_w is wellbore volume (m^3), θ is the inclination (°), v represents fluid flow rate m/s, ω stands for pipe rotation (s^{-1}), ρ_l is the density of drilling fluid (kg/m^3), ρ_s is the density of cuttings (kg/m^3), μ represents the viscosity of drilling fluid (Pa s), d_c is the cuttings size (m), D_h stands for the hydraulic diameter (m), C is the cuttings concentration, g stands for the gravity (N/kg).

In this simulation, size and density of cuttings, injected cuttings concentration, viscosity and density of drilling fluid are constant, therefore, π_3, π_5, and π_6 can be negligible, and Eq. (5) is rewritten by:

$$\frac{V_b}{V_W} = a_0 \pi_1^{a_1} \pi_2^{a_2} \pi_4^{a_3} \pi_7^{a_4} \tag{13}$$

Considering the effect of cuttings concentration, annular pressure drop can be defined as:

$$\frac{\Delta p}{\Delta L} = a_0 \pi_1^{a_1} \pi_2^{a_2} \pi_4^{a_3} \pi_7^{a_4} \left(\frac{V_b}{V_W}\right)^{a_5} \tag{14}$$

The least square method is applied to linearly regress for 12 groups data, and cuttings concentration and annular pressure drop for inclinations from 45° to 90° can be determined, respectively:

$$\frac{V_b}{V_W} = 0.01\left(\frac{\rho_l v D_h}{1,000\mu}\right)^{6.0493}\left(\frac{v^2}{gD_h}\right)^{-4.3133} \times (0.1\theta)^{-0.8977}\left(\frac{10\omega D_h}{v}\right)^{-1.3677} \tag{15}$$

$$\frac{\Delta p}{\Delta L} = 100\left(\frac{\rho_l v D_h}{1,000\mu}\right)^{1.0547}\left(\frac{v^2}{gD_h}\right)^{0.1376}(0.1\theta)^{-0.0740} \times \left(\frac{10\omega D_h}{v}\right)^{0.1435}\left(\frac{100 V_b}{V_W}\right)^{0.2877} \tag{16}$$

Conclusions

In this study, CFD simulations of the effects of drillpipe rotation on cuttings transport behavior in the complex structure well were performed using an Euler multiphase model. The following conclusions can be drawn as follows:

- Pipe rotation has an important effect on distribution of annular cuttings in inclined section and causes the asymmetrical distribution of cuttings along the circumference direction.
- Pipe rotation can significantly increase the tangential velocity of drilling fluid, which generates the drag force along the tangential direction, and the force can restrain cuttings bed development to some extent. In high rotation speeds, dragged effect is more pronounced, and the fully suspended flow is easy to be developed.
- At low or medium flow rates, increasing pipe rotation speed can significantly reduce cuttings volume and annular pressure drop. However, there exists a threshold point at high flow rates, after which no matter how much the pipe rotation speed is increased, the magnitudes of cuttings volume and annular pressure drop almost remain constant.
- Buckingham-π theorem was applied to define dimensionless groups, and eventually, user friendly empirical correlations about the cuttings concentration and pressure drop were developed using least square method based on the CFD simulations on cuttings transport.

Acknowledgments This work was supported by Program for National Science and Technology Major Project of China (No. 2011ZX05009-005) and the National Natural Science Foundation of China (Nos. 51174043, 51204056) are gratefully acknowledged.

References

Bassal AA (1996) The effect of drillpipe rotation on cuttings transport in inclined wellbores. MS Thesis, University of Tulsa

Bilgesu HI, Ali MW et al. (2002) Computational fluid dynamics (CFD) as a tool to study cutting transport. In: SPE Eastern Regional Meeting, Lexington

Bilgesu HI, Mishra N et al. (2007) Understanding the effects of drilling parameters on hole cleaning in horizontal and deviated wellbores using computational fluid dynamics. In: SPE Eastern Regional Meeting, Lexington

Duan MQ, Miska S et al (2008) Transport of small cuttings in extended reach drilling. SPE Drill Complet 23(3):258–265

Li J, Walker S (2001) Sensitivity analysis of hole cleaning parameters in directional wells. SPE J 6(4):356–363

Li M, Jv YF et al (2010) Simulation on the effect of drillstring rotation on hole cleaning for extended reach well. West-China Explor Eng 5:51–54

Mishra N (2007) Investigation of hole cleaning parameters using computational fluid dynamics in horizontal and deviated wells. MS Thesis, West Virginia University

Ozbayoglu ME, Saasen A et al. (2008). Effect of pipe rotation on hole cleaning for water-based drilling fluids in horizontal and deviated wells. In: IADC/SPE Asia Pacific drilling technology conference and exhibition, SPE, Jakarta, Indonesia

Peden JM, Ford JT et al. (1990) Comprehensive experimental investigation of drilled cuttings transport in inclined wells including the effects of rotation and eccentricity. In: European Petroleum Conference, SPE, Hague, Netherlands

Sanchez RA, Azar JJ et al (1999) Effect of drillpipe rotation on hole cleaning during directional-well drilling. SPE J 4(2):101–108

Shih TH, Liou WW et al (1995) A new κ–ε eddy viscosity model for high Reynolds number turbulent flows. Comput Fluids 24(3):227–238

Sifferman TR, Becker TR (1992) Hole cleaning in full-scale inclined wellbores. SPE Drill Eng 7(2):115–120

Sorgun M (2010) Modeling of Newtonian fluids and cuttings transport analysis in high inclination wellbores with pipe rotation. Ph. D Thesis, Middle East Technical University

Syamlal M, O'Brien TJ (1989) Computer simulation of bubbles in a fluidized bed. AIChE Symp Series 85:22–31

Wang ZM, Guo XL et al (2009) Effect of drillpipe rotation on borehole cleaning for extended reach well. J Hydrodyn 21(3):366–372

17

Wettability alteration to intermediate gas-wetting in low-permeability gas-condensate reservoirs

Shuo Zhang · Guan-Cheng Jiang · Le Wang · Wang Qing · Hai-Tao Guo · Xin-guo Tang · Dian-Gang Bai

Abstract Gas-condensate reservoirs experience significant productivity losses as reservoir pressure drops below the dew point due to condensate accumulation and water blocking and the subsequent reduction in gas relative permeability. One potential way to overcome this problem is to alter reservoir wettability to gas-wetting to reduce condensate accumulation and water blocking in the near wellbore and maintain high productivity. The major goal of this work was to study the mobility of the gas and liquid phase (both water and oil) before and after wettability alteration from strong liquid-wetting to intermediate gas-wetting. For this purpose, in addition to relative permeability measurements, we also conducted various other tests to demonstrate that liquid mobility can be improved significantly due to wettability alteration. As wettability modifiers, fluorinated polymers are capable of delivering a good level of oil and water repellency to the rock surface, making it intermediate gas-wet and alleviating such liquid blockage under high temperature. The contact angle analyses, capillary rise, flow tests, and imbibition spontaneous tests were used to estimate the effect of treatments on wettability. Experimental results demonstrated that the fluorinated polymers can alter the wettability of cores from strong liquid-wetting to gas-wetting, which could decrease the amount of water invaded and resided in the gas formation. Core flood test results demonstrated that the relative permeabilities of both the gas and the liquid phases were increased significantly after the wettability alteration to preferential gas-wetness. The residual liquid saturation was decreased, and the gas production was enhanced greatly due to the wettability alteration. These results imply that gas well deliverability may increase substantially when wettability is altered to intermediate gas-wetting. Efficiency in the extraction of natural gas is important to improve the productivity of gas-condensate reservoirs where liquid accumulates, which is beneficial for the economic and environmental considerations of the oil and gas industry.

Keywords Gas-condensate reservoir · Gas-wetting · Low-permeability gas reservoir · Liquid blockage · Wettability alteration

S. Zhang · G.-C. Jiang · L. Wang · W. Qing
State Key Laboratory of Petroleum Resource and Prospecting, Department of Petroleum Engineering, China University of Petroleum, Beijing 102249, China

H.-T. Guo · X. Tang · D.-G. Bai
Tan Hai Development Company, China National Petroleum Corporation, Tianjin 300289, China

Introduction

A sharp reduction in gas well deliverability is often observed in many low-permeability gas-condensate reservoirs even at very high reservoir pressure. The decrease in well deliverability is attributed to condensate accumulation and water blocking (Li and Firoozabadi 2000; Allen and Roe. 1950; Barnum. 1995; Wagner and Leach 1958). As the pressure drops below the dew point, liquid accumulates at the wellbore in high saturations reducing gas relative permeability (Tang and Firoozabadi 2003; Fahes and Firoozabadi 2005), the result is a decrease in the gas production rate. Gas deliverability often increases after removal of water and condensate phases from the near-wellbore region. Several techniques have been used to increase gas well deliverability after the initial decline. Hydraulic fracturing and solvent injection are implemented in order to remove the accumulated liquid (Tannich 1975; Holditch 1979). The increased rates were, however, only sustained for a period. The approach is not a permanent solution to the problem as the condensate bank will form again. On the other hand, when hydraulic fracturing is used by injecting aqueous fluids, the clean-up of water accumulation from the formation after fracturing is essential to obtain an increased productivity. Water is removed in two phases, immiscible displacement by gas, followed by vaporization by the expanding gas flow (Mahadevan and Sharma 2003). Due to the low permeability and the wettability characteristics, it takes a long time to perform the clean-up; in some cases, as little as 10–15 % of the water load could be recovered. In a new approach, Li and Firoozabadi examined the alteration of wettability to intermediate gas-wetting by chemical treatment of the rock. A major factor in liquid accumulation is the liquid's low mobility because of strong liquid-wetting (Walker et al. 2005; Zuluaga and Monsalve 2003). By altering the wettability of the rock from liquid-wetting to intermediate gas-wetting, an increase in liquid mobility is achieved, preventing the accumulation of liquid in high saturations and resulting in high rates of gas production (Boom and Zeelenberg 1996; Fahes and Firoozabadi 2005). The work of Firoozabadi and coworkers on wettability alteration may have a practical application in improving gas effective permeability due to the prevention of liquid accumulation in high saturations at the wellbore; the main requirement is the permanent alteration of wettability. Tang and Firoozabadi performed wettability alteration and measured the effect of wettability alteration on increased liquid mobility. The major goal of this work was to study the mobility of the gas and liquid phase (both water and oil) before and after wettability alteration from strong liquid-wetting to intermediate gas-wetting. For this purpose, in addition to relative permeability measurements, we also conducted various other tests to demonstrate that liquid mobility can be improved significantly due to wettability alteration. The alteration of wettability was achieved by treating the rock with the Zonyl8740 solution at 24 °C. This paper is organized along the following lines: First, the description of rocks, model reservoir fluids, chemicals used in treating the rocks, and the treatment procedure were presented. Then, the measurements of contact angle, imbibition, and flow testing were described, and in the section that follows, the results of these measurements in treated and untreated cores are presented. At the end, conclusions from the work were drawn. Note that the wettability alteration to gas-wetness may also be applied in drilling and completion to reduce the water loss to tight gas formation, in water shutoff in tight gas reservoirs with bottom water, and acid-fracturing to contain secondary formation damage in low-permeability gas reservoirs.

Table 1 Composition of the reservoir simulated brine

Composition	Content (mg/l)	Composition	Content (mg/l)
HCO_3^-	677	Ca^{2+}	172
Cl^-	5,527	Mg^{2+}	42
SO_4^{2-}	125	$Na^+ + K^+$	3,257

Table 2 Properties of rock samples

NO. 1	Length (cm)	Diameter (cm)	Porosity (%)	Permeability (md)
A-1	6.13	3.66	9.2	0.13
A-2	6.15	3.65	9.1	0.21
A-3	6.12	3.66	9.8	0.18
A-4	6.13	3.66	9.4	0.23
A-5	6.18	3.66	8.9	0.14

Experiment

Materials

The polymer used was Zonyl8740. This DuPont product is an aqueous fluorochemical that provides water and oil repellency on porous mineral surfaces. It is soluble in water and methanol; it is chemically bond to the surface of the rock at the molecular level, allowing durable results. At room temperature, it has a density of 1.095 g/cm^3 and a viscosity of less than 250 mPa s. The reservoir simulated brine with the concentration of 9,800 mg/L, and the simulated water was used. The composition is shown in Table 1.

Hexadecane is produced by Beijing Hengye company, China. Several core plugs with a diameter of about 3.7 cm sampled from the targeted gas-condensate reservoir (Changqing reservoir in China) were used in this study. As is shown in Table 2, the porosity was around 10 % and the permeability ranged from 0.13–0.23 md.

Wettability tests

The experimental work is designed to investigate wettability alteration and its effect on liquid mobility. Wettability is examined before and after treatment by contact angle measurements, capillary rise analyses, and spontaneous imbibition tests. The rock was saturated with 1.0 % Zonyl8740 solution and was aged to change the wettability at 24 °C. The aging time was about 8 h. After saturation with the chemical solution, the rock was evacuated for about an hour and was then dried in order to remove the extra liquid chemicals. A very small amount of the chemical was left on the rock surface; the surface energy decreased and the solid surface was altered to preferential gas-wettability as a result of adsorption. Liquid mobility, before and after wettability alteration, is measured by performing unsteady-state displacement tests.

Contact angles

Contact angle meter SL200 B (USA Kino Industry Co.) was used to test liquid contact angles on rock surface in air at ambient temperature. Liquid contact angle measurements were conducted using the sessile drop method. Droplets of liquid were carefully dropped onto the surface of each core. Then, liquid contact angles on the rock surface were obtained. The liquid volume is about 5×10^{-3} ml. The simulated water and hexadecane were selected for measuring the contact angles of rock cores. At the same time, the images of contact angle were obtained.

Spontaneous imbibition

The experimental apparatus used to conduct the spontaneous water imbibition tests was similar to that used by Li and Horne. The schematic of apparatus is shown in Fig. 1. The core sample was hung under an electronic balance and submerged in the liquid. Change in weight of the core versus time was recorded. Spontaneous water imbibition tests were conducted to alter the wettability from preferential liquid-wetness to gas-wetness in the core sample A-1 before and after 1.0 % Zonyl8740 solution treatments. Brine was made using the compositions of the formation water. The salinity of the brine was about 98,000 mg/l. The spontaneous water imbibition into the air-saturated rock was conducted before the chemical treatment. The rate of imbibition spontaneous was calculated using the data from the electronic balance. The decrease in the imbibition rate and recovery by spontaneous imbibition of liquid after chemical treatment give the extent of wettability alteration.

The glass capillary tube rise test

The glass capillary tube rise test was conducted to examine the effect of the fluoro-copolymer emulsion on wettability alteration. The experiment procedure (Li and Firoozabadi 2000) was described as follows. Firstly, fluorochemical solutions Zonyl8740 of different concentrations were prepared with distilled water. And then, the capillary tube was aged in the solutions of various concentrations for 8 h and dried at room temperature. After the chemical treatment, the capillary tube was inserted vertically in the liquid, and then, the liquid level (h) in the capillary tube was gained,

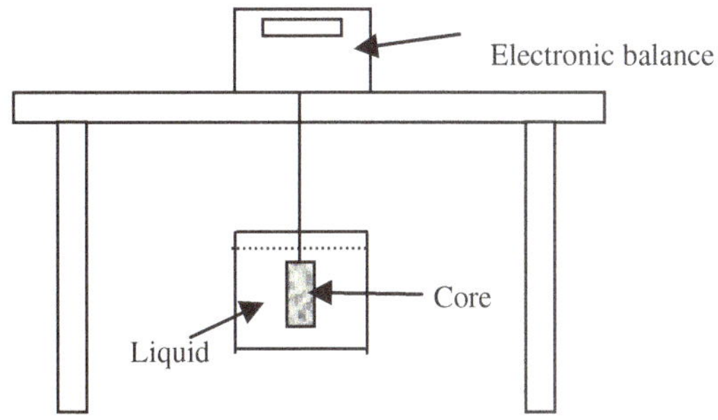

Fig. 1 Imbibition device

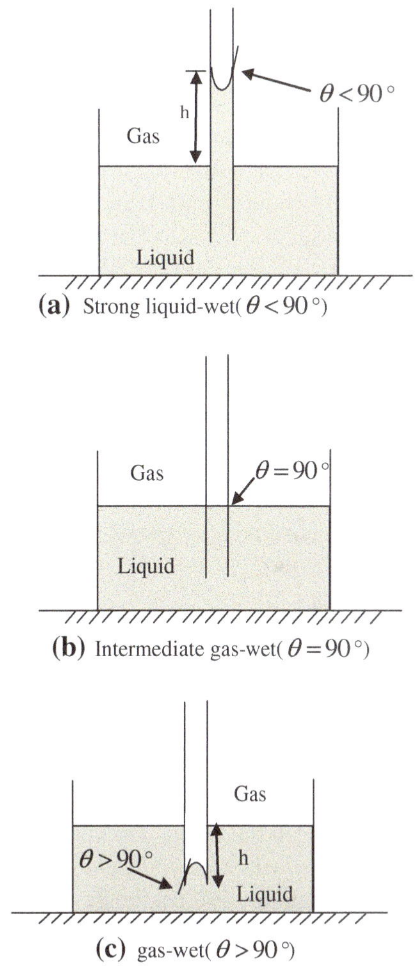

Fig. 2 Sketch of the apparatus for capillary rise tests

whether it rose or descended. The schematic diagram is shown in Fig. 2. According to the Young's equation,

$$\theta = \arccos\left(\frac{\rho g h r}{2\sigma}\right) \tag{1}$$

(where σ and ρ are surface tension and density of liquid, respectively, g is the acceleration of gravity, and r represents radius of capillary tube), the contact angle θ of the

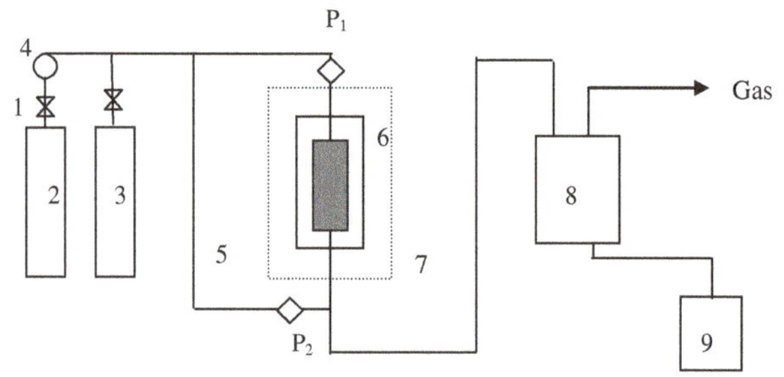

1- Valve 2-Gas 3-liquid 4- Regular 5- Pressure Transducer 6-Core Holder 7-Oven 8-Separator 9-Water

Fig. 3 Schematic of the apparatus for core flow measurements

capillary tube can be computed when σ, ρ, r, and h are available. The radius of the circular capillary tube is 0.37 mm.

If the liquid in the circular capillary tube rises ($h > 0$), the contact angle in gas–liquid system is <90°. If there is no liquid rise ($h = 0$) in the capillary, the contact angle is equal to 90°. And when the liquid in the circular capillary tube goes down ($h < 0$), the contact angle is larger than 90°. So one may define preferential liquid-wetting when $\theta < 90°$ and preferential gas-wetting when $\theta > 90°$. But for a capillary tube of equilateral triangle cross section, the gas–liquid interface will be flat when $\theta = 60°$, which suggests preferential gas-wetting for $\theta > 60°$. Actually, there is complex geometry in porous media of reservoirs. In this article, strong liquid-wetting is defined when $\theta < 60°$, neutral gas-wetting when $60° < \theta < 90°$, and preferential gas-wetting when $\theta > 90°$.

Core flooding

For measuring gas and liquid relative permeability before and after wettability alteration, the core was first saturated with oil (hexadecane) or the brine with a salinity of 9,800 mg/l. The absolute permeability of the core was measured. The air displacement in the core without the chemical treatment was then carried out at room temperature with a confining pressure of 500 psi, while the outlet was open to the atmosphere. Oil recovery, residual oil saturation, and gas-phase relative permeability were measured. The oil relative permeability was also calculated at low oil saturations. After the air displacement without chemical treatment, the core was cleaned, dried, and treated by 1.0 % Zonyl8740 solution. Fluorinated polymers are well known as low surface energy materials, which have oil and water repellent property (Huang et al. 2004). And the acrylic main-chains make the polymer adhere well to various materials (Zhang and Chen 2005). Following that, the chemical solution was removed from the core, and the core was saturated with oil. Air displacement was performed again. The core was then dried in order to remove the extra liquid chemicals. Oil recovery, residual oil saturation, and gas-phase relative permeability were then measured. The same experiments were repeated to measure the gas–water relative permeability after the rock was treated using the chemical solution to change the wettability. The schematic of the apparatus for core flow measurements is shown in Fig. 3.

Results and discussion

Contact angles

In order to demonstrate the wettability alteration by 1.0 % Zonyl8740 solution, droplets of water and hexadecane were placed on the surface of the rock before and after the chemical treatments (see Fig. 4). Water and hexadecane did not imbibe into the rock. Instead, it forms a ball-shape droplet. The contact angle of water and the core surface increases from 5.6° to 112.3°. The value of the measured contact angle is greater than 90°, which demonstrates that the wettability of the rock was altered to preferential gas-wetness by the chemical treatments. Figure 4b shows a droplet of hexadecane placed on the surface of the rock after the chemical treatments. The contact angle of hexadecane and the core surface increases from 35.3° to 85.7°. One can see from Fig. 4, and the value of the measured contact angle through the liquid phase that the wettability of the low-permeability rock was changed from liquid-wetness to preferential gas-wetness by the chemical treatments in the gas–liquid–rock system.

Capillary tube rise test

According to the Young equation,

$$P_c = 2\sigma \cos\theta / r \tag{2}$$

where P_c is the capillary pressure, σ is the surface tension, θ is the contact angle, and r is the diameter of capillary.

If the contact angle increases, then the capillary pressure will decrease. Hence, the liquid cannot be imbibed into the capillary tube and rock core after wettability alteration to gas-wetting. Therefore, the capillary rise and spontaneous imbibition tests are used to evaluate further the effect of the emulsion on rocks surface wettability. Based on Eq. (2), the lower the capillary rise and the higher the contact angle as the surface tension of liquid is constant, indicating a deviation from strong liquid-wetting.

Figure 5 shows that liquid rise versus the concentration of Zonyl8740 solution for both the gas/oil (Hexadecane) and gas/water systems. The liquid rise in the capillary tube decreases with an increase of the concentration, and the liquid level of capillary is the lowest when the concentration reaches around 1.0 % in both gas/water and oil/gas

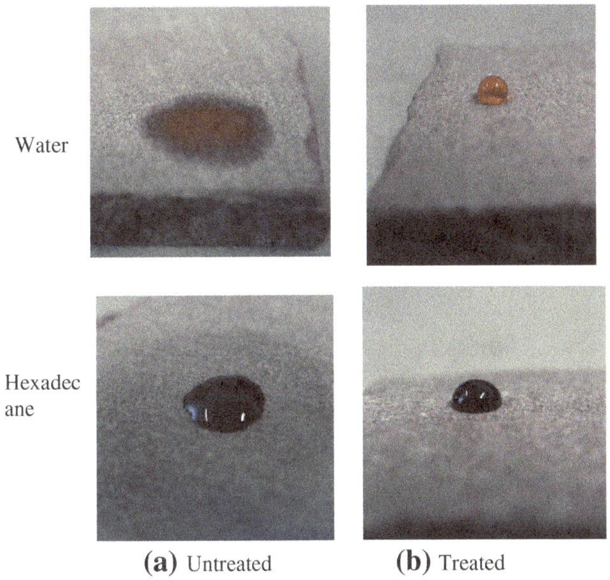

Fig. 4 Images of contact angle

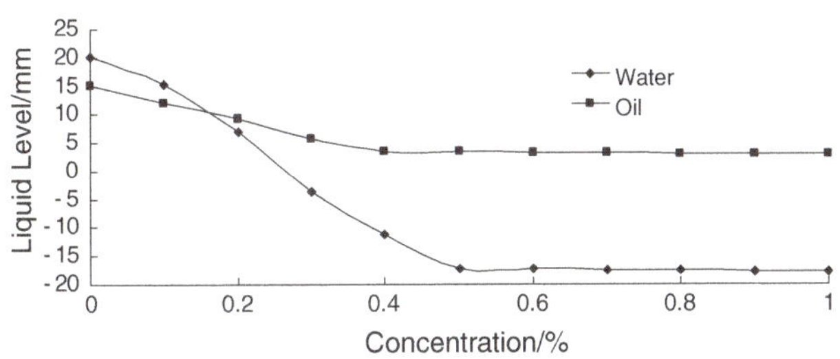

Fig. 5 Liquid level versus the concentration of Zonyl8740

systems. The contact angle can be calculated using Eq. (1). Figure 6 illustrates the computed contact angle versus the concentration of Zonyl8740 solution for both the gas/oil and gas/water systems. The contact angle is 25.2° in gas/oil system and 60.5° in gas/water system before treatment.

Figure 6 elucidates that the contact angle increases with the increase of the Zonyl8740 solution concentration in both the gas/oil and gas/water systems.

Imbibition and drainage

Spontaneous water imbibition tests were performed in the core samples A-2 and A-3 from the Changqing gas-condensate reservoir before and after the chemical treatments in this study. Spontaneous imbibition experiments were conducted at a room temperature to test the wettability effect of the low-permeability rock sampled from liquid-wetness to gas-wetness.

Figure 7 plots the gas recovery by both oil (hexadecane) and water spontaneous imbibitions in an air-saturated core before and after treatment. The gas recovery is represented by the fraction of gas originally in-place (GOIP). One can

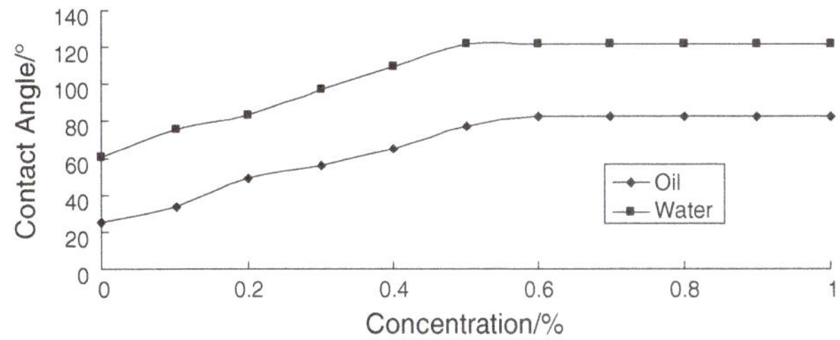

Fig. 6 Contact angle versus the concentration of Zonyl8740

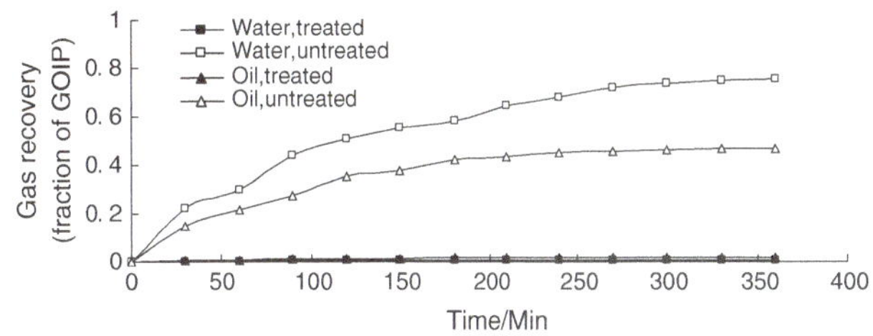

Fig. 7 Gas recovery by liquid imbibition into the air-saturated rock

see from Fig. 7 that the final recovery of the reservoir rock (A-2) by spontaneous water imbibition was about 75.6 % without the chemical treatment. After the chemical treatment, the amount of water imbibed into the reservoir rock was almost zero. The final recovery of the reservoir rock (A-3) by spontaneous oil imbibition was about 46.9 % without the chemical treatment. After the chemical treatment, the amount of oil imbibed into the reservoir rock was almost zero too. This demonstrates the wettability alteration and a brief explanation are given in the following. There are only two types of forces, gravity and capillary forces, that govern the spontaneous water imbibition shown in Fig. 1. In this case, capillary pressure is the driving force, while gravity is the resistance force. Gravity force did not change during the spontaneous water imbibition with and without the chemical treatments. So the only possibility was because of the change in capillary pressure to cause the reduction in recovery by spontaneous liquid imbibition. A simple definition of capillary pressure is adopted here:

$$P_c = \frac{\sigma \cos \theta_p}{\sqrt{\frac{k}{\phi}}} \tag{3}$$

where P_c is the capillary pressure, k and ϕ are permeability and porosity, respectively; δ is the surface tension and θ_p pseudo contact angle. We found that there was no change in surface tension and almost no change in porosity and permeability by the chemical treatment. Based on the above analysis and according to Eq. 2, it was the change in wettability represented by the contact angle to cause the reduction in recovery by spontaneous liquid imbibition

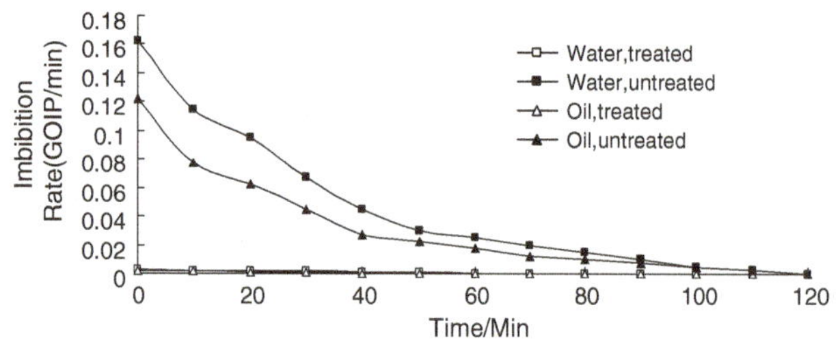

Fig. 8 Imbibition rate by liquid imbibition into the air-saturated rock

shown in Fig. 7. Therefore, the results shown in Fig. 7 demonstrate that the wettability of the very low-permeability rock alters from liquid-wetness to gas-wetness significantly.

A significant decrease in liquid imbibition rate with treatment is observed from Fig. 8. The spontaneous imbibition is related to wettability and gas recovery by spontaneous liquid imbibition decreases when the rock wettability changes from liquid-wetting to preferential gas-wetting. Therefore, these results demonstrate that the wettability of reservoirs rock in both gas/water and gas/oil systems has been altered from strong liquid-wetting to intermediate gas-wetting by the chemical treatment. These results are consistent with our previous contact angle measurement and capillary rise tests.

Core flooding test

Effect of wettability alteration on relative permeability

The gas and liquid relative permeability by gas flooding in the rock were calculated using the experimental data measured before and after the wettability alteration. The relative permeabilities were calculated by the JBN method (Johnson et al. 1959). Figure 9 shows the values of gas and water or oil (hexadecane) relative permeability measured in the core sample A-4 using an unsteady-state displacement approach before and after wettability alteration by the chemical treatments. After core flooding tests, a significant effect of wettability alteration on both gas and oil relative permeabilities was observed in Fig. 9a. Both the gas and water phase relative permeability increased after the wettability was altered from liquid-wetness to gas-wetness. The residual oil saturation by gas flooding decreased from 0.418 to 0.361 PV (PV = pore volume); the gas-phase relative permeability at the residual oil saturation increased about two times because of the wettability alteration from liquid-wetness to gas-wetness.

Gas and water relative permeabilities were measured in the core sample A-5 using the same unsteady-state displacement approach before and after wettability alteration by the chemical treatments. The effect of wettability alteration on gas and water relative permeabilities shows a trend similar to that for gas and oil relative permeabilities (see Fig. 9b). Residual water saturation decreased from 0.45 to 0.35 PV. Water saturation at the cross point, krg = krw, for treated cores was less than 0.4 PV, implying gas-wetting. This result is in line with the spontaneous imbibition measurements. These results imply that gas well deliverability may increase substantially when wettability is altered to intermediate gas-wetting.

Effect of wettability alteration on recovery

Figure 10 plots the values of recovery by gas flooding before and after the wettability alteration by the chemical treatments. The water recovery (see Fig. 10a) by gas flooding increased significantly after the wettability alteration. The final water recovery by gas flooding after wettability alteration increased from 50.5 to 72.9 %. The effect of wettability alteration on oil recovery shows a trend similar to that of water recovery (see Fig. 10b). The final oil (hexadecane) recovery by gas flooding after wettability alteration increased from 60.9 to 80.6 %. In summarizing, the experimental results shown above demonstrate that the effects of wettability alteration from liquid-wetness to gas-wetness on gas–water relative permeability and recovery are very significant in the rock with a low permeability.

Thermal stability

The thermal stability of the chemical treatment is very important in this study because the Changqing gas-

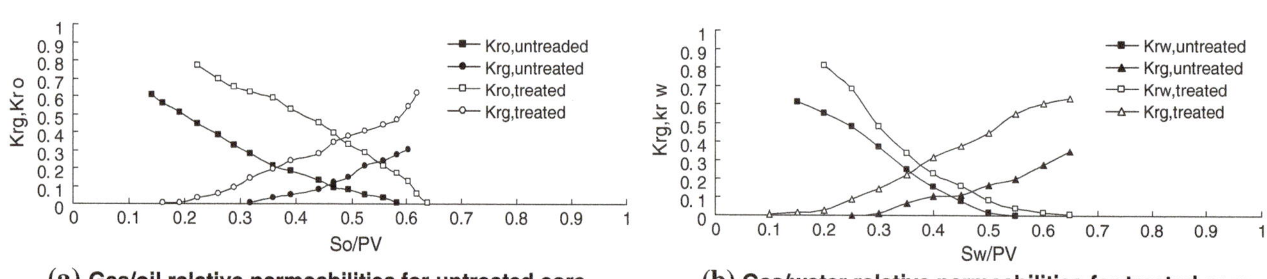

(a) Gas/oil relative permeabilities for untreated core **(b) Gas/water relative permeabilities for treated core**

Fig. 9 Effect of wettability alteration on relative permeability

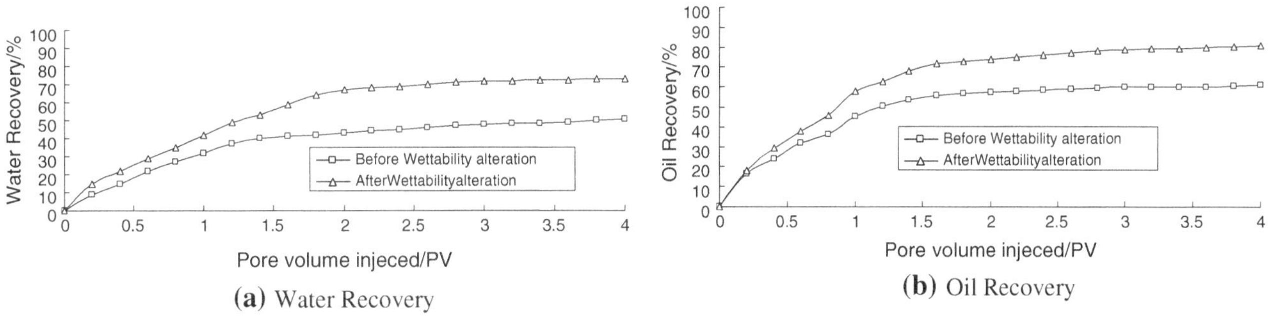

Fig. 10 Recovery measured before and after the wettability alteration to gas-wetness

condensate reservoir has a temperature of about 170 °C. To meet the requirement of high reservoir temperature, the thermal stability of the chemical Zonyl8740 was tested and verified. The experimental procedure and results are described briefly. The solution of Zonyl8740 solution (100 ml) was filled in a stainless steel cylinder and was pressurized to 5 MPa using nitrogen. Then, the cylinder with the solution was sealed and was put in an oven in which the temperature was raised gradually to 170 °C. The experimental results demonstrated that the chemical solution could still change the wettability of the low-permeability rock after 48 h high temperature aging.

In summarizing, the experimental results demonstrate that the effects of wettability alteration from liquid-wetness to gas-wetness on gas–liquid relative permeability and recovery are very significant in the rock with a low permeability of less than 1.0 md. One can expect that the wettability alteration from water-wetness to gas-wetness may enhance the gas well deliverability and the gas production notably in gas-condensate reservoirs.

Conclusions

1. The wettability of gas–liquid–rock systems can be altered from strong liquid-wetting to intermediate gas-wetting by Zonyl8740. The capillary rise test ensured the effectiveness of wettability alteration by Zonyl8740, and the result was in good agreement with contact angle measurement. It was evidenced by the spontaneous imbibition tests that Zonyl8740 could effectively suppress the imbibition of water and oil into rock.
2. Experimental results demonstrated that the relative permeabilities of both the gas and the liquid phases were increased significantly after the wettability alteration to preferential gas-wetness. The residual liquid saturation was decreased, and the gas production was enhanced greatly due to the wettability alteration. These results imply that gas well deliverability may increase substantially when wettability is altered to intermediate gas-wetting.
3. Note that the wettability alteration to gas-wetness may also be applied in drilling and completion to reduce the water loss to tight gas formation, in water shutoff in tight gas reservoirs with bottom water, and acid-fracturing to contain secondary formation damage in low-permeability gas reservoirs.

Acknowledgments This work is supported by the national science foundation of innovative research group project (Project NO. 51221003), the National Science and Technology Major Project(Project NO. 2011ZX05050)and the National Science Fund (Project NO.51074173).

References

Allen FH, Roe RP (1950) Performance characteristics of a volumetric condensate reservoir. Trans AIME 189:83

Barnum RS (1995) Gas-condensate reservoir behavior: productivity and recovery reduction due to condensation, paper SPE 30767 presented at the 1995 Annual Technical Conference and Exhibition, Dallas, 22–25 October

Boom W, Wit K, Zeelenberg JPW, Weeda HC, Maas JG (1996) On the use of model experiments for assessing improved gas-condensate mobility under near-wellbore flow conditions, paper SPE 36714 presented at the 1996 SPE annual technical conference and exhibition, Denver, Colorado, Oct. 6–9

Fahes M, Firoozabadi A (2005) Wettability alteration to intermediate gas wetting in gas condensate reservoirs at high temperatures, paper SPE 96184 presented at the SPE annual technical conference and exhibition held in Dallas, Texas, CA, 9–12 October 2005

Holditch SA (1979) Factors affecting water blocking and gas flow from hydraulically fractured gas wells, paper SPE 7561, SPE AIME, December 1979

Huang PY, Chao YC, Liao YT (2004) Preparation of fluoroacrylate nanocopolymer by miniemulsion polymerization used in textile finishing. J Appl Polym Sci 94:1466–1472

Johnson EF, Bossler DP, Naumann VO (1959) Calculation of relative permeability from displacement experiments. Trans AIME 216:370–372

Li K, Firoozabadi A (2000) Phenomenological modeling of critical-condensate saturation and relative permeabilities in gas-condensate systems, paper SPE 56014 available from SPE, Richardson, Texas

Mahadevan J, Sharma MM (2003) Clean-up of water blocks in low permeability formations, paper SPE 84216 presented at the SPE annual technical conference and exhibition held in Denver, Colorado, USA, 5–8 October 2003

Tang G, Firoozabadi A (2003) Wettability alteration to intermediate gas wetting in porous media at elevated temperatures. Transp Porous Media 52:185–211

Tannich JD (1975) Liquid removal from hydraulically fractured wells. J Pet Tech 1309–1317

Wagner OR, Leach RO (1958) Improving oil displacement efficiency by wettability adjustment, paper SPE 1101-G presented at the 1958 Annual Meeting, Houston, 5–8 October

Walker JG, Pope GA, Sharma MM, Wang P (2005) Use of solvents to improve the productivity of gas condensate wells, paper SPE 62935 presented at the SPE Annual Technical Conference and Exhibition held in Dallas, Texas, CA, 1–4 October 2005

Zhang CC, Chen YJ (2005) Investigation of fluorinated polyacrylate latex with coreshell structure. Polym Int 54:1027–1033

Zuluaga E, Monsalve JC (2003) Water vaporization in gas reservoirs, paper SPE 84829 presented at the SPE Eastern Regional/AAPG Eastern Section Joint Meeting held in Pittsburgh, Pennsylvania, USA, 6–10 September 2003

18

Particle agglomeration in sheared fluids

David Dogon · Michael Golombok

Abstract Orthokinetic agglomeration is a candidate mechanism for selectively increasing proppant size in large fractures during subsurface flow. The rate of agglomeration shows a maximum at shear rates of 275 s^{-1} suggesting that the ability to selectively grow particles depends on the hydrodynamics. Growth rates are shown to increase with increasing ion availability per unit surface area. The availability of solution ions and the average shear rates are the key parameters determining agglomeration. Precipitated species act as bridging "glue" between particles. Reduced growth rates are found for higher particle concentrations and also for larger particle size. This result has possible applications for shear-selective mechanical blocking which can be applied to solve the problem of high conductivity paths in fractured reservoirs for both traditional oil recovery as well as geothermal heat mining from fractured basement.

Keywords Fracture flow · Proppants · Blocking · Agglomeration · Shear

List of symbols

d	Particle diameter
D	Impeller diameter
h	Fracture aperture
k	Agglomeration rate constant
n	Number density in size class
N	Total number density
N_P	Power number
P	Pressure
r	Rate
w	Wall
x	Fracture axis

Greek

α	Supersaturation rate
β	Collision rate constant
γ	Shear rate
μ	Viscosity
ϕ	Sticking frequency
Ω	Stirrer speed

Subscripts

agg	Aggregation
col	Collision
N	Average overall particle sizes

D. Dogon · M. Golombok
Faculty of Mechanical Engineering, Technische Universiteit Eindhoven, Den Dolech 2, 5600 MB Eindhoven, The Netherlands

D. Dogon · M. Golombok
Shell Global Solutions International B.V., Kessler Park 1, 2288 GS Rijswijk, The Netherlands

Introduction

Particle suspensions in fluids are used in various aspects of petroleum production. These include drilling muds in wells and proppants in fractures. The latter category includes a number of options for various stages of the production process. Currently the best known application is in unconventional gas where the role of particles is to keep fractures open following high-pressure hydraulic fracturing. In oil recovery the problem is pre-existing rather than stimulated fractures, which provide a high conductivity path for secondary production injected water to break through to the producer well. This limits the amount of oil that can be swept in the permeable matrix. In this case proppants are used to reduce leak-off via fractures. The same effects are also observed in enhanced geothermal systems where following stimulation it is desirable to block high conductivity paths (i.e. large fractures) so that the heat is more efficiently extracted (Brown et al. 1999).

The common element in all these applications is the introduction of small particles suspended in the injection water. Proppant materials are pumped in suspensions into fractured unconventional shale gas reservoirs to prop open fractures and maintain conductivity (Smith and Shlyapubersky 2000). Silica micro-particles are introduced in fractured oil reservoirs to block pores and reduce fluid leak-off. In oil reservoirs larger fractures divert a disproportionately large amount of the injected water and prevent efficient matrix penetration and permeable reservoir sweep. These fluid diversions are known as subsurface "short-circuits". They are also a feature of fractured enhanced geothermal systems. Most geothermal reservoirs occur in fractured basement such as granite where permeable flow is impossible. Short circuits mean that heat is not efficiently recovered, and high flows of unheated water are obtained rather than a smaller flow with higher quality heat.

In both oil and geothermal cases, fluid diversion thus results in non-optimal use of the energy resource stored in the reservoir (Chilingarian et al. 1989). Managing such short-circuits becomes a significant challenge for reservoir engineers and operators. The classic reservoir techniques for fracture shut-off applied to oil/gas reservoirs, such as cementing or polymer injection, have the problem that they are not fracture-size selective (Schechter 1992).

In this paper, we explore the application of orthokinetic agglomeration of particles which could be used to selectively shut off larger fractures. This lowers fluid losses by fracture leak-off and could also be applied to selectively increase proppant size in large fractures. We couple a flow characteristic of large fractures to cause selective material blocking in them and not in small fractures. The questions are: (1) are the shear rates at which this agglomeration occurs relevant for subsurface flow of injected fluids and (2) what are the rates and are they fast enough? "Background" outlines the physical processes we seek to exploit. An experimental description follows in "Experiment" and the results are discussed and analysed in "Results and discussion".

Background

In a reservoir with fractured connectivity there are a multitude of fractures and none of them will be simple single-dimensioned conductive routes. Nevertheless, the size selective shut-off problem can be reduced conceptually to a comparison of a small and large fracture pair extending

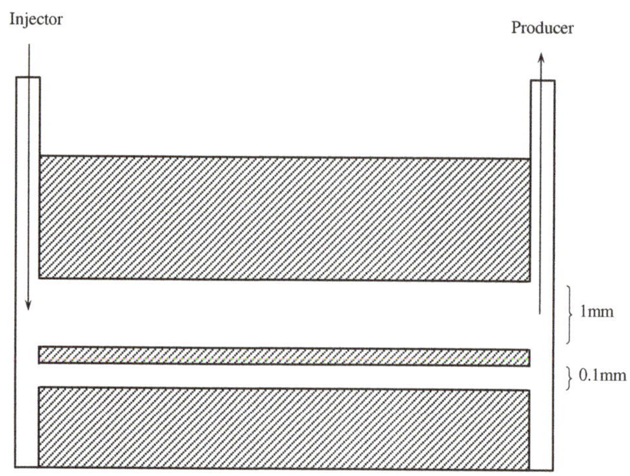

Fig. 1 Schematic of dual fracture system for competitive flow in fractured connectivity systems

between injector and producer. The basic principle of a low and high resistance path can thus be represented as in Fig. 1. There is a fixed pressure drop between the two ends of the fracture network—the injection well at one end and the producer at the other. If the large fracture aperture is 10 times that of the small fracture, then 1,000 times as much water passes through the large fracture compared to the small one due to the cubic dependence of flow rate on fracture height given by the Hagen–Poiseuille relationship. In a hydrocarbon reservoir, injected water is diverted away from optimally sweeping the reservoir permeable zones and driving oil through the porous matrix to the production well (Zimmerman and Bodvarsson 1996).

In the oil reservoir we need a fluid property related to the fracture size which can be coupled to particle growth. Such a property is the fluid shear to which two properties can be coupled:

- The shear is proportional to fracture width and therefore is larger in large fractures than in smaller ones.
- Suspensions of small particles agglomerate faster under high shear conditions in the process of "orthokinetic agglomeration".

The first observation follows from the Hagen–Poiseuille equation. For flow within fractures the shear rate at the wall γ_w is determined by the fracture aperture height h and the pressure drop dP between the injection and producer wells separated by a distance dx,

$$\gamma_W = \frac{h}{2\mu}\frac{dP}{dx} \qquad (1)$$

The second observation relates to shear-induced agglomeration processes. In general, particle agglomeration can be attributed to two physical causes. Brownian motion results in isokinetic agglomeration which is shear independent. In orthokinetic agglomeration, particle collision rates increase with shear: at higher shear rates, layers of fluid (with suspended particles), move faster past one another (Elimelech et al. 1995). Equation 1 shows that the shear is larger in larger fractures. Orthokinetic processes thus have potential to mechanically block or narrow larger fractures selectively because of the faster agglomeration processes in them due to larger shear. This applies to both particles agglomerating with each other or agglomerating with the fracture wall.

Smoluchowski (1917) first described the rate of collision as dependent on the shear rate γ, particle number concentration n for particle i and j, respectively, and particle size d:

$$r_{\mathrm{col}} = \beta_{ij} n_i n_j, \tag{2}$$

where the collision rate constant β_{ij}, between particle i and j, is given by:

$$\beta_{ij} = \frac{1}{6}(d_i + d_j)^3 \gamma \tag{3}$$

Not every collision leads to an agglomeration event, the rate at which two colliding particles (i and j) agglomerate is therefore different from the collision rate:

$$r_{\mathrm{agg}} = k_{ij} n_i n_j \tag{4}$$

The fraction of collisions which lead to agglomeration (either with another particle or with a wall) is defined by the sticking frequency, ϕ,

$$\phi = \frac{r_{\mathrm{agg}}}{r_{\mathrm{col}}} \tag{5}$$

The validity of this theory has been investigated both numerically and experimentally (Balakin et al. 2012; Yates et al. 2008) and has been shown to describe the relationship between particle growth and collision rates. At moderate shear rates this dependence is linear (Eq. 3), however, at high shear rates where tensile forces are large, particle disruption occurs resulting in breakup and reduced agglomeration rates, which ultimately fall to zero (Hollander et al. 2003; Colomer et al. 2005). Mumtaz and Hounslow (2000) showed that for batch and continuous experiments of the crystallization and growth of calcium oxalate, the agglomeration rate constant which best fits the data is particle size-independent. This convenient finding allows the agglomeration rates described in Eq. 4 to be simplified in terms of N, the total particle concentration, as follows:

$$r_{\mathrm{agg}} = \frac{1}{2} k_N N^2 \tag{6}$$

Furthermore, Mumtaz et al. (1997) employed this simplification to develop a model to quantify agglomeration rate constants. Using the same assumptions, the collision rate constant β can be estimated using the average particle diameter for the size-independent system and is given by a simplification of Eq. 3 for the monodisperse case:

$$\beta_N = \frac{4}{3} \bar{d}^3 \bar{\gamma} \tag{7}$$

The size-independent agglomeration rate constant k_N and collision rate constant β_N can therefore be related through total number concentration and mean particle diameter—which can be determined experimentally.

These are all physical effects. However, the final result is dependent on the sticking efficiency and this is a largely chemical effect. It is determined by solution ionic strength and counterion supersaturation (Mumtaz and Hounslow 2000; Ilievski and Livk 2006; Livk and Ilievski 2007). Whereas shear forces promote collisions, it is the interaction energy that determines sticking between particles. The ionic nature of the solution influences these forces: ions determine the surface charge and the ability of free ions to shield surfaces charges. Mumtaz and Hounslow (2000) describe supersaturation as providing a "bridging cement" between particles upon collision due to the precipitation of mineral salts on the surface. Since the degree of supersaturation directly influences the rate of precipitation, the sticking efficiency is expected to be positively correlated with it. Huang et al. (1991) showed that in $CaCO_3$ suspensions Ca^{2+} ions were the zeta potential determining ion—ions that are free to move from the surface to the bulk solution—and that surface charge increased with increased Ca^{2+} concentration. They also showed that increases in surface charge resulted in increased particle–particle repulsion and reduced agglomeration. Rashchi et al. (1998) showed that Ca^{2+} adsorption on silica was again the zeta potential determining ion. However, in the presence of carbonate ions, $CaCO_3$ is precipitated, suggesting the importance of supersaturation and precipitation as a way to overcome the effects of high Ca^{2+} adsorption and high surface charges. Our experiments examine a combination of the features above, in order to assess applicability to the subsurface fracture environment.

Experiment

Stirred cell

Inside a fracture, there is a distribution of shear rates present. Shear is actually a tensor rather than vector quantity and Eqs. 3 and 7 refer to the absolute magnitude of the shear. The smallest value is $\gamma = 0$ in the middle of the fracture. Referring to the fracture pair Fig. 1, then for a pressure gradient of 10 mbar/m in a 1 mm fracture, the wall shear is 500 s^{-1} with an average of 250 s^{-1}.

Similarly there is a distribution of shear rates present in stirred cells which are used in this study to shear fluids containing suspended particles. In a stirred cell, the stirrer speed, Ω, is used to vary the average shear rate according to (Sanchez Perez et al. 2006):

$$\bar{\gamma} = \sqrt{\frac{4N_P \rho D^2 \Omega^3}{\pi \mu}}, \tag{8}$$

where N_P is the power number, D is the paddle diameter, Ω the stirrer speed and μ is the fluid viscosity. This expression for shear rate is derived from the total energy dissipation rate. Spatial averages will bias the results, and the shear experienced by particles will differ between their positions in the cell (Hollander et al. 2003). Errors due to concentration gradients in the stirred cell can be overcome by sampling from numerous points in the batch system.

Batch tests in a stirred cell provide a controlled environment to study particle systems maintained in suspension. The tests were performed in a glass cylindrical cell of diameter 100 mm and height 140 mm with a square stirrer paddle of 85 mm × 85 mm, providing a well-defined average shear rate. Particles were first sheared in deionised water at a rate of 1,800 s^{-1} for 1 h to break up existing agglomerates and were then measured for size. After this, particles were sheared for 8 h at 100, 150, 200, 275 and 500 s^{-1} with an IKA-WERKE Eurostar digital electric stirrer. Experiments were carried out under laminar conditions as determined by Reynolds numbers for stirred tanks.

Materials

Polydisperse industrial grade $CaCO_3$ particles ("Baracarb") were obtained from Halliburton Europe in size classes with volume median diameters of 2, 5 and 25 μm. (The number given after the name "Baracarb" designates the size class based on this median diameter.) The particles are produced from ground marble, resist size reduction beyond their specified range and have a specific gravity of 2.7 g/cm^3. They were chosen because of existing application in the oil and gas industry.

Reagent grade $CaCl_2$ and Na_2CO_3 provided by Sigma Aldrich were used to make up two 4.2 mM solutions which were then added to the particle suspension to obtain the required degree of Ca^{2+} and $CO_3{}^{2-}$ supersaturation α defined by.

$$\alpha = \frac{[Ca^{2+}]_{SS} \times [CO_2^{2-}]_{SS}}{[Ca^{2+}]_{Eq} \times [CO_2^{2-}]_{Eq}} \tag{9}$$

The solubility product in water at equilibrium is $K_{sp} = 5 \times 10^{-9}$ giving a nominal solubility of 7×10^{-5} mol/l. Final solutions were therefore made up to $\alpha = 5$, 10 and 20 times this concentration and preliminary tests showed that no spontaneous nucleation and precipitation occurred. These solutions were then sheared for 8 h. In this way, the effect of low and high shear rates on agglomeration under different conditions of ionic concentration, particle concentration and particle size was investigated.

Particle size analysis

Particle size was measured using a Galai CIS-100 Particle sizer. A rotating prism scans a laser beam across the particle suspension. As the laser beam encounters a particle, the beam is obscured for a time and measured by a detector. The time of obscuration is measured and since the rotary speed of the laser is known, the particle diameter can be determined using the Laser Obscuration Time (LOT) principle (Moes and Flash 2013). Due to direct particle size measurement, in comparison to other optical techniques, LOT is independent of material optical properties and solution chemistry. The technique also does not assume spherical particles. Particle sizes can be measured in the range 0.1–300 μm and data are analysed for particle size distribution, number (NMD) and volume (VMD) mean diameters and particle concentration. Samples were taken every 2 h and analysed in duplicate for size distribution.

Our analysis is based on Eq. 6. Integrating over the time of shearing we obtain:

$$\frac{1}{N_t} - \frac{1}{N_{\text{initial}}} = \frac{1}{2} k_N t \tag{10}$$

Since the initial and final particle concentrations as well as the average particle size are measured and the time known, we can calculate the average agglomeration rate, k_N from Eq. 10 and β_N from Eq. 7. Using the empirical values for k_N and β_N the sticking efficiency can be obtained from Eq. 4.

Results and discussion

Growth and distribution

Since the aim of this study was to apply particle agglomeration in water or brines of similar carrying power, it was important to first test the fluid's suspension ability at the required shear rates. Initial visual tests showed Baracarb-2 and -5 to be maintained in suspension even at the lowest shear rates. Baracarb-25 settled out at shear rates below 275 s^{-1}—a critical value for characterising behaviour—see below. As a result these large particles were excluded from further investigation in this study as we only deal here with aqueous carrying power as outlined above. The

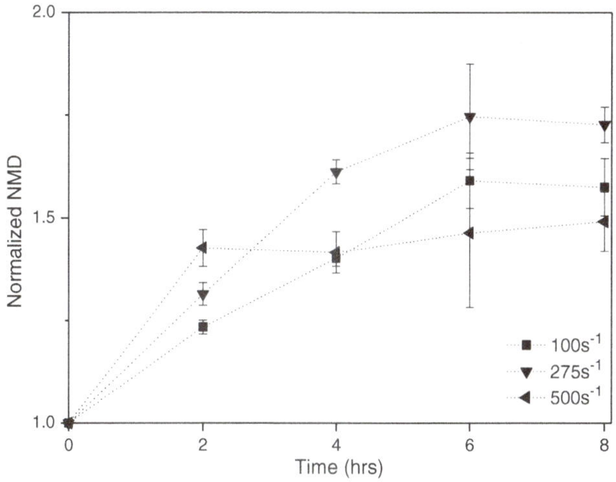

Fig. 2 Particle agglomeration for Baracarb-5 at initial concentration of 0.25 g/l and solution supersaturation $\alpha = 10$

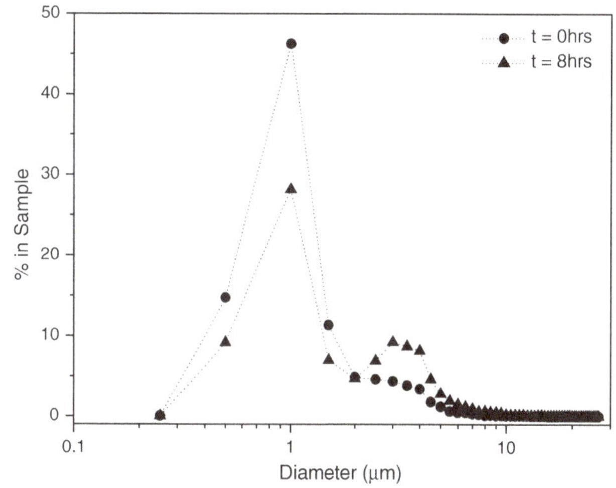

Fig. 3 Particle size distribution for 0.25 g/l of Baracarb-5 before and after shearing for 8 h at $\alpha = 10$ and $\gamma = 275\ \mathrm{s}^{-1}$

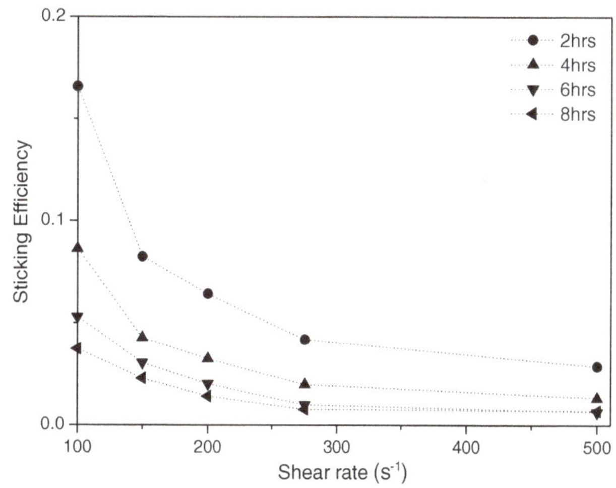

Fig. 4 Sticking efficiency as a function of shear rate γ for different time durations of test

suspension ability of the fluid is governed by factors that include viscosity, particle size, and fluid velocity. There is therefore potential for applications of larger particles at higher shear rates or viscosified aqueous media. This is the focus of future research and is not germane to the study described here which focuses on small particle shear-induced agglomeration potential in water-like media, i.e. aqueous viscosities.

Figure 2 shows particle size evolution for Baracarb-5 over 8 h for three shear rates tested in the stirred cell. Results are reported as a normalized number mean diameter (NMD), $\bar{d}(t)/\bar{d}(0)$, since it is the number of particles on which the agglomeration equation is based rather than for example other standards such as volume or Sauter mean diameters. The initial rates over the first 0–2 h support Smoluchowski's theory where the 500 s^{-1} shear case has the highest initial growth and 100 s^{-1} the lowest. Increase in size is also a linear function of shear rate at this stage. These rates decrease as the solution becomes less saturated and other disruptive mechanisms develop. This growth rate decrease is seen most notably for the highest shear case, 500 s^{-1}, where rapid growth over the first 2 h falls to nearly zero equilibrating at 1.5 times initial size—the lowest value for all cases tested. This is because larger agglomerates are unstable as high shearing forces break them apart.

For all cases, final particle size did not exceed 2 times the initial number mean diameter. This indicates that the smaller rather than larger particles in the size distribution agglomerate. This is supported by Fig. 3 which shows a particle size distribution for Baracarb-5 before and after shearing at 275 s^{-1} for 8 h. The initial size distribution is unimodal with a second peak developed after shearing. The initial peak, centred on 1 μm, is reduced, indicating a significant reduction in the number of particles in the smaller size classes. The percentage of particles above 5 μm remains relatively unchanged suggesting that the second peak is composed of agglomerates of the smaller particles.

Sticking probability

The sticking efficiency ϕ (Eq. 5) can be calculated by comparing the measured agglomeration rate (Eq. 6) and predicted shear-induced collision rate (Eq. 7). Figure 4 shows that the sticking efficiency ϕ, i.e. the ratio of observed mean agglomeration to orthokinetic collision rate derived from Eq. 4, is negatively affected by fluid shear and the duration thereof. These results agree with Mumtaz and Hounslow (2000) in so far that the number of collisions resulting in sticking is much lower than predicted and

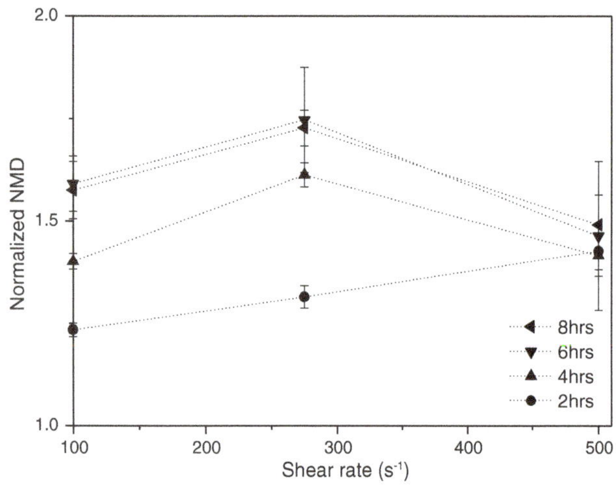

Fig. 5 Normalized number mean diameter at different shear rates and experimental time for 0.25 g/l of Baracarb-5 at $\alpha = 10$

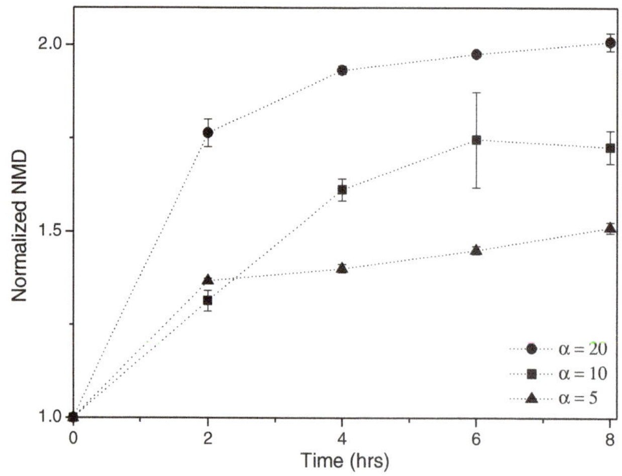

Fig. 6 Normalized number mean diameter for 0.25 g/l of Baracarb-5 at different solution supersaturations after shearing for 8 h at $\gamma = 275\ s^{-1}$

decreases for increasing shear. This can be explained by hydrodynamic forces and collision contact time having a significant effect in maintaining a newly formed "doublet" of two sticking particles. For higher shear, contact time is reduced and sufficient precipitation cannot occur. In addition, the strength of the bond depends on bridging chemistry and electrostatic forces (Hounslow and Andreassen 2004) and, as time elapses, bridging ions are depleted. The sticking efficiency decreases with shear rate (Fig. 4) but also decreases with the time duration of shearing.

Even though sticking efficiency decreases with increasing stirrer speed, we still observe increasing growth as predicted by Smoluchowski. Figure 5 shows particle NMD versus shear rate at different experimental times. Over the first 2 h the agglomeration rate increases linearly with shear rate. This means that the increase in collision rates as a result of higher shear rates is enough to overcome the lower sticking efficiency. As time progresses a peak develops for the intermediate shear rate, 275 s^{-1}. This is due to lower collision rates in the low shear regime and breakage in the high shear regime as discussed previously. There is also a relatively larger drop in sticking efficiency with time for the lower shear rates compared to higher ones (see Fig. 4). This indicates that growth will slow even more rapidly for low shear rates as time progresses. These findings have positive implications for application in size selective fracture blocking: lower wall shear rates (as found in smaller fractures) will result in both lower collision rates (Eq. 7) and a more significant drop in sticking efficiency as shown in Fig. 4.

DLVO theory describes electrochemical sticking by adding the repulsive forces, arising from a common particle surface charge called the zeta potential, and the attractive forces, arising from Van der Waals interactions between particles of the same type (Elimelech et al. 1995). Both the sign and magnitude of the zeta potential is a function of the solid material and solution pH. Yates et al. (2008) found that hetero-agglomeration, between two different mineral types, resulted in larger particles than homo-agglomeration studied here. This is due to the differences in magnitude and sign of the zeta potential between two different mineral types which lead to improved sticking efficiency. A similar phenomenon can be seen in dishwashers where lime-scale deposits on glassware. Due to the opposite zeta potential surface charges of $CaCO_3$ and SiO_2 in the dishwasher environment (Fuerstenau 1976), electrochemical attractions promote sticking of precipitated $CaCO_3$ on glass resulting in the white coating often seen. The manipulation of these charges in subsurface environments through changes in the pH and ionic concentrations can be used to force particle–wall sticking. This is applicable to fractured granite reservoirs, composed of mostly silicates, where a zeta potential difference would exist between the fracture surface and a $CaCO_3$ particle.

Supersaturation/precipitation

Figure 6 shows the effect of varying the solution supersaturation. At higher supersaturations the particle growth is faster and reaches larger final particle sizes. This is due to the availability of the solvated Ca^{2+}/CO_3^{2-} in the supersaturated solution which forms the "glue" for enabling colliding particles to stick. At a higher supersaturation there are more ions in solution. However, the particle mass concentration remains the same. There is therefore more "glue" per unit surface area in the form of ions available to precipitate out and form interparticle connecting bridges. For the same particle concentration, but with more ions in the solution, the mass of solute depositing from solution

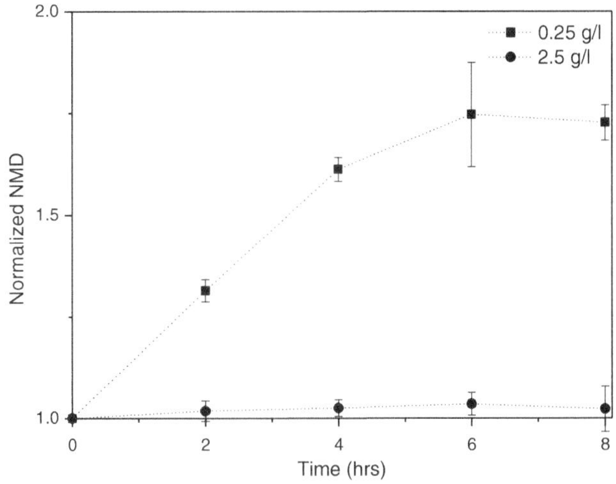

Fig. 7 Normalized number mean diameter for different concentrations of Baracarb-5 at 0.25 and 2.5 g/l and $\alpha = 10$

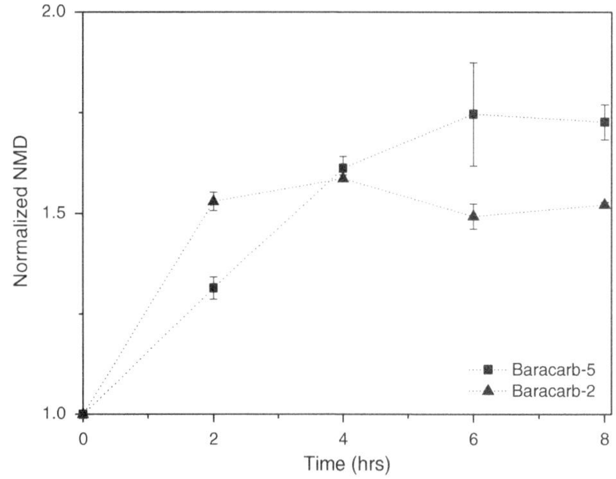

Fig. 8 Normalized number mean diameter for Baracarb-2 and Baracarb-5 showing the effect of particle size on agglomeration at 0.25 g/l and $\alpha = 10$

has a correspondingly larger effect. The same effect is found for lower particle number concentrations at constant supersaturation. Figure 7 shows Baracarb-5 particle growth for two different mass concentrations, 0.25 and 2.5 g/l. In this case the number of particles in the 2.5 g/l system is 10 times higher than that in the 0.25 g/l. The growth for lower concentration is faster and reaches a larger final size due to the higher ion to surface area ratio.

If we consider the effect of size on particle agglomeration using Eqs. 2 and 3, for the same mass in solution, the rate of collision of particle size d_i differs by a factor of $\left(\frac{d_j}{d_i}\right)^3$ compared to that for particle size d_j. This is because of the higher number concentration of smaller particles. To test this 0.25 g/l of Baracarb-2 and -5 were sheared separately at the same supersaturation condition, $\alpha = 10$. These results are shown in Fig. 8 where initially the Baracarb-2 agglomerates at a faster rate due to the higher collision rate. As time progresses however Baracarb-5 results in greater agglomeration compared to Baracarb-2. This is due to the reduced surface area-to-mass ratio for larger particles compared to smaller ones which leads to a higher ion to surface area ratio.

When applied to reservoir applications, these findings indicate that manipulation of the chemical environment can be a means to enhance particle–particle and particle–wall sticking. Future investigation into solid surface charges through ion concentration coupled with pH manipulation is expected to further improve sticking.

Conclusion

1. Smaller size distributions agglomerate faster than larger ones.
2. The hydrodynamic effects formalized by Smoluchowski are supported below a critical shear rate of 275 s^{-1} after which breakage dominates. (The shear process arising from stirring increases the collision rate.)
3. Low particle number-concentration suspensions agglomerate faster than higher ones when considering the critical role of ion precipitation. This contradicts Smoluchowski's theory which does not take agglomeration chemistry into consideration.
4. The sticking efficiency is a function of the particle/fluid chemistry. Factors such as surface charge and supersaturation govern these effects.
5. Selective agglomeration controlled by shear rate and chemistry manipulations can thus be induced in particles flowing in the subsurface. The rates we found were too low and final particle size too small for immediate practical use.
6. Hetero-agglomeration (i.e. between two different mineral types) would result in larger particles than the homo-agglomeration studied here. This is the subject of current investigations as exemplified by application of dishwasher phenomena discussed above.

References

Balakin B, Hoffman A, Kosinski P (2012) The collision efficiency in a shear flow. Chem Eng Sci 68:305–312

Brown D, DuTeaux R, Kruger P, Swenson D, Tsutomu Y (1999) Fluid circulation and heat extraction from engineered geothermal reservoirs. Geothermics 28:553–572

Chilingarian G, Robertson J, Kumar S (1989) Surface operations in petroleum production. In: Chilingarian G (ed) Developments in petroleum science. Elsevier, Amsterdam

Colomer J, Paters F, Marrase C (2005) Experimental analysis of coagulation of particles under low-shear flow. Water Res 39:2994–3000

Elimelech M, Gregory J, Jia X, Williams RA (1995) Particle deposition and aggregation. Butterworth-Heinemann, Oxford

Fuerstenau MC (1976) Flotation, vol 1. American Institute of Mining, Metallurgical and Petroleum Engineers Inc, New York

Hollander E, Derksen JJ, Bruinsma OSL, van den Akker HEA, van Rosmalen GM (2003) A numerical study on orthokinetic agglomeration in stirred tanks. Powder Tech 130:169–173

Hounslow M, Andreassen JP (2004) Growth and aggregation of vaterite in seeded-batch experiments. AIChE J 50(11):2772–2782

Huang Y, Fowkes F, Lloyd T, Sanders N (1991) Adsorption of calcium ions from calcium chloride solutions onto calcium carbonate particles. Langmuir 7:1742–1748

Ilievski D, Livk I (2006) An agglomeration efficiency model for gibbsite precipitation in a turbulently stirred vessel. Chem Eng Sci 61:2010–2022

Livk I, Ilievski D (2007) A macroscopic agglomeration kernel model for gibbsite precipitation in turbulent and laminar flows. Chem Eng Sci 62:3787–3797

Moes H, Flash R (2013) Advances in particle size analysis by laser obscuration time. http://www.rktech.hu/dokumentaciok/AmbiValue/Poster/EyeTech_Poster.pdf

Mumtaz H, Hounslow M (2000) Aggregation during precipitation from solution: an experimental investigation using Poiseuille flow. Chem Eng Sci 55:5671–5681

Mumtaz H, Hounslow M, Seaton N, Paterson W (1997) Orthokinetic aggregation during precipitation: a computational model for calcium oxalate monohydrate. Trans IChemE 75(A):152–159

Sanchez Perez J, Rodriguez Porcel E, Casas Lopez J, Fernandez Seville JCY (2006) Shear rate in stirred tank and bubble column bioreactors. Chem Eng J 124:1–5

Rashchi F, Xu ZFJ, Finch JA (1998) Adsorption on silica in Pb- and Ca–SO_4–CO_3 systems. Colloids Surf A: Physiochem Eng Aspects 132:159–171

Schechter R (1992) Oil well stimulation. Prentice Hall, Eaglewood Cliffs

Smith W, Shlyapubersky J (2000) Basics of hydraulic fracturing. In: Economides M, Nolte K (eds) Reservoir stimulation. Wiley, Chichester

Smoluchowski M (1917) Mathematical theory of the coagulation of colloidal solutions. Z Phys Chem 92:129–168

Yates P, Franks G, Jameson G (2008) Orthokinetic heteroaggregation with nanoparticles: effect of particle size ratio on aggregate properties. Colloids Surf A: Physiochem Eng Aspects 326:83–91

Zimmerman R, Bodvarsson G (1996) Hydraulic conductivity of rock fractures. Transp Porous Med 23:1–30

19

An improved method to study CO_2–oil relative permeability under miscible conditions

F. F. Li · S. L. Yang · H. Chen · X. Zhang · D. D. Yin · L. P. He · Z. Wang

Abstract Relative permeability curves are crucial parameters for reservoir engineer and reservoir commercial simulator to predict reservoir performance throughout the life of a reservoir, but meet difficulties in laboratory to obtain reliable data under miscible conditions due to the lack of proper testing and formulation methods. Up to now, most relative permeability curves are measured in short core segments by core flooding, which can hardly display miscible flooding features for early gas breakthrough and insufficient contacting time between CO_2 and oil. In addition, the commonly used analytical and semi-analytical data processing methods are not suitable for miscible flooding for ignoring the mechanism of vaporizing and dissolving mechanism. In this study, slim tubes (101 and 1,528 cm in length) and long composite cores (74.46 cm in length) instead of conventional core segments were used to acquire reliable experimental data of CO_2 flooding under miscible or near miscible condition. Then, using improved empirical Corey model which assumes shape defining factor b_{og} is a function of displacement pressure P combined with history-matching method to calculate relative permeability curves under near miscible and miscible conditions. Results indicated core length is another important parameter to simulate miscible flooding other than pressure, temperature and oil composition, and using long composite cores and improved data processing method more reliable data can be obtained compared with conventional measured method. It is found residual oil saturation in short slim tube is 16.25 % higher than that of long slim tube and CO_2 relative permeability is lower in short slim tube/core segment than in long slime tube/long composite cores.

Keywords Long core flow tests · Miscible conditions · Relative permeability · Improved method

F. F. Li · S. L. Yang · H. Chen · X. Zhang · D. D. Yin · L. P. He
Key Lab of Petroleum Engineering of MOE, China University of Petroleum, Beijing 102249, People's Republic of China

Z. Wang
Technology Department, Petro-China International Iraq, Beijing 100724, People's Republic of China

Introduction

CO_2 EOR is the first option to implement CO_2 geological sequestration because of the additional economic benefit, existing infrastructure and knowledge of petroleum engineering and operations (Holt et al. 2000). Experimental studies showed that oil recovery factor after CO_2 injection improved up to 60 % of original oil in place while this value was only 44 % for water flooding (Hadlow 1992; Taber et al. 1997). The increase in oil recovery is mainly due to swelling of crude oil by CO_2, which help to reduce interfacial tension and crude oil's viscosity. By approaching the miscible conditions, these changes can be introduced in relative permeability curves in multi-contact miscible displacement (Al-Wahaibi et al. 2005, 2006). Unfortunately, most of the recent relative permeability data under miscible or near miscible conditions can not be directly used in commercially available compositional simulators due to the uncertainty of measurement and lack of proper formulation (Prieditis and Brugman 1993; Dria et al. 1993).

Firstly, there is no proper way of measuring CO_2–oil relative permeability curves under miscible condition due to the slow development of dynamic miscibility. In CO_2

miscible flooding process, miscible displacement between oil and CO_2 is caused by the extraction of light and intermedium hydrocarbons into CO_2 and dissolution of CO_2 into oil (Okuno and Xu 2013; Wang and Pope 2001). CO_2 and oil are not miscible with each other during their first contact but will become miscible at CO_2 front gradually after multiple contact process. Component mass transfer between the CO_2 phase and the oil phase will generate a miscible zone, and distances are required for the development of dynamic miscibility (Shtepani 2007). It is concluded by some researchers (Omole and Osoba 1989; Ekundayo and Ghedan 2013) that core lengths in range of 0.5 to 6 m might well be below the minimum length required for the development of dynamic miscibility. Therefore, for better performance and accuracy of recovery process under miscible flooding, it is recommended to measure relative permeability curves with long composite cores. Unfortunately, most current relative permeability curves were obtained in conventional core segments which can hardly simulate real miscible flooding process (Shtepani 2007; Prieditis and Brugman 1993; Ren et al. 2011). In general, conventional core segments are 3 to 7 cm in length, and short core segment frequently led to early CO_2 breakthrough and inadequate time for vaporizing before it enriched enough to get miscible with reservoir oil (Ekundayo and Ghedan 2013). Therefore, direct application of relative permeability curves generated from conventional core segments may induce significant errors in predicting reservoir performance in miscible condition.

Furthermore, improper interpretation of laboratory experimental data may also add further errors to reservoir simulation. In principle, three major approaches—analytical methods, semi-analytical methods and history-matching methods can be used to determine two-phase relative permeability from unsteady state displacement experiments (Hussain and Cinar 2010). However, analytical methods and semi-analytical methods failed to represent miscible flooding relative permeability for their ignorance of capillary pressure effects and vaporizing mechanism (Gu and Oliver 2005). History-matching methods iterate the laboratory core flooding pressure and production data by changing the relative permeability values. Unlike analytical and semi-analytical methods pointed defects can be removed easily. This possibility encourages us to use this inverse modeling technique to calculate CO_2 flooding relative permeability curves under near miscible and miscible conditions (Watson et al. 1984; Reynolds et al. 2004; Eydinov et al. 2007). However, most current relative permeability representation models have nothing to do with injection pressure or interfacial tension (IFT) which is crucial scaling factor for miscible displacements.

To provide representative laboratory measured relative permeability data for reservoir simulators, this paper discusses how to obtain more reliable data of CO_2 miscible flooding in laboratory and how these data could be used in an improved empirical Corey model which incorporates injection pressure as a scaling factor.

Experimental studies

Four sets of relative permeability experiments were conducted to obtain reliable data for CO_2 flooding. Since distances are required for the development of dynamic miscibility and existence difficulties to conduct a long real core flooding due to low permeability, long experiment time and severe sleeve corrosion, two experiments were conducted in slim tubes to study the effect of core length on miscible process and displacement efficiency. Then based on natural cores and reservoir fluid of Jilin oil field, long composite core and short core segument experiments of CO_2 flooding were preformed above MMP in the presence of irreducible water saturation to study CO_2–oil relative permeability in natural cores.

Materials

The experimental design adopts unsteady state procedure under reservoir temperature and reservoir pressure in which effluent production from the core sample was recorded by liquid meter and gasometer. The injected fluid CO_2 with purity of 99.95 % and its physical properties have been calculated with PVTi simulator. The crude oil used in the experiments is recombined according to GOR of the stock-tank oil and gas from the ground separator of a well in Block H79 of Jilin oil field. The reservoir description and fluid properties are exhibited in Tables 1 and 2.

The effect of core length experiments was conducted in slim tubes of different lengths. The slim tube used in this study was made of Hastelloy C tubing. The coiled tubing was packed with 120 mesh sand and mounted horizontally inside an air bath to alleviate the influence of gravity. The physical properties of the two slim tubes are shown in Table 3.

The long real core flooding experiments were performed on the low-permeable natural core samples from Block H79 in Jilin oil field. Ten natural cores were selected and ordered according to the length and permeability based on Bradley's law. End effect was eliminated by adding filter paper between the natural cores. The total length of core group is 74.45 cm, and the average porosity and permeability are 15.28 % and 6.06×10^{-3} μm^3, respectively. The parameters for the cores utilized in the tests are shown in Table 4.

Table 1 Block Hei-79 reservoir descriptions and its fluid properties

Average reservoir burial depth (m)	2,400
Gross sand layers (m)	Range from 11.2 to 18.2 m
Average porosity (%)	12.7
Average permeability (10^{-3} μm^2)	3.5
Reservoir temperature (°C)	98.9
Original reservoir pressure (MPa)	24.15
Oil viscosity at the reservoir condition (mPa s)	Vary from 1.85 to 2.18 mPa s
Oil density at the reservoir condition (kg/m^3)	761.5
Averaged formation water salinity (mg/l)	14,607

Table 2 Live oil properties and oil compositions

Density (kg/cm^3)	Viscosity (mPa s)	Saturation pressure (MPa)	GOR (m^3/m^3)
(a) Live oil properties			
761.5	1.85	7.1	30

Components	C1	C2	C3	C4–C7	C8–C20	C20+
(b) Live oil compositions						
Mol fraction (%)	25.353	3.818	1.670	6.040	38.087	25.096

Table 3 Properties of the slim tubes

Slim tubes	Length (cm)	Diameter (mm)	Porosity (%)	Permeability (10^{-3} μm^2)
No. 1	1,528	4	31.4	5
No. 2	101	4.4	34.17	7

Table 4 Properties of the cores used in long core experiment

Order	No	Diameter (cm)	Length (cm)	Porosity (%)	Permeability (10^{-3} μm^2)
1	H-1	2.5	8.422	11.43	6.10
2	H-60-2	2.5	8.104	12.06	5.63
3	H-35	2.5	8.066	13.77	6.95
4	H-24-3	2.5	7.64	18.29	4.60
5	H-34-3	2.5	7.87	13.32	7.80
6	H-60-1	2.5	5.984	23.58	4.51
7	H-10	2.5	6.386	19.06	8.72
8	H-8	2.5	8.05	12.47	4.01
9	H-60-5	2.5	7.32	17.23	8.746
10	H-26-2	2.5	6.61	14.95	9.33
Total length	74.46	Average porosity	15.28	Average permeability	6.06

Apparatus and experimental procedure

The unsteady displacement experiments are performed in CFS-200 core flooding system, which consists of an injection system, four piston accumulators, a TEMCO BP-100 back pressure regulator, a long core holder/slim tubes, a confining pressure system, three differential pressure measurement system, an oil–gas separator, a gasometer and computer system for data acquisition and process control. Figure 1 provides a schematic illustration of the experimental apparatus used to measure relative permeability curves.

Lead is used as a sleeve material to avoid the diffusion and penetration of CO_2 through the rubber core sleeves. The ductility of lead sleeve allowed a confining external overburden pressure to be transferred to the core in a radial mode to simulate reservoir overburden pressure. The steel core holder can simulate reservoir confining pressures of up to 50 MPa. This pressure is applied by filling the annular space between the core sleeve and the core holder with non-damaging saline brine. Test fluids were stored in titanium accumulators that were connected to a variable-rate injection pump. The CO_2 cells were kept outside the oven with a temperature-controlled air bath for safety reasons. However, the injected CO_2 was initially passed through stainless steel coils to heat up to reservoir temperature before it entered the core sample. The simulated reservoir temperature was regulated by a high-quality thermocouple. Pressure readings across the core sample were instantly recorded by high-precision pressure transducers that were digitally synchronized to a laboratory PC.

The experiments were performed as follows: (1) clean the core or slim tube with solvent, dry with hot nitrogen and evacuate, and then saturate the core with the formation water. (2) Assemble the cores in long core holder and upload confining pressure and back pressure, generally confining pressure is 2–3 MPa higher than back pressure, heat the thermostatic oven to reservoir temperature at 98.9 °C and keep it constant. (3) Saturate the core or slim tube with live oil, measure the oil-phase permeability. (4) Inject CO_2 at a setting constant rate of 0.2 cc/min; record the injection volume, pressure, instantaneous oil and gas production data dynamically.

Mathematical model

End effect elimination

The construction of long cores caused strong interface between core segments when capillary contact was not maintained between contiguous segments, which led to strong effect on the saturation distributions. One of the

Fig. 1 The schematic diagram of core flood laboratory

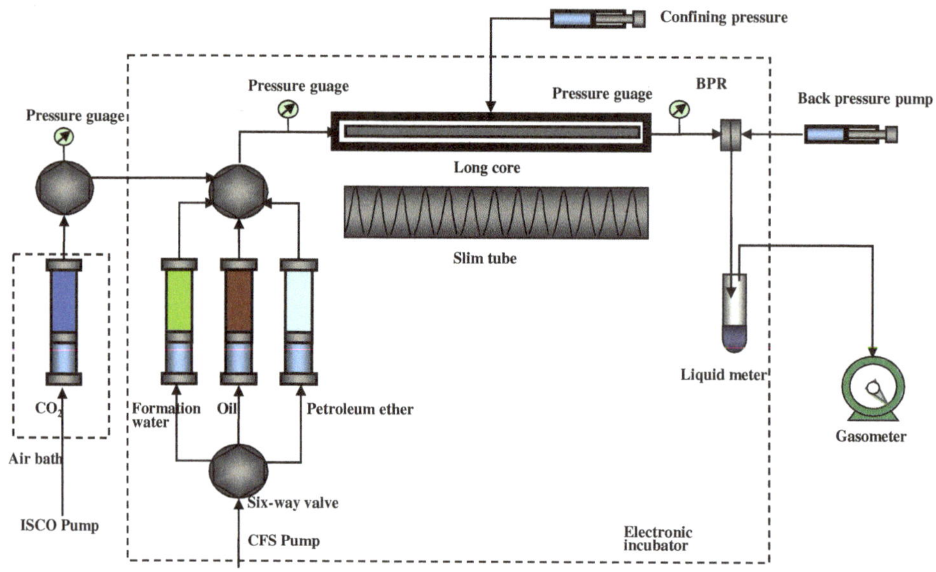

effective ways to decrease the end effect caused by capillary pressure is to put a filter paper between the core segments, the other is increase the flow rate or differential pressure. Rapoport formula shown in Eq. 1 is used to determine the displacement pressure to eliminate end effect. It can be calculated from interfacial tension σ, permeability k and porosity ϕ as shown in Eq. 1.

$$\frac{\sigma \times 10^{-3}}{\Delta p\sqrt{k/\phi}} \leq 0.6 \tag{1}$$

Fluid volume calibration

The gas production volume was measured at ambient temperature and pressure, which contains not only produced CO_2 but also light hydrocarbons escaped from oil, therefore, the instantaneous gas production needs to be calibrated to get the gas production volume under experimental pressure and temperature. According to GOR and the EOS for real gas, the CO_2 production can be calculated using Eq. 2

$$V_{gi} = V_{gi-1} + \frac{2Z_i P_a T}{(P_{i1} + P_{i2} + 2P_a) \cdot (\Delta V_{gi} - \mathrm{GOR} \cdot \Delta V_{oi}) T_a} \tag{2}$$

where V_{gi-1} and V_{gi} are accumulated gas production at time $i-1$ and i; ΔV_{oi} is accumulated oil production at time i; P_{i1} and P_{i2} are inlet and outlet pressures of the core holder; P_a and T_a are ambient pressure and temperature; T is experimental temperature, Z_i is CO_2 compressibility factor at temperature T and average pressure $\frac{P_{i1}+P_{i2}}{2}$; and GOR is gas–oil ratio.

In addition, oil production was measured at ambient temperature and pressure with no dissolved gas, which should be calibrated according to oil volume factor at testing temperature P and testing pressure T. As shown in Eq. 3.

$$V_{oi} = V_{oi-1} + B_o \cdot \Delta V_{oi} \tag{3}$$

Methods to obtain relative permeability for CO_2 miscible flooding

After more reliable experimental data have gained, we need proper interpretation method to get representative relative permeability curves. In general, there are two types of components to determine relative permeability curves: endpoints and shape factors (Eydinov et al. 2007). Endpoints could be measured directly by core flooding, while shape factors are depending on the fluid/fluid interactions and usually obtained by empiric. Up to now, the proper formulation of relative permeability curves under miscible flooding is mostly based on empirical Corey model (Lee and Seinfeld 1987; Yang and Watson 1991). Therefore, in this paper, Corey model combined with history-matching method is used in this paper to obtain CO_2–oil relative permeability curves.

For the oil/gas system:

$$k_{rg} = a_g\left(\frac{S_g - S_{gc}}{1 - S_{org} - S_{wc} - S_{gc}}\right)^{b_g} \tag{4}$$

$$k_{rog} = a_{og}\left(\frac{1 - S_{org} - S_{wc} - S_g}{1 - S_{org} - S_{wc} - S_{gc}}\right)^{b_{og}} \tag{5}$$

where k_{rg} and k_{rog} are relative permeability of gas and oil phase, respectively; a_g and a_{og} are gas relative permeability at $S_g = 1 - S_{org} - S_{wc}$ and the oil relative permeability at $S_g = S_{gc}$, respectively; S_g is gas saturation; S_{org} is residual oil saturation; b_g and b_{og} are exponential factors.

Every parameter in Corey model can be obtained directly by laboratory measurements except b_g and b_{og}, they are exponential factors defining shapes of relative permeability curves. It is confirmed that phase transition has no concern with gas relative permeability, thus b_g is a constant parameter with empirical value of 3.5. To get the value of b_{og}, interfacial tension is usually used as scaling factor to estimate the miscible flooding performance (Shtepani 2007). However, high-temperature and high-pressure interfacial tension test could be very costly, while this value calculated from PVT modeling needs high quality of laboratory PVT data and procedure (Ren et al. 2011). Since interfacial tension is a function of temperature, pressure and composition of target reservoir oil. For a specific oil reservoir, reservoir temperature and oil composition are considered to be constant, then interfacial tension between CO_2 and oil only depends on reservoir pressure. Thus, we could use pressure as a scaling factor to estimate the miscible flooding performance. Generally, b_{og} equals to 1 represents miscible flooding and 3 represents immiscible flooding, suppose the exponential factor b_{og} is between 1 and 3 (Li et al. 2012), and according to linear interpolation function, exponential factor b_{og} can be written as follows:

$$b_{og} = \begin{cases} 1 & P \geq P_{mmp} \\ \frac{2P + P_{nm} - 3P_{mmp}}{P_{nm} - P_{mmp}} & P_{nm} < P < P_{mmp} \\ 3 & P < P_{nm} \end{cases} \quad (6)$$

where P_{mmp} is minimum miscibility pressure, P_{nm} is critical immiscible pressure. Critical immiscible pressure means the maximum pressure for an immiscible flooding and can be determined by IFT. It is proved when injected gas and oil interfacial tension is higher than 7 dyne/cm, gas injection can be considered to immiscible flooding (Guo et al. 2000; Ren et al. 2011).

Results and discussion

Determination of MMP and IFT

Laboratory slim tube experiments and field experience have shown that oil recovery of CO_2 flooding depends on pressure, the higher the reservoir pressure the greater the oil recovery because dynamic miscibility condition is more likely to be obtained. The miscibility between CO_2 and oil decreases interfacial tension and capillary forces, which could help recover essentially all of the remaining oil in theory. MMP is used to define as the threshold pressure to achieve miscibility in situ between injected gas and reservoir oil. To measure MMP in laboratory, grain-packed coil of considerably large length and relatively small diameter is recommended as it helps to avoid unfavorable effects of fingering, transition zone length and transverse compositional variations (Elsharkawy et al. 1992; Randall and Bennion 1988). A 1,528-cm length and 4-mm diameter slim tube, Jilin recombined oil and pure CO_2 are used to determine MMP of the aimed field; the ultimate oil recovery was obtained after 1.2 hydrocarbon pore volume (HCPV) injection in different pressure steps. The plot of oil recovery versus pressure is shown in Fig. 2. The point in which the trend of recovery plot has been changed sharply is MMP which is 22.8 MPa in this case, indicating CO_2 injection can achieve miscible flooding at initial reservoir pressure in theory.

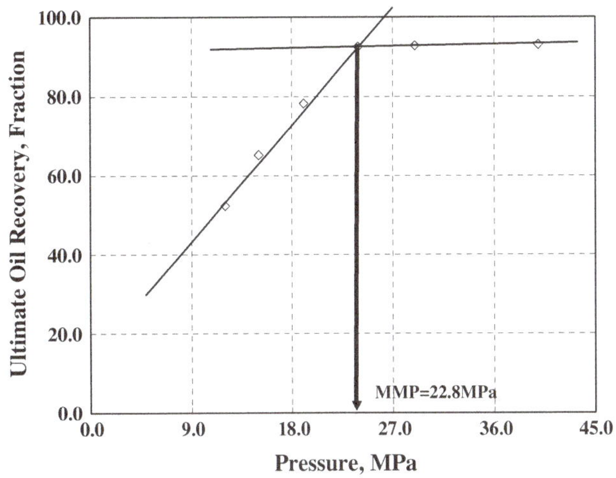

Fig. 2 Oil recovery versus pressure in slime tube test

Table 5 showed the measured interfacial tension between CO_2 and crude oil under different pressures. It was measured with DSA100 contact angle meter at reservoir temperature. As can be seen from Table 5, at testing pressure of 9 MPa, IFT is 7.35 mN/m and demonstrated Pnm of this system is about 9 MPa. Thus, when injection pressure is lower than 9 MPa, CO_2 flooding is under immiscible state. When pressure increases from 19 MPa to MMP, IFT decreases sharply from 1.54 to 0.12 mN/m which demonstrated the miscibility between CO_2 and oil. When pressure varies from Pnm to P_{mmp}, IFT is approximately a linear function of pressure. When pressure is higher than MMP, because of the miscibility of CO_2 and oil, the contacting surface disappears, and IFT is nearly 0.

The effects of core length to oil recovery

Figure 3 is oil recovery in slim tubes under the same back pressure 24.5 MPa. As can be seen, oil recovery variation curves are quite different in the two slim tubes. In short slim tube with length of 101 cm, gas breakthrough early at 0.82 PV, then oil recovery increases slightly and ultimate oil recovery is 74.20 % at 1.2 PV. In long slim tube with length of 1,528 cm, gas breakthrough later at 0.92 PV and

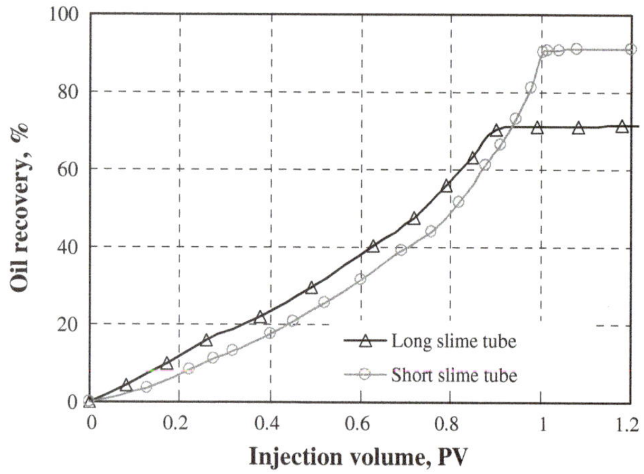

Fig. 3 Oil recovery versus injection volume in long and short slim tube tests

Table 5 IFT between CO_2 and oil at different pressures

Pressure (MPa)	IFT between CO_2 and oil (mN/m)
9	7.35
12	5.86
15	3.67
19	1.54
24	0.12
29	0.006
40	0

ultimate oil recovery is 90.45 % at 1.2 PV. In addition, the maximum pressure differential in long slim tube is 3.35 MPa, while it is 2.53 MPa in short slim tube. The higher pressure differential is caused by CO_2 compression, which supplies opportunity of extraction of light and medium hydrocarbons from oil and dissolution of CO_2 into the oil. In short slim tube flooding, although the injection is done above MMP, dynamic miscibility seems not to be achieved due to early gas breakthrough, insufficient contact time and unfavorable IFT (Table 5).

The effects of core length on relative permeability curves in slim tubes

With experimental data of endpoints, calculated shape factor b_{og} and empirical shape factor b_g, the initial guess of CO_2–oil relative permeability curves was measured, then using history-matching method we reversed the curves by matching production data. The final results of CO_2–oil relative permeability curves are shown in Fig. 4. As can be seen in Fig. 4, CO_2–oil relative permeability curves in long slim tube are nearly two intersecting straight lines, and endpoint of gas relative permeability at residual oil saturation is 1. However, these features in short slim tube are different: (1) oil-phase relative permeability decreases faster and gas phase relative permeability increases slower with gas saturation, (2) crossing point is lower and endpoint of oil relative permeability at residual oil saturation is 0.473, (3) residual oil saturation is 16.25 % higher.

It seems that CO_2 flooding in short slim tube is not under miscible flooding but near miscible flooding although the pressure is higher than MMP. With CO_2 injected into pores, CO_2 contacted and dissolved into the oil while extracting light hydrocarbons into CO_2, in this way CO_2 got enriched. The enriched CO_2 gathered at the leading edge of displacement, however due to the insufficient length of core, CO_2 breakthrough early from production well and generated a communication path before it enriched enough to achieve fully miscible flooding. Inadequate time of extracting and vaporizing light and medium hydrocarbons caused low rate of mass transfer. Consequently, oil recovery, endpoints and oil-phase relative permeability are affected.

Length determination of long composite cores

Trapping and mobilization of fluids in a porous medium can be described by complex interactions between viscous, gravity and capillary forces in a complex way. Slim tube test has merit of simulating miscible process via extremely long contacting distance, but fails to simulate natural core structure and wettability. Besides, slim tube test is often conducted without irreducible water saturation. But the presence of irreducible water would expect to reduce oil recovery and affect endpoints by increasing capillary force and decreasing flow channels. Therefore, natural core is recommended.

However, implementing CO_2 miscible flooding process in several meters long composite cores under HTHP is very difficult. At first, the core was designed to 200 cm in length, but the extremely high injection pressure and severe corrosion problem made the attempt a failure. After several times testing experiment, the core length was decided to 74 cm which is still ten times longer than conventional core segments. Although the length of the core may not be long enough to develop dynamic miscible flooding, it is a great improvement over the conventional core segments.

The effects of core length on relative permeability curves in natural cores

CO_2 flooding was then conducted in natural cores with lengths of 74 and 7.2 cm to study the effects of core length on relative permeability curves in real cores. The experimental conditions and data processing method used are the same as in slim tube. Figure 4 shows the history-matching

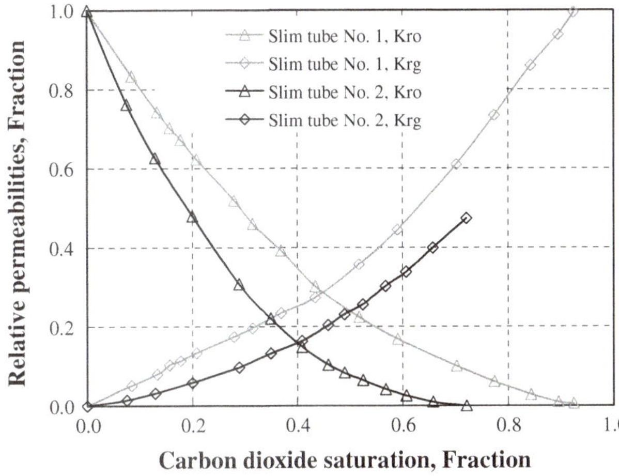

Fig. 4 Relative permeability curves in long and short slim tube tests

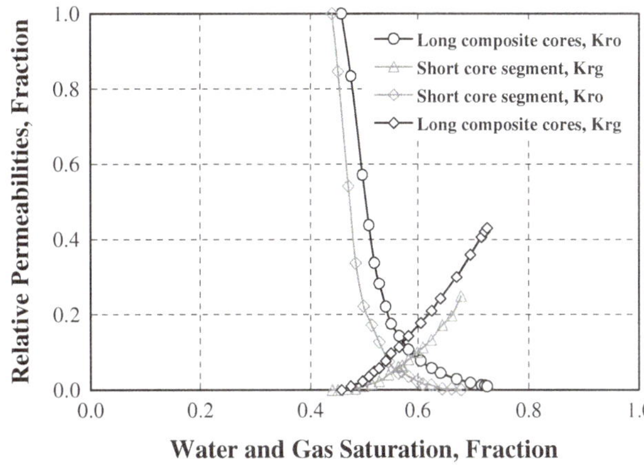

Fig. 6 Relative permeability curves in long and short real cores

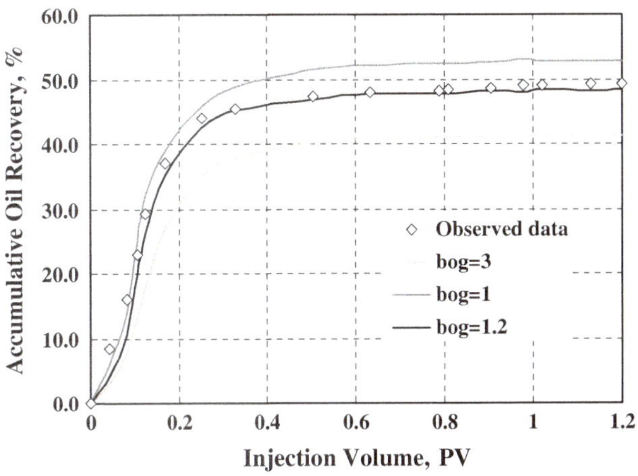

Fig. 5 History-matching production data with different b_{og} at back pressure 24.5 MPa in core test

data at long core test with different b_{og} values using Eqs. 4–6. As can be seen from Fig. 4, the calculated data fits well with reference production history at b_{og} equals to 1.2 which indicated CO_2 did not develop into full miscible flooding in 74-cm length natural cores. This is consistent with the results in slim tubes. In addition, it can be found in Fig. 4, oil recovery is affected by relative permeability shape factors and deviation of simulated oil recovery was occurred at the middle and late stage of core flooding.

Figure 5 provides the CO_2–oil relative permeability curves in long composite cores and short core segment. As can be seen from the figure, both CO_2–oil relative permeability curves in long and short real cores are not two interacting lines and endpoints of CO_2 relative permeability are not equal to 1. In addition, relative permeability endpoints are different. $K_{rg}(S_{org})$ is 0.431 in long composite cores, while this value is 0.249 in short core segment, and S_{org} is 0.275 in long composite cores, while this value is 0.333 in short core segments (Fig. 6).

Obviously, with new data processing method and long composite cores, the relative permeability curves for CO_2 flooding obtained in this paper are more reliable. Firstly, combined with improved Corey model and history-matching method effects of capillary pressure and vaporizing mechanism can be incorporated in relative permeability curves. Secondly, short core segment failed to simulate mass transfer mechanism due to early gas breakthrough and more oil remained in pores. Thirdly, longer contacting time between CO_2 and oil strengthens the vaporizing and dissolution mechanism which helps to decrease oil viscosity and IFT. However, it is should be noticed, long composite cores and new data processing method in this paper improved CO_2–oil relative permeability curves obtained in laboratory, but it is still failed to simulate the fully miscible flooding in natural cores.

Conclusions

Slim tubes and natural cores in different lengths were used to simulate strong extraction and vaporization mechanism between CO_2 and oil during the CO_2 injection process above the MMP. This is to provide reliable basic data for CO_2–oil relative permeability data process under miscibility condition. Then a new method launched to yield relative permeability for miscible flooding which considered the shape factor b_{og} as a function of pressure as a first guess and then inverted the final relative permeability curves by history-matching method.

1. Experimental results demonstrated that the early gas breakthrough and inadequate contacting time in conventional core segment led to weak mass transfer and phase transition between CO_2 and oil, and this is reason why current relative permeability curves failed to predict miscible flooding performance. Therefore,

long composite core is recommended to implement CO_2 and oil relative permeability under miscible or near miscible flooding.
2. By considering shape factor b_{og} as a function of displacement pressure, CO_2 flooding relative permeability could be determined under different degrees of miscible conditions via classic Corey model. However, due to the insufficient core length and influence of irreducible water, relative permeability curves should be corrected by history-matching method by matching measured production data.
3. The CO_2–oil relative permeability curves obtained in slim tubes and natural core in different core lengths showed core length is another important factor to achieve miscible flooding other than pressure, temperature and oil composition. Both CO_2–oil relative permeability curves in 101-cm length slim tube and 74-cm length long composite cores depicted near miscible flooding features, while CO_2–oil relative permeability curves in 1,528-cm length slim tube depicted miscible flooding features in CO_2 flooding test above MMP.
4. Although CO_2–oil relative permeability curves obtained in long composite core failed to simulate fully miscible condition, it is more reliable than that of short core segment. It is showed CO_2 and oil-phase relative permeability increased and residual oil saturation decreased in long composite cores due to stronger vaporizing and dissolution mechanism and lower capillary force.

Acknowledgments This study was supported by National Basic Research Program of China, "Fundamental research of carbon dioxide emission reduction, geological storage and utilization (FCGU) 2011CB707300". The authors would like to thank Jilin oil field for its encouragement and assistance throughout the project and preparation of this paper.

References

Al-Wahaibi YM, Muggeridge AH, Grattoni CA (2005) Experimental and numerical studies of gas/oil multicontact miscible displacements in homogeneous porous media. SPE reservoir simulation symposium, 31 January–2 February 2005, The Woodlands, Texas

Al-Wahaibi YM, Muggeridge AH, Grattoni CA (2006) Gas-oil non-equilibrium in multicontact miscible displacements within homogeneous porous media. SPE/DOE symposium on improved oil recovery, 22–26 April 2006, Tulsa, Oklahoma, USA

Dria DE, Pope GA, Sepehrnoori K(1993) Three-phase gas/oil/brine relative permeabilities measured under CO_2 flooding conditions. SPE Reserv Eng 8(2):143–150

Ekundayo JM, Ghedan SG (2013) Minimum miscibility pressure measurement with slim tube apparatus-how unique is the value? SPE reservoir characterization and simulation conference and exhibition, 16–18 September, Abu Dhabi, UAE, SPE-165966-MS

Elsharkawy AM, Poettmann FH, Christiansen RL (1992) Measuring minimum miscibility pressure: slim-tube or rising-bubble method. SPE Paper 24114, 8th symposium on enhanced oil recovery, Tulsa, USA, 22–24 April 1992

Eydinov D, Gao G, Li G, Reynolds AC (2007) Simultaneous estimation of relative permeability and porosity/permeability fields by history matching production data. Paper CIPC 2007-143 presented at the 8th Canadian international petroleum conference, Calgary, 12–14 June 2007

Gu Y, Oliver DS (2005) History matching of the PUNQ-S3 reservoir model using the ensemble Kalman filter. SPE J 10(2):217–224. SPE-89942-PA

Guo P, Sun L, Sun L et al (2000) Influences of injection gas on physical behavior of crude. J Southwest Pet Inst 22(3):57–64

Hadlow RE (1992) Update of industry experience with CO_2 injection. SPE annual technical conference and exhibition, 4–7 October 1992, Washington, DC. SPE 24928

Holt T, Lindeberg EGB, Taber JJ (2000) Technologies and possibilities for larger-scale CO_2 separation and underground storage. SPE annual technical conference and exhibition, 1–4 October 2000, Dallas, Texas, SPE 63103-MS

Hussain F, Cinar Y (2010) Comparison of methods for drainage relative permeability estimation from displacement tests. SPE improved oil recovery symposium, 24–28 April 2010, Tulsa, Oklahoma, USA. SPE 129678-MS

Lee TY, Seinfeld JH (1987) Estimation of absolute and relative permeabilities in petroleum reservoirs. Inverse Probl Eng 3(4):711–728

Li H, Chen S, Yang D et al (2012) Estimation of relative permeability by assisted history matching using the ensemble-Kalman-filter method. J Can Pet Technol 51(3):205–214

Okuno R, Xu Z (2013) Mass transfer on multiphase transitions in low-temperature CO Floods. SPE annual technical conference and exhibition, 30 September–2 October, New Orleans, Louisiana, USA. SPE-166345-MS

Omole O, Osoba JS (1989) Effect of column length on CO_2-crude oil miscibility pressure. J Can Pet Technol 28(4):97–102

Prieditis J, Brugman RJ (1993) Effects of recent relative permeability data on CO_2 flood modeling. SPE annual technical conference and exhibition, 3–6 October 1993, Houston, Texas, SPE 26650-MS

Randall TE, Bennion DB (1988) Recent developments in slim tube testing for hydrocarbon-miscible flood (HCMF) solvent design. J Can Pet Technol 27:33–44

Ren M, Wang Z, Xu Y (2011) An improved relative permeability model and its application for CO_2 injection EOR process. J China Univ Pet 35(4):108–112

Reynolds AC, Li R, Oliver DS (2004) Simultaneous estimation of absolute and relative permeability by automatic history matching of three-phase flow production data. J Can Pet Technol 43(3):37–46

Shtepani E (2007) Experimental and modeling requirements for compositional simulation of miscible CO_2-EOR processes. SPE/EAGE reservoir characterization and simulation conference, 28–31 October 2007, Abu Dhabi, UAE, SPE 111290-MS

Taber JJ, Martin FD, Seright RS (1997) EOR screening criteria revisited—part 1: introduction to screening criteria and enhanced recovery field projects. SPE Reserv Eng 12(3). SPE 39234-PA

Wang P, Pope GA (2001) Proper use of equations of state for compositional reservoir simulation. J Pet Technol 53(7). SPE 69071-MS

Watson AT, Gavalas GR, Seinfeld JH (1984) Identifiability of estimates of two-phase reservoir properties in history matching. SPE J 24(6):697–706. SPE-12579-PA

Yang PH, Watson AT (1991) A Bayesian methodology for estimating relative permeability curves. SPE Reserv Eng 6(2):259–265. SPE-18531-PA

Permissions

The contributors of this book come from diverse backgrounds, making this book a truly international effort. This book will bring forth new frontiers with its revolutionizing research information and detailed analysis of the nascent developments around the world.

We would like to thank all the contributing authors for lending their expertise to make the book truly unique. They have played a crucial role in the development of this book. Without their invaluable contributions this book wouldn't have been possible. They have made vital efforts to compile up to date information on the varied aspects of this subject to make this book a valuable addition to the collection of many professionals and students.

This book was conceptualized with the vision of imparting up-to-date information and advanced data in this field. To ensure the same, a matchless editorial board was set up. Every individual on the board went through rigorous rounds of assessment to prove their worth. After which they invested a large part of their time researching and compiling the most relevant data for our readers. Conferences and sessions were held from time to time between the editorial board and the contributing authors to present the data in the most comprehensible form. The editorial team has worked tirelessly to provide valuable and valid information to help people across the globe.

Every chapter published in this book has been scrutinized by our experts. Their significance has been extensively debated. The topics covered herein carry significant findings which will fuel the growth of the discipline. They may even be implemented as practical applications or may be referred to as a beginning point for another development. Chapters in this book were first published by Springer; hereby published with permission under the Creative Commons Attribution License or equivalent.

The editorial board has been involved in producing this book since its inception. They have spent rigorous hours researching and exploring the diverse topics which have resulted in the successful publishing of this book. They have passed on their knowledge of decades through this book. To expedite this challenging task, the publisher supported the team at every step. A small team of assistant editors was also appointed to further simplify the editing procedure and attain best results for the readers.

Our editorial team has been hand-picked from every corner of the world. Their multi-ethnicity adds dynamic inputs to the discussions which result in innovative outcomes. These outcomes are then further discussed with the researchers and contributors who give their valuable feedback and opinion regarding the same. The feedback is then collaborated with the researches and they are edited in a comprehensive manner to aid the understanding of the subject.

Apart from the editorial board, the designing team has also invested a significant amount of their time in understanding the subject and creating the most relevant covers. They scrutinized every image to scout for the most suitable representation of the subject and create an appropriate cover for the book.

The publishing team has been involved in this book since its early stages. They were actively engaged in every process, be it collecting the data, connecting with the contributors or procuring relevant information. The team has been an ardent support to the editorial, designing and production team. Their endless efforts to recruit the best for this project, has resulted in the accomplishment of this book. They are a veteran in the field of academics and their pool of knowledge is as vast as their experience in printing. Their expertise and guidance has proved useful at every step. Their uncompromising quality standards have made this book an exceptional effort. Their encouragement from time to time has been an inspiration for everyone.

The publisher and the editorial board hope that this book will prove to be a valuable piece of knowledge for researchers, students, practitioners and scholars across the globe.

List of Contributors

C. Gao
Sinopec Shengli Oil Field Company, Dongying, China

L. Aniefiok
Department of Chemical/Petroleum Engineering, University of Uyo, Uyo, Akwa-Ibom State, Nigeria

O. Boniface
Federal University of Technology, Owerri, Imo State, Nigeria

R. Larpudomlert, V. A. Torrealba, Z. T. Karpyn and P. Halleck
John and Willie Leone Family Department of Energy and Mineral Engineering, EMS Energy Institute, The Pennsylvania State University, University Park, PA 16802-5000, USA

B. Wei and L. Romero-Zeroón
Department of Chemical Engineering, University of New Brunswick, Fredericton, NB E3B 5A3, Canada

D. Rodrigue
Chemical Engineering Department, Universite´ Laval, Pavillon Adrien-Pouliot, 1065, Avenue de la Me´decine, Local 3546, Quebec, QC G1V 0A6, Canada

I. A. Purwasena
School of Life Science and Technology, Bandung Institute of Technology, Ganesha 10, Bandung 40132, West Java, Indonesia

Y. Sugai and K. Sasaki
Department of Earth Resources Engineering, Faculty of Engineering, Kyushu University, 744 Motooka, Nishi-ku, Fukuoka 819-0395, Japan

O. Ogiesoba and U. Hammes
Bureau of Economic Geology, Jackson School of Geosciences, The University of Texas at Austin, Austin, USA

R. Ahmadi and M. Masihi
Department of Chemical and Petroleum Engineering, Sharif University of Technology, Tehran, Iran

M. R. Rasaei
Institute of Petroleum Engineering, Tehran University, Tehran, Iran

K. Eskandaridalvand
DeGolyer and MacNaughton Petroleum Consulting Company, Dallas, TX, USA

R. Shahalipour
Petroiran Development Company, Tehran, Iran

J. M. Carcione and S. Picotti
Instituto Nazionale di Oceanografia e di Geofisica Sperimentale (OGS), Borgo Grotta Gigante 42c, 34010 Sgonico, Trieste, Italy

J. E. Santos
CONICET, Instituto del Gas y del Petro´leo, Facultad de Ingenierı´a, Universidad de Buenos Aires, Buenos Aires, Argentina

J. E. Santos
Universidad Nacional de La Plata, La Plata, Argentina

J. E. Santos
Department of Mathematics, Purdue University, West Lafayette, IN 47907-2067, USA

A. Qadrouh and H. S. Almalki
SAC, KACST, PO Box 6086, Riyadh 11442, Saudi Arabia

M. Alsouki and O. Al Hamad
Syrian Petroleum Company, Damascus, Syria

R. Taifour
Geology Department, Sciences Faculty, University of Damascus, Damascus, Syria

K. Ling, J. He, J. Ge and P. Pei
University of North Dakota, Grand Forks, TX, USA

Z. Shen
Texas A&M University, College Station, ND, USA

M. Ihtsham Hashmi and B. Ghosh
Petroleum Engineering Department, The Petroleum Institute, 2533, Abu Dhabi, UAE

C. Gao
The Petroleum Institute, SINOPEC SLOF, Dongying, China

S. Shokrolahzadeh-Behbahani and M.-A. Ahmadi
Department of Petroleum Engineering, Ahwaz Faculty of Petroleum Engineering, Petroleum University of Technology, Ahwaz, Iran

M. Zeinali-Hasanvand
Research Institute of Petroleum Industry (RIPI), Tehran, Iran

M. A. Safarzadeh
Research Institute of Petroleum Industry, Tehran, Iran

M. Motealleh
Computer Aided Process Engineering Group, Tehran, Iran

J. Moghadasi
Petroleum University of Technology, Ahwaz, Iran

H. Akhondzadeh
Faculty of Petroleum and Renewable Energy Engineering, Universiti Teknologi Malaysia, No 63, Lane 13, Jahanara St., Shiraz, Iran

A. Fattahi
Faculty of Petroleum and Renewable Energy Engineering, Universiti Teknologi Malaysia, No 10, level 2, Sanei Bon bast, Fifth Ave, Fatemi St., Tehran, Iran

X. Sun, K. Wang, T. Yan and S. Shao
College of Petroleum Engineering, Northeast Petroleum University, Daqing 163318, China

J. Jiao
Tianjin Branch of CNPC Offshore Engineering Co. Ltd, Tianjin 300450, China

S. Zhang, G.-C. Jiang, L. Wang and W. Qing
State Key Laboratory of Petroleum Resource and Prospecting, Department of Petroleum Engineering, China University of Petroleum, Beijing 102249, China

H.-T. Guo, X. Tang and D.-G. Bai
Tan Hai Development Company, China National Petroleum Corporation, Tianjin 300289, China

D. Dogon and M. Golombok
Faculty of Mechanical Engineering, Technische Universiteit Eindhoven, Den Dolech 2, 5600 MB Eindhoven, The Netherlands
Shell Global Solutions International B.V., Kessler Park 1, 2288 GS Rijswijk, The Netherlands

F. F. Li, S. L. Yang, H. Chen, X. Zhang, D. D. Yin and L. P. He
Key Lab of Petroleum Engineering of MOE, China University of Petroleum, Beijing 102249, People's Republic of China

Z. Wang
Technology Department, Petro-China International Iraq, Beijing 100724, People's Republic of China